金仓数据库 KingbaseES DBA 实践

曾庆峰 杜 胜 冯 玉 编著

清华大学出版社
北京

内 容 简 介

本书的主要内容包括 KingbaseES 数据库管理的各方面：KingbaseES 数据库单机环境安装、客户端工具、KingbaseES 数据库体系结构、实例管理、用户数据管理、事务与并发控制、数据库日常运行监控、数据库性能问题诊断工具、SQL 语句执行计划、物理备份与逻辑备份、闪回技术，以及主备集群。本书采用 TPC-H 生成的测试数据集，以实战的方式，帮助读者理解和掌握 KingbaseES 数据库运维中涉及的原理和知识点。

本书适合作为 KingbaseES 数据库管理员的参考书，也可作为高等院校计算机大类高年级本科生或研究生理解数据库管理系统体系结构的实现、数据库系统管理等方面的实验实践教材。

版权所有，侵权必究。举报：010-62782989，beiqinquan@tup.tsinghua.edu.cn。

图书在版编目（CIP）数据

金仓数据库 KingbaseES DBA 实践 / 曾庆峰，杜胜，冯玉编著. -- 北京：清华大学出版社，2024.12. -- ISBN 978-7-302-67784-0

Ⅰ. TP311.132.3

中国国家版本馆 CIP 数据核字第 2024R4C258 号

责任编辑：张　玥　薛　阳
封面设计：常雪影
责任校对：王勤勤
责任印制：刘　菲

出版发行：清华大学出版社
网　　址：https://www.tup.com.cn，https://www.wqxuetang.com
地　　址：北京清华大学学研大厦 A 座
邮　　编：100084
社 总 机：010-83470000
邮　　购：010-62786544
投稿与读者服务：010-62776969，c-service@tup.tsinghua.edu.cn
质量反馈：010-62772015，zhiliang@tup.tsinghua.edu.cn
课件下载：https://www.tup.com.cn，010-83470236

印 装 者：北京鑫海金澳胶印有限公司
经　　销：全国新华书店
开　　本：185mm×260mm
印　　张：23
字　　数：560 千字
版　　次：2024 年 12 月第 1 版
印　　次：2024 年 12 月第 1 次印刷
定　　价：75.00 元

产品编号：106183-01

前 言

  金仓数据库 KingbaseES 是由中电科金仓(北京)科技股份有限公司(简称"电科金仓")研发的一款面向大规模并发交易处理的企业级关系数据库,融合了电科金仓在数据库领域几十年产品研发与企业级应用的实践经验,可满足各行业用户多种场景的数据处理需求。KingbaseES 遵循严格的事务 ACID 特性、结合多核架构的极致性能、行业最高的安全标准、完备的高可用方案,以及可覆盖迁移、开发及运维管理全使用周期的智能便捷工具,可为用户带来更为极致的使用体验。金仓数据库 KingbaseES 广泛服务于电子政务、能源、金融、电信等 60 余个重点行业和关键领域,累计装机部署超过 100 万套,入选国务院国资委发布十项国有企业数字技术典型成果。

  关系数据库仍然是目前数据存储和管理主要方式,并且主流的商业化和开源关系数据库产品主要运行在 UNIX/Linux 平台。本书的 KingbaseES 数据库使用的数据库版本是 KingbaseES V9。本书包含了大量有趣的实验,它们基于 TPC-H 生成的数据集,以实践的方式,帮助读者理解和掌握 KingbaseES 数据库运维中涉及的原理和知识点。本书既可作为数据库管理人员的参考书,也可作为高等院校计算机相关专业的学生理解数据库管理系统体系结构的实现、数据库系统管理等方面的实验/实践教材。

  全书共 14 章,内容由浅入深,层次清晰,通俗易懂。第 1 章的主要内容是为读者安装一个简单规范的 KingbaseES 数据库实战环境,并导入本书的测试数据集;第 2 章的主要内容是关于 KingbaseES 客户端工具 ksql 的用法介绍;第 3 章的内容是关于 KingbaseES 数据库的体系结构,对于数据库管理员 DBA 来说,非常重要,建议读者反复重点学习,直到完全掌握;第 4 章的内容是管理 KingbaseES 数据库实例和数据库集簇;第 5 章的内容是管理用户与会话连接,包括管理配置 sys_hba.conf 和 sys_ident.conf 这两个配置文件;第 6 章是管理用户数据,包括表空间管理、用户数据存储规划、管理表中的数据;第 7 章的内容是关于事务与并发控制的原理与管理实践,包括自动(手动)VACUUM 操作、冻结事务号、配置数据库运行在归档日志模式等;第 8 章是数据库的日常运行监控,包括监控操作系统的内存、I/O、CPU 和文件系统,监控 KingbaseES 数据库的运行维护日志,数据库运行监控,DBA 如何接手一个 KingbaseES 数据库;第 9 章介绍电科金仓的数据库性能问题诊断工具 KWR、KSH 和 KDDM;第 10 章的内容是关于 SQL 语句的执行计划;第 11 章和第 12 章是关于 KingbaseES 的物理备份和逻辑备份;第 13 章介绍 KingbaseES 的数据库闪回技术;第 14 章的内容是 KingbaseES 主备集群的安装、运维问题的简单处理和备份恢复。

  本书具有以下特点:

  (1) 以 KingbaseES 数据库体系结构为核心,全书的实践均围绕体系结构进行。

（2）介绍 KingbaseES 数据库的特性。

（3）注重 KingbaseES 数据库管理的规范性。

（4）本书提供配套的资源文件，包括每章的命令和 SQL 脚本、用于实战的 VMware Workstation 虚拟机备份。读者可以使用这些资源从任何一章开始学习，并获得与本章内容一致的输出。

本书由曾庆峰、杜胜、冯玉共同编写。本书大纲由杜胜、冯玉拟制，第 1、2、4～6、8、9、11～14 章由曾庆峰执笔，第 3、7 和第 10 章由冯玉执笔，最后由冯玉统稿。电科金仓的窦培、靳国军、刘贺、王胜利等提供了金仓数据库的相关资料，并解答了本书撰写过程中遇到的很多问题。

在编写过程中，还参阅了电科金仓公司、甲骨文（Oracle）、PostgreSQL 开源数据库等相关的数据库文档、联机帮助和教学培训成果，也吸取了国内外相关参考书的精髓，对这些作者的贡献表示由衷的感谢。本书在出版过程中，得到了中国人民大学王珊教授的支持和帮助；还得到了清华大学出版社张玥编辑的大力支持，在此表示诚挚的感谢。

由于作者水平有限，书中难免有不妥和疏漏之处，恳请各位专家、同仁和读者不吝赐教。

作　者

2024 年 6 月于北京

目　录

第 1 章　KingbaseES 数据库安装部署 ········· 1
1.1　KingbaseES 数据库简介 ········· 1
1.2　准备安装 KingbaseES 数据库的软硬件环境 ········· 1
1.2.1　准备安装 KingbaseES 数据库的服务器 ········· 1
1.2.2　安装 Linux 操作系统 ········· 2
1.2.3　准备安装 KingbaseES 数据库的存储空间 ········· 3
1.3　安装 KingbaseES 数据库 ········· 6
1.3.1　KingbaseES 数据库软件安装包文件和许可证 ········· 6
1.3.2　KingbaseES 数据库安装前的准备工作 ········· 7
1.3.3　安装 KingbaseES 数据库软件 ········· 9
1.3.4　KingbaseES 数据库安装后的操作 ········· 16
1.4　本书的实验环境 ········· 17
1.4.1　配置实验环境 ········· 17
1.4.2　导入 TPC-H 测试数据集 ········· 18
1.4.3　命令行提示符约定 ········· 21
1.4.4　本书提供的资源文件 ········· 22
1.4.5　制作自己的虚拟机备份 ········· 22

第 2 章　KingbaseES 数据库客户端工具 ········· 24
2.1　安装 KingbaseES 数据库客户端 ········· 24
2.1.1　在 Linux 操作系统上安装 KingbaseES 数据库客户端 ········· 24
2.1.2　在 Windows 操作系统上安装 KingbaseES 数据库客户端 ········· 27
2.2　客户端与服务器的连接方式 ········· 28
2.2.1　使用命令行选项 ········· 28
2.2.2　使用服务名 ········· 28
2.2.3　数据库用户密码文件.kbpass ········· 30
2.3　客户端程序 ksql ········· 30
2.3.1　ksql 的命令行选项 ········· 31
2.3.2　ksql 的元命令 ········· 32
2.3.3　ksql 初始化文件.ksqlrc ········· 39

2.4 图形客户端程序 KStudio ·· 40

第 3 章 KingbaseES 数据库体系结构 ·· 43

3.1 KingbaseES 数据库服务器 ·· 43
3.2 KingbaseES 数据库的进程结构 ·· 44
   3.2.1 KingbaseES 数据库主进程 ·· 45
   3.2.2 KingbaseES 数据库服务进程 ·· 45
   3.2.3 KingbaseES 数据库后台进程 ·· 45
3.3 KingbaseES 数据库的内存结构 ·· 48
   3.3.1 系统全局区 ·· 49
   3.3.2 程序全局区 ·· 49
   3.3.3 内存参数初始优化建议 ·· 50
3.4 KingbaseES 数据库逻辑结构 ·· 50
   3.4.1 数据库集簇 ·· 50
   3.4.2 数据库 ·· 51
   3.4.3 模式 ·· 54
   3.4.4 数据库对象 ·· 57
3.5 KingbaseES 数据库物理结构 ·· 57
   3.5.1 数据文件 ·· 58
   3.5.2 控制文件 ·· 60
   3.5.3 WAL 文件 ·· 63
   3.5.4 配置文件 ·· 64
3.6 系统表与系统视图 ·· 65
   3.6.1 系统表与系统视图的类别 ·· 65
   3.6.2 查询数据库对象信息 ·· 66
3.7 连接和会话 ·· 68
3.8 SQL 语句的执行过程 ·· 69

第 4 章 管理 KingbaseES 数据库实例 ·· 70

4.1 数据库实例的启动与关闭 ·· 70
   4.1.1 启动数据库 ·· 70
   4.1.2 关闭数据库 ·· 72
   4.1.3 kingbase 服务 ·· 74
4.2 系统配置参数 ·· 74
   4.2.1 系统配置参数概述 ·· 75
   4.2.2 设置系统配置参数 ·· 78
   4.2.3 查看系统配置参数值的设置来源 ·· 85
4.3 管理数据库的扩展插件 ·· 86
4.4 管理数据库的软件许可证 ·· 88

4.5　数据库实例与数据库集簇 ·················· 90

第 5 章　管理用户与会话连接 ·················· 92
5.1　用户管理 ························· 92
    5.1.1　创建数据库用户 ················· 92
    5.1.2　删除数据库用户 ················· 95
    5.1.3　查询用户信息 ·················· 96
5.2　管理连接会话 ······················ 97
    5.2.1　设置与会话连接相关的系统配置参数 ······· 97
    5.2.2　查看会话连接信息 ················ 98
    5.2.3　处理有问题的连接会话 ············· 101
5.3　配置文件 sys_hba.conf ················· 102
5.4　配置文件 sys_ident.conf ················ 105

第 6 章　管理用户数据 ···················· 107
6.1　管理表空间 ······················ 107
    6.1.1　创建表空间 ·················· 107
    6.1.2　删除表空间 ·················· 112
6.2　用户数据的存储规划 ·················· 116
    6.2.1　表空间的分类 ················· 116
    6.2.2　表空间的规划 ················· 116
6.3　管理表中数据 ····················· 118
    6.3.1　表的填充因子 ················· 118
    6.3.2　生成批量测试数据 ··············· 119
    6.3.3　使用 COPY 命令导入/导出数据库表 ······· 120
    6.3.4　使用 sys_bulkload 装载大表 ··········· 121
    6.3.5　大表模糊查询 ················· 123
    6.3.6　数据存储空间查询 ··············· 124

第 7 章　事务与并发控制 ··················· 126
7.1　事务的基本概念 ···················· 126
7.2　事务处理模型 ····················· 127
    7.2.1　显式事务与隐式事务 ·············· 127
    7.2.2　DDL 语句与事务 ················ 129
    7.2.3　事务隔离级别 ················· 130
    7.2.4　事务并发控制机制 ··············· 132
7.3　MVCC ························ 133
    7.3.1　事务号与事务状态 ··············· 133
    7.3.2　元组的结构 ·················· 134

	7.3.3	元组的增、删、改 ············	135
	7.3.4	元组的访问 ················	139
	7.3.5	元组的并发更新 ·············	143
7.4	管理元组的多版本 ··················		147
	7.4.1	手工清理无效元组 ············	147
	7.4.2	自动清理无效元组 ············	150
7.5	管理事务号 ·························		151
	7.5.1	自动冻结事务号 ·············	151
	7.5.2	手工冻结事务号 ·············	153
7.6	数据库锁 ···························		158
	7.6.1	表级锁 ····················	158
	7.6.2	事务锁 ····················	164
	7.6.3	死锁 ······················	166
7.7	故障恢复机制 ······················		168
	7.7.1	故障恢复概述 ···············	168
	7.7.2	日志系统组件 ···············	168
	7.7.3	WAL 文件 ··················	169
	7.7.4	检查点机制 ·················	171
	7.7.5	配置 WAL 文件 ··············	173
	7.7.6	归档日志模式 ···············	174

第 8 章 数据库日常运行监控 ·················· 176

8.1	数据库服务器的运行维护日志 ········		176
	8.1.1	启动日志文件 startup.log ·····	176
	8.1.2	运行日志文件 ···············	177
8.2	数据库服务器的操作系统监控 ········		181
	8.2.1	本节实验环境说明 ············	181
	8.2.2	监控服务器内存 ·············	183
	8.2.3	监控服务器磁盘 I/O ··········	184
	8.2.4	监控服务器 CPU ·············	185
	8.2.5	监控服务器文件系统 ··········	186
8.3	数据库的运行监控 ··················		188
	8.3.1	监控会话 ···················	188
	8.3.2	监控长时间的活动事务 ········	190
	8.3.3	监控长时间运行的 SQL 语句 ····	191
	8.3.4	监控锁 ·····················	192
	8.3.5	监控 vacuum 操作 ············	194
	8.3.6	监控事务 ID 回卷风险 ·········	195
8.4	接手一个生产数据库 ·················		196

第 9 章　数据库性能问题诊断工具 ·············· 198
9.1　性能问题诊断工具概述 ·············· 198
9.2　KWR ·············· 199
9.2.1　KWR 的使用场景 ·············· 199
9.2.2　配置 KWR ·············· 200
9.2.3　创建 KWR 快照 ·············· 202
9.2.4　查看 KWR 快照 ·············· 202
9.2.5　删除 KWR 快照 ·············· 203
9.2.6　生成 KWR 报告 ·············· 205
9.2.7　生成 KWR 运行期对比报告 KWR DIFF ·············· 205
9.3　KSH ·············· 206
9.3.1　KSH 的使用场景 ·············· 206
9.3.2　配置 KSH ·············· 207
9.3.3　查看 KSH 数据 ·············· 208
9.3.4　生成 KSH 报告 ·············· 208
9.4　KDDM ·············· 210
9.4.1　KDDM 的使用场景 ·············· 210
9.4.2　生成 KDDM 用户报告 ·············· 211
9.4.3　获取数据库配置参数的建议值 ·············· 211
9.5　性能诊断工具 KWR 实战 ·············· 212

第 10 章　SQL 语句执行计划 ·············· 216
10.1　SQL 语句的执行过程 ·············· 216
10.2　查看 SQL 语句的执行计划 ·············· 218
10.3　阅读 SQL 语句的执行计划 ·············· 219
10.3.1　单表查询 ·············· 219
10.3.2　多表连接查询 ·············· 221
10.3.3　分组聚集查询 ·············· 225
10.3.4　子查询 ·············· 226
10.4　影响 SQL 语句的执行计划 ·············· 228
10.4.1　更新数据库的统计信息 ·············· 228
10.4.2　创建合适的索引 ·············· 229
10.4.3　影响执行计划的配置参数 ·············· 232
10.4.4　使用查询提示 SQL hint ·············· 234

第 11 章　物理数据库备份与恢复 ·············· 237
11.1　数据库备份与恢复的基本概念 ·············· 237
11.1.1　逻辑备份与物理备份 ·············· 237
11.1.2　冷备份和热备份 ·············· 238

　　　　11.1.3　全量备份、差异备份和增量备份 ·································· 239
　　　　11.1.4　数据库恢复 ·· 240
　11.2　为生产系统引入备份恢复测试机 ·· 241
　　　　11.2.1　准备数据库备份恢复测试机 ·· 241
　　　　11.2.2　为数据库服务器配置 rsync 服务 ··· 242
　　　　11.2.3　配置主机间的无密码 ssh ·· 243
　11.3　数据库脱机冷备份与恢复 ··· 244
　　　　11.3.1　数据库脱机冷备份 ··· 244
　　　　11.3.2　物理数据库脱机冷备份恢复 ·· 248
　11.4　数据库联机热备与恢复 ··· 248
　　　　11.4.1　sys_rman 备份恢复工具简介 ··· 248
　　　　11.4.2　配置 sys_rman ··· 250
　　　　11.4.3　使用 sys_rman 备份数据库 ·· 252
　　　　11.4.4　管理 sys_rman 备份集 ·· 253
　　　　11.4.5　在生产数据库上执行完全恢复 ··· 256
　　　　11.4.6　异机恢复 KingbaseES 数据库 ·· 259
　　　　11.4.7　不完全恢复到指定时间点 ·· 260
　　　　11.4.8　不完全恢复到指定事务号 ·· 263

第 12 章　逻辑备份与恢复 ··· **267**
　12.1　sys_dump 和 sys_restore ·· 267
　　　　12.1.1　数据库的逻辑备份与恢复 ·· 267
　　　　12.1.2　模式的逻辑备份与恢复 ·· 271
　　　　12.1.3　表的逻辑备份与恢复 ··· 273
　　　　12.1.4　逻辑备份用户和表空间定义 ·· 275
　12.2　Oracle 兼容的 exp 和 imp ··· 276
　　　　12.2.1　导出/导入数据库 ··· 276
　　　　12.2.2　导出/导入用户模式 ··· 277
　　　　12.2.3　导出/导入表 ··· 277

第 13 章　闪回查询与闪回表 ··· **280**
　13.1　配置 KingbaseES 数据库的闪回功能 ··· 280
　13.2　闪回回收站 ··· 280
　　　　13.2.1　清空闪回回收站 ·· 281
　　　　13.2.2　删除闪回回收站中的一个表 ·· 281
　　　　13.2.3　从闪回回收站恢复被误删除的表 ·· 283
　13.3　闪回查询 ·· 283
　　　　13.3.1　基于时间戳的闪回查询 ·· 284
　　　　13.3.2　基于 CSN 的闪回查询 ··· 285

13.4 闪回表 ······ 286
13.5 闪回技术的限制 ······ 287

第 14 章 KingbaseES 主备集群 ······ 288

14.1 KingbaseES 主备集群简介 ······ 288
    14.1.1 KingbaseES 主备集群的拓扑结构 ······ 288
    14.1.2 KingbaseES 主备集群的组件 ······ 289

14.2 安装 KingbaseES 主备集群 ······ 291
    14.2.1 规划一个 KingbaseES 主备集群 ······ 291
    14.2.2 准备安装主备集群的服务器 ······ 291
    14.2.3 安装主备集群的准备工作 ······ 293
    14.2.4 安装主备集群 ······ 300
    14.2.5 部署完主备集群后的操作 ······ 301

14.3 管理 KingbaseES 主备集群 ······ 301
    14.3.1 获取集群信息的指令 ······ 301
    14.3.2 停止主备集群 ······ 305
    14.3.3 启动主备集群 ······ 305
    14.3.4 让节点重新加入集群 ······ 306
    14.3.5 集群主备切换 ······ 307
    14.3.6 重做备用节点 ······ 308
    14.3.7 为集群添加新的备节点 ······ 309
    14.3.8 从集群中删除备节点 ······ 311
    14.3.9 从集群中删除见证节点 ······ 312
    14.3.10 为集群添加见证节点 ······ 313

14.4 主备集群 sys_rman 备份实战 ······ 314
    14.4.1 REPO 外部部署 ······ 314
    14.4.2 REPO 内部部署 ······ 318

附录 A 安装 CentOS 7 操作系统 ······ 323

A.1 准备服务器硬件 ······ 323
A.2 下载 CentOS 7 ······ 328
A.3 安装 CentOS 7 ······ 328
A.4 安装 Google Chrome 浏览器 ······ 339
A.5 删除逻辑卷 centos/toBeDeleted ······ 339
A.6 备份 CentOS 虚拟机文件 ······ 341

附录 B 安装 KingbaseES 单机数据库的最佳实践 ······ 342

B.1 最佳实践安装规划 ······ 342
B.2 基础安装 ······ 342

B.3 KingbaseES 数据库优化配置 ··· 343
 B.3.1 创建数据库用户 kingbase ·· 343
 B.3.2 优化 WAL 日志 ·· 344
 B.3.3 配置 KingbaseES 数据库工作在归档模式 ······················ 344
 B.3.4 初步优化 KingbaseES 数据库的系统参数 ······················ 345
 B.3.5 控制文件多元化 ·· 346
 B.3.6 设置默认表空间和临时表空间 ·································· 346
 B.3.7 配置服务名 ·· 347
B.4 导入测试数据集 ··· 347
B.5 CentOS 7 操作系统安全加固 ·· 347
 B.5.1 隐藏 GNOME 登录界面中的用户名 ····························· 347
 B.5.2 在 GNOME 登录界面显示文本信息 ····························· 348
 B.5.3 禁止普通用户关机 ·· 348
 B.5.4 禁用 Ctrl+Alt+Del 组合键 ··· 348
 B.5.5 禁止系统休眠 ·· 349
B.6 备份 CentOS 虚拟机文件 ··· 349

附录 C 生成 TPC-H 测试数据集 ··· 350

参考文献 ··· 354

第 1 章

KingbaseES 数据库安装部署

 ## 1.1 KingbaseES 数据库简介

KingbaseES(Kingbase Enterprise Server)是由电科金仓公司自主研发的通用关系数据库管理系统(database management system,DBMS)。作为一款拥有自主知识产权的创新数据库产品,KingbaseES 体现了电科金仓公司对技术创新的持续投入和努力。

KingbaseES 数据库主要面向事务处理类应用,同时也兼顾各类数据分析类应用,其适用范围涵盖管理信息系统、业务及生产系统、决策支持系统、多维数据分析、全文检索、地理信息系统、图片搜索等,在多领域应用中表现出色,具备强大的承载能力。

KingbaseES 数据库支持多种操作系统和硬件平台,如 Linux、Windows 和国产 Kylin、统信、中科方德、深之度 openEuler 等操作系统,以及通用 x86_64 和国产龙芯、飞腾、申威等中央处理器(central processing unit,CPU)硬件体系架构。

 ## 1.2 准备安装 KingbaseES 数据库的软硬件环境

1.2.1 准备安装 KingbaseES 数据库的服务器

可以把 KingbaseES 数据库部署在物理服务器或者云服务器上。如果决定使用物理服务器直接部署 KingbaseES 数据库,建议服务器的配置如下。

(1) **CPU**:KingbaseES 数据库支持 Intel 公司或 AMD 公司生产的 x86_64 架构处理器,也支持国产的 Kunpeng64、AArch64、MIPS64、LoongArch64 等架构的处理器。

(2) **物理内存**:KingbaseES 数据库服务器(KingbaseES database server)的物理内存,一般需要根据用户的数据总量(未来 5 年甚至 8 年后的情况)、用户客户端连接(connection)的数量等因素进行容量规划。由于当前服务器的内存并不昂贵,因此一定要为 KingbaseES 数据库服务器配置足够容量的物理内存,避免内存成为 KingbaseES 数据库服务器的性能瓶颈。

(3) **存储**:KingbaseES 数据库服务器一般会有内置的独立磁盘冗余阵列(redundant arrays of independent disks,RAID)控制器(特别提醒要为内置 RAID 控制器配置尽量大的高速缓存(cache),这样可以获得更好的性能)。

① 如果有外置的 RAID 用于存储 KingbaseES 数据库的数据,可以为服务器配置 4 块

内置磁盘：2 块磁盘做 raid1，用于安装操作系统；1 块磁盘用于 raid 的热冗余（hot spare）；1 块磁盘用于存储 KingbaseES 数据库备份。

② 如果没有外置的 RAID 用于存储 KingbaseES 数据库的数据，那么服务器应该配置至少 12 块内置磁盘：2 块磁盘做 raid1，用于安装操作系统；8 块磁盘做 raid10，用于存储 KingbaseES 数据库的数据；1 块磁盘用于 raid 的热冗余；1 块磁盘用于存储 KingbaseES 数据库备份。

（4）**网络**：建议为 KingbaseES 数据库服务器配置多个 10Gb/s 以太网卡，或者性能更好的 25Gb/s、40Gb/s、56Gb/s、100Gb/s 以太网卡，以满足大量客户端访问 KingbaseES 数据库服务器的需要。

如果 KingbaseES 数据库的数据库管理员（DBA）决定在私有云或者公有云上部署 KingbaseES 数据库服务器，相对于直接使用物理服务器，DBA 只需要设置 CPU 的类型和数量、磁盘存储的规格和容量、物理内存的容量、网络带宽的要求。DBA 可以根据 2～3 年内的数据和访问情况来规划这些需求。

DBA 通常会使用一个属于个人的 KingbaseES 数据库测试环境，一般在自己的个人电脑上，使用 VMware Workstation 软件创建 CentOS 操作系统的虚拟机，在虚拟机上安装 KingbaseES 数据库测试环境。

1.2.2 安装 Linux 操作系统

KingbaseES 数据库是运行在操作系统之上的系统软件，操作系统的配置直接影响 KingbaseES 数据库的正常运行。本书以 CentOS 7 操作系统为例，其他 Linux 操作系统同理。

安装 CentOS 7 操作系统需要注意以下几点。

（1）安装过程的提示语言选择英语。这个建议基于过往项目经验，因为部分软件对中文支持不完善，容易出现使用问题，所以推荐选择英语。

（2）软件安装选项，建议初学者选择安装 Server with GUI 软件组，并选择安装所有软件包。

（3）建议采用逻辑卷管理（logical volume manager，LVM）技术来管理文件系统，这样做的好处是未来某个文件系统存储空间不足时，可以通过扩大该文件系统所在逻辑卷的大小来扩展这个文件系统。

（4）基于项目经验，建议安装 CentOS 操作系统时使用 XFS 文件系统，并采用如下原则设置文件系统的大小。

① 基于**够用就行（限制最大值）**的原则，设置文件系统或分区为/boot 文件系统、/（根目录）文件系统、/usr 文件系统、/tmp 文件系统。

② 基于**尽量大**的原则（相对于系统盘的总大小），设置文件系统或分区为交互区（swap）、/opt 文件系统、/var 文件系统、/home 文件系统。

附录 A 给出了在 VMware Workstation 17 Pro 软件下创建一个 CentOS 7 操作系统虚拟机的详细过程。示例将主机名设置为 dbsrv，将对外服务网卡的 IP 地址设置为 192.168.100.22/24。使用一块 900GB 的磁盘作为操作系统盘，在该磁盘上创建名为 centos 的卷组，文件系统的大小建议设置如下。

（1）/boot 文件系统，大小为 10GB。

（2）/（根目录）文件系统，大小为 10GB。

（3）交换区（swap），大小为 64GB。

（4）/usr 文件系统，大小为 20GB。

（5）/opt 文件系统，大小为 40GB。

（6）/var 文件系统，大小为 40GB。

（7）/tmp 文件系统，大小为 10GB。

（8）/home 文件系统，大小为 40GB。

（9）/toBeDeleted，剩余空间，示例大小为 666GB。

1.2.3　准备安装 KingbaseES 数据库的存储空间

为了数据的安全和方便管理，KingbaseES 数据库中的数据文件、控制文件、预写式日志（write ahead log，WAL）文件、数据库备份通常使用不同的逻辑卷组来保存。可以按照如下规划来创建逻辑卷组和逻辑卷。

（1）在卷组 dbvg 上创建多个逻辑卷，如 u00lv、u01lv、u02lv 和 u03lv，用于存放 KingbaseES 数据库安装后的二进制介质、KingbaseES 数据库集簇（database cluster）的数据文件、控制文件、日志文件，并需要在其上创建多个逻辑卷，用于存放不同应用或用户的数据文件、日志文件、控制文件等。

（2）在卷组 archvg 上创建逻辑卷 u04lv，用于存放 KingbaseES 数据库的归档日志。

（3）在卷组 bakvg 上创建逻辑卷 baklv，用于存放 KingbaseES 数据库的备份。

下面以有 4 块硬盘的服务器为例，介绍创建数据库使用的逻辑卷组的过程。如果无法获得物理服务器，可以使用 VMware Workstation 软件来仿真这个环境，并按照附录 A 安装 CentOS 7 操作系统。安装完成后，为虚拟机增加 3 块大小为 1200GB 的虚拟磁盘。

（1）磁盘/dev/sda：大小为 900GB，CentOS 操作系统盘，在上面创建了名字为 centos 的卷组。

（2）磁盘/dev/sdb：大小为 1200GB，计划用于创建卷组 dbvg。

（3）磁盘/dev/sdc：大小为 1200GB，计划用于创建卷组 archvg。

（4）磁盘/dev/sdd：大小为 1200GB，计划用于创建卷组 bakvg。

执行下面的命令，可以查看当前有哪些卷组和逻辑卷：

```
[root@dbsvr ~]# vgs
  VG     #PV #LV #SN Attr   VSize    VFree
  centos   1   7   0 wz--n- <890.00g <666.00g
[root@dbsvr ~]# lvs
  LV    VG     Attr       LSize  Pool Origin Data%  Meta%  Move Log Cpy%Sync Convert
  home  centos -wi-ao---- 40.00g
  opt   centos -wi-ao---- 40.00g
  root  centos -wi-ao---- 10.00g
  swap  centos -wi-ao---- 64.00g
  tmp   centos -wi-ao---- 10.00g
  usr   centos -wi-ao---- 20.00g
```

```
  var   centos -wi-ao---- 40.00g
[root@dbsvr ~]# df -h
Filesystem               Size  Used Avail Use% Mounted on
devtmpfs                 7.8G     0  7.8G   0% /dev
tmpfs                    7.8G     0  7.8G   0% /dev/shm
tmpfs                    7.8G   13M  7.8G   1% /run
tmpfs                    7.8G     0  7.8G   0% /sys/fs/cgroup
/dev/mapper/centos-root   10G   84M   10G   1% /
/dev/mapper/centos-usr    20G  5.8G   15G  29% /usr
/dev/mapper/centos-var    40G  1.4G   39G   4% /var
/dev/mapper/centos-opt    40G   33M   40G   1% /opt
/dev/mapper/centos-tmp    10G   33M   10G   1% /tmp
/dev/mapper/centos-home   40G   33M   40G   1% /home
/dev/sda1                 10G  187M  9.9G   2% /boot
tmpfs                    1.6G   24K  1.6G   1% /run/user/0
[root@dbsvr ~]#
```

从输出可以看到,安装 CentOS 7 操作系统时,使用 /dev/sda 创建了 LVM 卷组 centos,在卷组 centos 上创建了逻辑卷 home、opt、root、swap、tmp、usr 和 var,它们被挂载在 CentOS 操作系统的文件系统上。

按照之前的规划,完成如下任务。

(1) 在磁盘 /dev/sdb 上创建卷组 dbvg,并在其上创建逻辑卷 u00lv、u01lv、u02lv 和 u03lv。

(2) 在磁盘 /dev/sdc 上创建卷组 archvg,并在其上创建逻辑卷 u04lv。

(3) 在磁盘 /dev/sdd 上创建卷组 bakvg,并在其上创建逻辑卷 baklv。

(4) 在新创建的所有逻辑卷上创建 XFS 文件系统,并将它们挂载到 CentOS 操作系统的文件系统上。

具体操作步骤如下。

(1) 将磁盘初始化为物理卷(physical volume,PV):

```
[root@dbsvr ~]# pvcreate /dev/sdb
[root@dbsvr ~]# pvcreate /dev/sdc
[root@dbsvr ~]# pvcreate /dev/sdd
```

(2) 使用物理卷创建卷组,并在卷组上创建逻辑卷:

```
[root@dbsvr ~]# # 创建卷组 dbvg,并在卷组 dbvg 上创建逻辑卷 u00lv、u01lv、u02lv 和 u03lv
[root@dbsvr ~]# vgcreate dbvg /dev/sdb
[root@dbsvr ~]# lvcreate -L 200G -n u00lv dbvg
[root@dbsvr ~]# lvcreate -L 200G -n u01lv dbvg
[root@dbsvr ~]# lvcreate -L 200G -n u02lv dbvg
[root@dbsvr ~]# lvcreate -L 200G -n u03lv dbvg
[root@dbsvr ~]# # 创建卷组 archvg,并在卷组 archvg 上创建逻辑卷 u04lv
[root@dbsvr ~]# vgcreate archvg /dev/sdc
[root@dbsvr ~]# lvcreate -L 800G -n u04lv archvg
```

```
[root@dbsvr ~]# # 创建卷组 bakvg,并在卷组 bakvg 上创建逻辑卷 baklv
[root@dbsvr ~]# vgcreate bakvg /dev/sdd
[root@dbsvr ~]# lvcreate -L 800G -n baklv bakvg
```

(3) 在步骤(2)中创建的逻辑卷上创建 XFS 文件系统:

```
[root@dbsvr ~]# mkfs.xfs /dev/dbvg/u00lv
[root@dbsvr ~]# mkfs.xfs /dev/dbvg/u01lv
[root@dbsvr ~]# mkfs.xfs /dev/dbvg/u02lv
[root@dbsvr ~]# mkfs.xfs /dev/dbvg/u03lv
[root@dbsvr ~]# mkfs.xfs /dev/archvg/u04lv
[root@dbsvr ~]# mkfs.xfs /dev/bakvg/baklv
```

(4) 创建文件系统挂载的目录,挂载在逻辑卷上创建的 XFS 文件系统:

```
[root@dbsvr ~]# mkdir /u00      # 用于挂接上面创建的 XFS 文件系统/dev/dbvg/u00lv
[root@dbsvr ~]# mkdir /u01      # 用于挂接上面创建的 XFS 文件系统/dev/dbvg/u01lv
[root@dbsvr ~]# mkdir /u02      # 用于挂接上面创建的 XFS 文件系统/dev/dbvg/u02lv
[root@dbsvr ~]# mkdir /u03      # 用于挂接上面创建的 XFS 文件系统/dev/dbvg/u03lv
[root@dbsvr ~]# mkdir /u04      # 用于挂接上面创建的 XFS 文件系统/dev/archvg/u04lv
[root@dbsvr ~]# mkdir /dbbak    # 用于挂接上面创建的 XFS 文件系统/dev/bakvg/baklv
```

(5) 将逻辑卷上的 xfs 文件系统挂载到 CentOS 操作系统的文件系统上:
以用户 root 的身份使用 vi 编辑器,编辑文件/etc/fstab,在文件末尾添加以下内容:

```
/dev/mapper/dbvg-u00lv      /u00        xfs     defaults    0 0
/dev/mapper/dbvg-u01lv      /u01        xfs     defaults    0 0
/dev/mapper/dbvg-u02lv      /u02        xfs     defaults    0 0
/dev/mapper/dbvg-u03lv      /u03        xfs     defaults    0 0
/dev/mapper/archvg-u04lv    /u04        xfs     defaults    0 0
/dev/mapper/bakvg-baklv     /dbbak      xfs     defaults    0 0
```

以用户 root 的身份执行下面的命令,将这些逻辑卷挂载到 CentOS 操作系统的文件系统上:

```
[root@dbsvr ~]# mount -a
[root@dbsvr ~]# df -h
Filesystem                  Size    Used Avail Use% Mounted on
--省略了一些输出
/dev/mapper/dbvg-u00lv      200G    33M  200G   1% /u00
/dev/mapper/dbvg-u01lv      200G    33M  200G   1% /u01
/dev/mapper/dbvg-u02lv      200G    33M  200G   1% /u02
/dev/mapper/dbvg-u03lv      200G    33M  200G   1% /u03
/dev/mapper/archvg-u04lv    800G    33M  800G   1% /u04
/dev/mapper/bakvg-baklv     800G    33M  800G   1% /dbbak
[root@dbsvr ~]#
```

1.3 安装 KingbaseES 数据库

按照 1.2 节准备好逻辑卷和文件系统后，就可以开始安装 KingbaseES 数据库了。

1.3.1 KingbaseES 数据库软件安装包文件和许可证

可以通过电科金仓官网下载对应平台的 KingbaseES 数据库软件的试用版本，也可以联系电科金仓公司的销售人员、售后支持人员或代理商，获取 KingbaseES 数据库软件的安装包文件（ISO 文件）及其 MD5 和 SHA1 校验值。

例如，从电科金仓官网下载安装包文件 KingbaseES_V009R001C001B0025_Lin64_install.iso，并且可以在官网可以查到这个安装包文件的 MD5 和 SHA1 校验值。

（1）MD5 校验值为 DF45575DFB4CEA1E2AD4A8F2905118D8。

（2）SHA1 校验值为 9B270E0C948E6523208C058EB7F9166999210DD7。

使用用户 root，创建目录 /opt/media，用于存放 KingbaseES 数据库软件安装包文件：

```
[root@dbsvr ~]# mkdir -p /opt/media
```

将 KingbaseES 数据库安装包文件 KingbaseES_V009R001C001B0025_Lin64_install.iso 上传到服务器 dbsvr 的目录 /opt/media 下，执行下面的命令，查看其 MD5 校验值：

```
[kingbase@dbsvr ~]$ cd /opt/media/
[kingbase@dbsvr media]# md5sum KingbaseES_V009R001C001B0025_Lin64_install.iso
df45575dfb4cea1e2ad4a8f2905118d8  KingbaseES_V009R001C001B0025_Lin64_install.iso
[kingbase@dbsvr media]#
```

执行下面的命令，查看其 SHA1 校验值：

```
[kingbase@dbsvr media]$ sha1sum KingbaseES_V009R001C001B0025_Lin64_install.iso
9b270e0c948e6523208c058eb7f9166999210dd7  KingbaseES_V009R001C001B0025_Lin64_install.iso
[kingbase@dbsvr media]$
```

从输出可以看到，在服务器 dbsvr 上根据安装包文件重新计算的 MD5 和 SHA1 校验值，与官方公布的值是一致的，说明安装包文件是完整和安全的。

如果重新计算的 MD5 和 SHA1 校验值与官方公布的值不一样，则说明安装文件可能不完整，安装过程可能无法正常完成，或者介质已经被篡改，即使完成安装也不安全。

许可证（license）是电科金仓公司给用户提供的授权文件，其中会对数据库有效日期、发布类型、最大并发连接数、MAC 地址（media access control address）等信息进行设置。运行 KingbaseES 数据库需要许可证，如果在许可证文件中信息与安装环境中的相关信息不匹配，数据库将无法启动。

用户购买 KingbaseES 数据库后，会获得相应的商业版软件许可证（文件）。如果只是想临时测试 KingbaseES 数据库，从电科金仓官网可以下载用于测试的开发版软件许可证或企业版软件许可证。开发版软件许可证的最大连接数被限制为 10 个连接，有效期为 365 天；企业版软件许可证不限制连接数，但有效期为 90 天。

本书的实验环境基于电科金仓官网下载的软件许可证(开发版和企业版)。下载后，上传到服务器 dbsvr 的目录/opt/media 下：

```
[root@dbsvr ~]# cd /opt/media/
[root@dbsvr media]# ls -l
total 2245408
-rw-r--r--. 1 root root 2299289600 Feb 16 14:51 KingbaseES_V009R001C001B0025_
Lin64_install.iso
-rw-r--r--. 1 root root       2735 Feb 16 14:51 license_企业版.zip
-rw-r--r--. 1 root root       2730 Feb 16 14:51 license_开发版.zip
[root@dbsvr media]#
```

1.3.2 KingbaseES 数据库安装前的准备工作

按如下的方案来安装 KingbaseES 数据库。
(1) KingbaseES 数据库的二进制程序将被安装在目录/u00/Kingbase/ES/V9/kingbase 下。
(2) KingbaseES 数据库的数据目录位于目录/u00/Kingbase/ES/V9/data 下。
在安装 KingbaseES 数据库之前，还需要完成以下准备工作。
(1) 创建安装 KingbaseES 数据库的用户组 dba 和用户 kingbase。
使用用户 root 执行下面的命令，创建用户组 dba 和用户 kingbase，并将用户 kingbase 的密码设置为 kingbase123(在生产环境中需要设置为强度更高的用户密码)：

```
[root@dbsvr ~]# groupadd dba -g 3000
[root@dbsvr ~]# groupadd kingbase  -g 3001
[root@dbsvr ~]# useradd   kingbase  -g 3000 -G 3001 -u 3001
[root@dbsvr ~]# echo "kingbase123"|passwd --stdin kingbase
[root@dbsvr ~]# id kingbase
[root@dbsvr ~]# userdel -r test
```

上面的最后一条命令删除了 CentOS 操作系统安装过程中创建的用户 test。
(2) 配置 IP 地址-主机名映射文件/etc/hosts。
使用用户 root 执行下面的命令，在文件/etc/hosts 中添加以下三条记录：

```
[root@dbsvr ~]# cat >>/etc/hosts<<EOF
192.168.100.22     dbsvr
192.168.100.19     dbtest
192.168.100.18     dbclient
EOF
[root@dbsvr ~]#
```

(3) 创建安装 KingbaseES 数据库需要的目录。
使用用户 root 执行下面的命令，创建下面的目录并修改目录的属主和属组，KingbaseES 数据库将使用这些目录：

```
[root@dbsvr ~]# mkdir -p /u00/Kingbase/ES/V9/kingbase
                          # KingbaseES 数据库二进制程序
```

```
[root@dbsvr ~]# mkdir -p /u00/Kingbase/ES/V9/data
                                       # KingbaseES 数据库的数据目录、控制文件
[root@dbsvr ~]# mkdir -p /u01/Kingbase/ES/V9/data
                                       # KingbaseES 数据库的用户表空间、控制文件
[root@dbsvr ~]# mkdir -p /u02/Kingbase/ES/V9/data
                                       # KingbaseES 数据库的用户表空间、控制文件
[root@dbsvr ~]# mkdir -p /u03/Kingbase/ES/V9/data
                                       # KingbaseES 数据库的用户表空间
[root@dbsvr ~]# mkdir -p /u04/Kingbase/ES/V9/archivelog
                                       # KingbaseES 数据库的归档日志
[root@dbsvr ~]# mkdir -p /dbbak/pbak    # KingbaseES 数据库的物理备份
[root@dbsvr ~]# mkdir -p /dbbak/lbak    # KingbaseES 数据库的逻辑备份
[root@dbsvr ~]# chown -R kingbase:dba /u0?/Kingbase
[root@dbsvr ~]# chown -R kingbase:dba /dbbak
```

说明：在1.3.3节安装 KingbaseES 数据库时，只会用到目录/u00/Kingbase/ES/V9，其他的目录将在附录 B"安装 KingbaseES 单机数据库的最佳实践"中使用。

（4）修改操作系统内核参数。

使用用户 root 修改 Linux 操作系统的内核参数（文件系统、共享内存、信号量和网络等）：

```
[root@dbsvr ~]# cat>>/etc/sysctl.conf<<EOF
fs.aio-max-nr= 1048576
fs.file-max= 6815744
kernel.shmall= 2097152
kernel.shmmax= 17179869184
kernel.shmmni= 4096
kernel.sem= 250 32000 100 128
net.ipv4.ip_local_port_range= 9000 65500
net.core.rmem_default= 262144
net.core.rmem_max= 4194304
net.core.wmem_default= 262144
net.core.wmem_max= 1048576
EOF
```

（5）修改 kingbase 用户的资源限制参数。

以用户 root 的身份执行下面的命令，配置用于安装 KingbaseES 数据库的用户 kingbase 的堆栈限制、打开文件数限制、用户的最大进程数限制：

```
[root@dbsvr ~]# cat>>/etc/security/limits.conf<<EOF
kingbase soft nofile    1048576
kingbase hard nofile    1048576
kingbase soft nproc     131072
kingbase hard nproc     131072
kingbase soft stack     10240
kingbase hard stack     32768
kingbase soft core      unlimited
kingbase hard core      unlimited
EOF
```

（6）修改 RemoveIPC 参数。

systemd-logind 服务中引入了一个特性：当用户退出系统后，会删除所有与这个用户有关的进程间通信（interprocess communication，IPC）对象。该特性由 /etc/systemd/logind.conf 文件中的 RemoveIPC 参数控制。在某些操作系统中会默认打开该参数，可能会造成程序信号丢失的问题，因此需要关掉这个参数，使用 vi 编辑器，编辑文件 /etc/systemd/logind.conf，将

```
#RemoveIPC=no
```

修改为

```
RemoveIPC=no
```

（7）使用用户 root 执行 reboot 命令，重新启动 KingbaseES 数据库服务器。

1.3.3 安装 KingbaseES 数据库软件

KingbaseES 数据库软件安装开始前，需要首先挂载 KingbaseES 数据库安装包文件。使用用户 kingbase（密码是 kingbase123）登录到 CentOS 7 操作系统的图形用户界面（graphical user interface，GUI），打开一个 Linux 操作系统的终端，执行如下命令，将 KingbaseES 数据库的安装包文件挂载到文件系统目录 /mnt/iso 上：

```
[kingbase@dbsvr ~]$ su -
Password: # 输入用户 root 的密码 "root123"
[root@dbsvr ~]# mkdir -p /mnt/iso                #创建挂接点目录 /mnt/iso
[root@dbsvr ~]# cd /opt/media
[root@dbsvr media]#mount -t iso9660 -o loop KingbaseES_V009R001C001B0025_Lin64_install.iso /mnt/iso
```

继续执行下面的命令，解压缩 KingbaseES 数据库的开发版软件许可证：

```
[root@dbsvr media]# unzip license_开发版.zip
[root@dbsvr media]# cd license_34151
[root@dbsvr license_34151]# cp license_34151_0.dat /home/kingbase
```

打开另外一个终端，以用户 kingbase 的身份启动 KingbaseES 数据库安装程序：

```
[kingbase@dbsvr ~]$ sh /mnt/iso/setup.sh
```

稍等片刻，将出现启动安装 KingbaseES 数据库的对话框，单击 OK 按钮，打开 KingbaseES 数据库安装欢迎界面，如图 1-1 所示。单击 Next 按钮出现如图 1-2 所示的接受 KingbaseES 数据库许可协议的界面，勾选 I accept the terms of License Agreement 复选框后，单击 Next 按钮，将出现如图 1-3 所示的 KingbaseES 数据库的安装类型选择界面，其中：

（1）Full 选项安装所有的组件，包括数据库和客户端；

（2）Client 选项只安装客户端组件；

（3）Custom 选项由用户自己决定安装哪些组件。

如图 1-3 所示，选中 Full 单选按钮，安装所有的 KingbaseES 数据库组件，单击 Next 按钮。

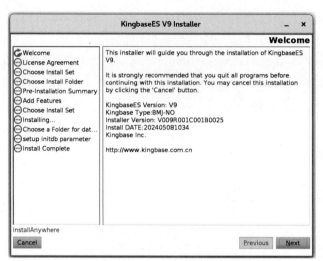

图 1-1　KingbaseES 数据库安装欢迎界面

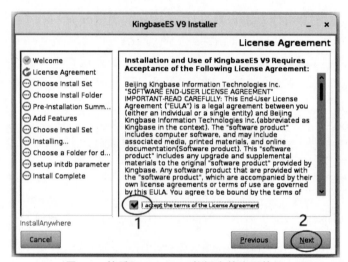

图 1-2　接受 KingbaseES 数据库的许可协议

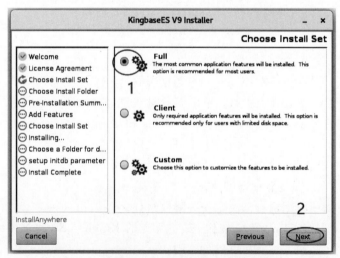

图 1-3　选择 KingbaseES 数据库的安装类型

接下来按照如图 1-4～图 1-6 所示进行操作，导入 KingbaseES 数据库的软件许可证文件。

图 1-4　导入软件许可证界面 1

图 1-5　导入软件许可证对话框

如图 1-7 所示，按照 1.3.2 节的规划，将 KingbaseES 数据库的二进制软件安装到目录 /u00/Kingbase/ES/V9/kingbase 下，单击 Next 按钮。

接下来将出现如图 1-8 所示的界面，显示 KingbaseES 数据库的安装配置信息。确定配置没有问题，可以单击 Install 按钮，开始安装 KingbaseES 数据库，如图 1-9 所示。

KingbaseES 数据库安装结束，将出现如图 1-10 所示的界面，开始进行 KingbaseES 数据库的初始化。输入 KingbaseES 数据库数据目录的位置信息（目录 /u00/Kingbase/ES/V9/data），单击 Next 按钮，设置 KingbaseES 数据库的一些安装配置信息。

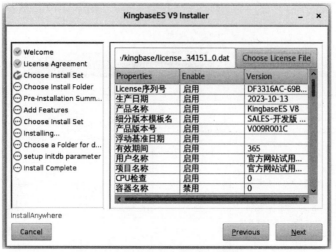

图 1-6　导入软件许可证界面 2

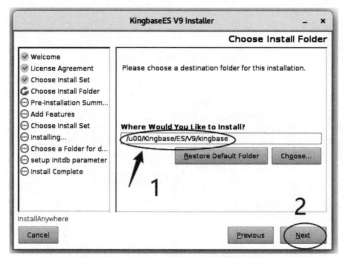

图 1-7　KingbaseES 数据库的安装目录

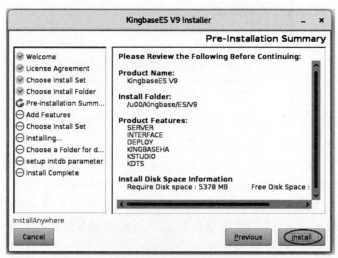

图 1-8　KingbaseES 数据库安装配置信息小结

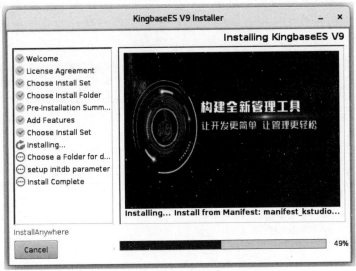

图 1-9　开始安装 KingbaseES 数据库

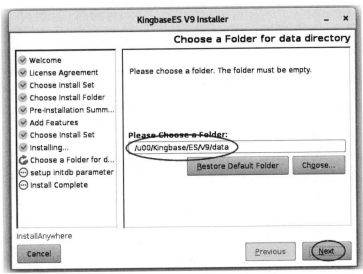

图 1-10　KingbasES 数据库的数据目录

（1）为数据库用户 system 设置密码。
（2）设置数据库的编码方式。
（3）设置数据库的兼容模式。
（4）设置字母大小写是否敏感。
（5）设置数据库块的大小。

如图 1-11 所示，将数据库用户 system 的密码设置为 Passw0rd（注意，此处把小写字母 o，用数字 0 进行了替换），密码需要输入 2 次。

如图 1-12 所示，设置数据库的编码方式。当前 KingbaseES 数据库有三种编码选项：UTF8、GBK 和 GB18030。本书选择数据库默认选项 UTF8。设置数据库的兼容模式。本书选择 ORACLE 选项，即在 KingbaseES 数据库中使用与 Oracle 数据库兼容的语法和数据字典视图。设置数据库的字母大小写是否敏感。对于 Oracle 和 PostgreSQL 数据库，字母大小写是敏感的；对于 SQL Server 和 MySQL 数据库，字母大小写是不敏感的。本书启用字母大小写敏感。

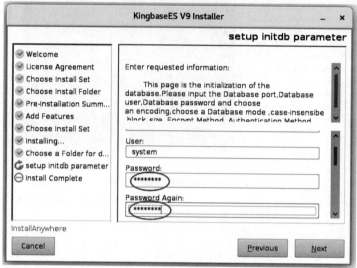

图 1-11　设置用户 system 的密码

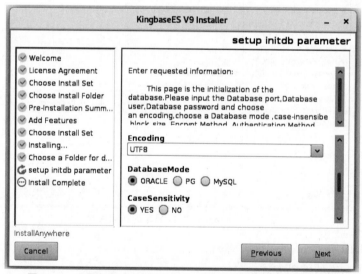

图 1-12　设置数据库的编码方式、兼容模式和字母大小写敏感

如图 1-13 所示，设置数据库的块大小。数据库块的大小有三种选择：8KB、16KB 和 32KB。如果数据库用于联机事务处理（online transaction processing，OLTP），可以使用默认值 8KB；如果数据库用于联机分析处理（online analytical processing，OLAP），数据库块的大小可以为 16KB 或 32KB。本书选择数据库块的大小为 8KB，并设置加密方法为 sm4，鉴权方法为 scram-sha-256。参数设置完成后，单击 Next 按钮，开始创建 KingbaseES 数据库集簇（database cluster）。创建完数据库集簇后，出现如图 1-14 所示的界面，说明已经成功完成 KingbaseES 数据库的安装。

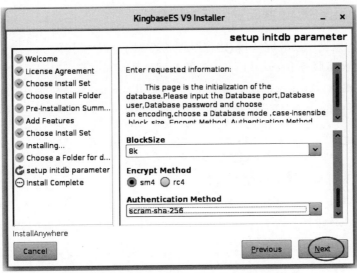

图 1-13　设置数据库块的大小、加密和鉴权方法

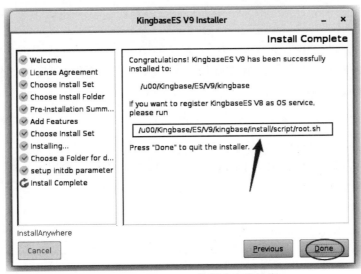

图 1-14　KingbaseES 数据库安装结束

如果想把 KingbaseES 数据库配置为操作系统服务，可以使用用户 root 执行脚本 /u00/Kingbase/ES/V9/kingbase/install/script/root.sh：

```
[root@dbsvr ~]# /u00/Kingbase/ES/V9/kingbase/install/script/root.sh
Starting KingbaseES V9:
waiting for server to start.... done
server started
KingbaseES V9 started successfully
[root@dbsvr ~]#
```

执行完 root.sh 脚本后，单击如图 1-14 所示的 Done 按钮，退出 KingbaseES 数据库的安装程序。

1.3.4 KingbaseES 数据库安装后的操作

KingbaseES 数据库安装完成后，建议继续执行如下配置操作。

(1) 打开防火墙服务 firewalld.service 的数据库访问端口 54321。

由于 CentOS 7 操作系统的防火墙服务 firewalld.service 默认是启用的，因此需要使用用户 root 执行下面的命令，打开防火墙服务 firewalld.service 的数据库访问端口 54321：

```
[root@dbsvr ~]# firewall-cmd --zone=public --add-port=54321/tcp --permanent
[root@dbsvr ~]# firewall-cmd --reload
```

(2) 为用户 kingbase 配置环境变量。

为了便于管理 KingbaseES 数据库，需要 DBA 使用 vi 编辑器，在 Bash shell 的初始化文件 /home/kingbase/.bashrc 中为用户 kingbase 添加以下环境变量：

```
export KINGBASE_HOME=/u00/Kingbase/ES/V9/kingbase
export PATH=$KINGBASE_HOME/Server/bin:$PATH
export PATH=$KINGBASE_HOME/ClientTools/bin:$KINGBASE_HOME/ClientTools/guitools/KStudio:$PATH
export LD_LIBRARY_PATH=$KINGBASE_HOME/Server/lib:$LD_LIBRARY_PATH
export KINGBASE_DATA=/u00/Kingbase/ES/V9/data
export KINGBASE_PORT=54321
```

为了使添加的环境变量生效，需要让用户 kingbase 退出系统，然后重新登录系统，或者在终端上执行下面的命令：

```
[kingbase@dbsvr ~]$ source /home/kingbase/.bashrc
```

至此，KingbaseES 单机数据库安装完成。

使用用户 kingbase，验证 KingbaseES 数据库是否已经正确安装。

(1) 执行下面命令，查看 KingbaseES 数据库的运行状态：

```
[kingbase@dbsvr ~]$ sys_ctl status
sys_ctl: server is running (PID: 14751)
/u00/Kingbase/ES/V9/kingbase/KESRealPro/V009R001C001B0025/Server/bin/kingbase "-D" "/u00/Kingbase/ES/V9/data"
[kingbase@dbsvr ~]$
```

(2) 执行下面命令，关闭 KingbaseES 数据库：

```
[kingbase@dbsvr ~]$ sys_ctl stop
waiting for server to shut down.... done
server stopped
[kingbase@dbsvr ~]$
```

(3) 执行下面的命令，查看 KingbaseES 数据库的运行状态：

```
[kingbase@dbsvr ~]$ sys_ctl status
sys_ctl: no server running
[kingbase@dbsvr ~]$
```

从输出可以看到，目前 KingbaseES 数据库没有运行。

(4) 执行下面命令，启动 KingbaseES 数据库：

```
[kingbase@dbsvr ~]$ sys_ctl start
waiting for server to start....2024-02-29 13:31:02.411 GMT [15417] WARNING:  max_connections should be less than or equal than 10 (restricted by license)
--忽略了一些输出
server started
[kingbase@dbsvr ~]$ sys_ctl status
sys_ctl: server is running (PID: 15417)
/u00/Kingbase/ES/V9/kingbase/KESRealPro/V009R001C001B0025/Server/bin/kingbase
[kingbase@dbsvr ~]$
```

从输出可以看到，目前 KingbaseES 数据库正在运行中。

(5) 执行下面 ksql 命令和 SQL 语句：

```
[kingbase@dbsvr ~]$ ksql -d test -U system
Password for user system:                   --输入数据库用户 system 的密码 Passw0rd
Type "help" for help.
test=# SELECT  version();
                    version
-----------------------------------------------------------------
KingbaseES V009R001C001B0025 on x86_64-pc-linux-gnu,
compiled by gcc (GCC) 4.8.5 20150623 (Red Hat 4.8.5-28), 64-bit
(1 row)                                     --已经对输出进行了排版美化
test=# \q
[kingbase@dbsvr ~]$
```

从输出可以看到，KingbaseES 数据库已经安装成功。

1.4 本书的实验环境

为了方便完成本书后面章节的 KingbaseES 数据库实战任务，请完成本节的配置任务。

1.4.1 配置实验环境

(1) 执行下面的命令，创建 KingbaseES 数据库用户密码文件 /home/kingbase/.kbpass：

```
[kingbase@dbsvr ~]$ cat >/home/kingbase/.kbpass<<EOF
192.168.100.22:54321:*:system:Passw0rd
localhost:54321:*:system:Passw0rd
EOF
[kingbase@dbsvr ~]$ chmod 600 /home/kingbase/.kbpass
```

创建完 KingbaseES 数据库用户密码文件后，使用 ksql 命令以用户 system 的身份登录到 KingbaseES 数据库时，不会再提示输入用户 system 的密码：

```
[kingbase@dbsvr ~]$ ksql -d test -U system
Type "help" for help.
test=# \q
```

```
[kingbase@dbsvr ~]$
```

(2) 配置 ksql 的提示符。

为了方便在 ksql 中执行 SQL 语句时,可以看到本会话(session)连接的用户名和数据库名,以用户 kingbase 的身份,使用 vi 编辑器创建 ksql 的初始化环境文件/home/kingbase/.ksqlrc,内容如下:

```
\set PROMPT1 '%n@%/%R%# '
```

文件保存之后,执行 ksql 命令访问 KingbaseES 数据库时,ksql 的提示符如下:

```
[kingbase@dbsvr ~]$ ksql -d test -U system
Type "help" for help.
system@test=# \q
[kingbase@dbsvr ~]$
```

从输出可以看到,ksql 的提示符被设置为"登录的数据库用户名@登录的数据库名=#"的形式。

(3) 创建工具脚本 kbps。

为了方便查看 KingbaseES 数据库中的所有进程,以用户 root 的身份使用 vi 编辑器创建一个工具脚本命令文件/usr/bin/kbps,内容如下:

```
#/bin/bash
ps -ef |grep $(head -1 /u00/Kingbase/ES/V9/data/kingbase.pid) |grep -v grep
```

创建完工具脚本命令文件后,使用用户 root 执行下面的命令,让脚本命令文件拥有可执行权限:

```
[root@dbsvr ~]# chmod 755 /usr/bin/kbps
```

之后,只要 KingbaseES 数据库在服务器上运行,就可以使用用户 kingbase 执行 kbps 命令,查看 KingbaseES 数据库的所有进程,如图 1-15 所示。

```
[kingbase@dbsvr ~]$ kbps
kingbase  15417     1  0 21:31 ?        00:00:00 /u00/Kingbase/ES/V9/kingbase/KESRealPro/V009R001C001B0025/Server/bin/kingbase
kingbase  15418 15417  0 21:31 ?        00:00:00 kingbase: logger
kingbase  15420 15417  0 21:31 ?        00:00:00 kingbase: checkpointer
kingbase  15421 15417  0 21:31 ?        00:00:00 kingbase: background writer
kingbase  15422 15417  0 21:31 ?        00:00:00 kingbase: walwriter
kingbase  15423 15417  0 21:31 ?        00:00:00 kingbase: autovacuum launcher
kingbase  15424 15417  0 21:31 ?        00:00:00 kingbase: stats collector
kingbase  15425 15417  0 21:31 ?        00:00:00 kingbase: kwr collector
kingbase  15426 15417  0 21:31 ?        00:00:00 kingbase: ksh writer
kingbase  15427 15417  0 21:31 ?        00:00:00 kingbase: ksh collector
kingbase  15428 15417  0 21:31 ?        00:00:00 kingbase: logical replication launcher
[kingbase@dbsvr ~]$
```

图 1-15 KingbaseES 数据库的进程

1.4.2 导入 TPC-H 测试数据集

本书的实验使用了 TPC-H(Transaction Processing Performance Council benchmark H)数据集。TPC-H 是事务处理性能委员会(Transaction Processing Performance Council,TPC)设计的用于评估关系数据库管理系统的决策支持型工作负载的基准测试。它模拟了一个复杂的商业决策支持系统场景,通过执行一系列复杂的 SQL 语句来测试数据库管理系统的性能。

TPC-H 的数据库模式(schema)由 Part(零件)表、Supplier(供应商)表、Partsupp(零件供应商联系)表、Customer(顾客)表、Nation(国家)表、Region(地区)表、Orders(订单)表和 Lineitem(订单明细)表 8 个基本表组成。这 8 个表的关系如图 1-16 所示,图中的箭头表示表之间的参照完整性约束。

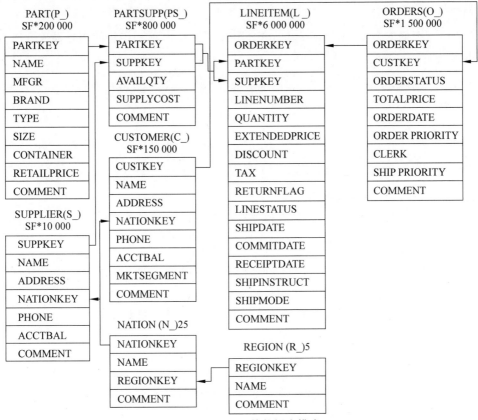

图 1-16　TPC-H 的数据库模式

TPC-H 的表中元组的数量由比例因子(scale factor,SF)定义,SF 描述了数据库规模的缩放比例。TPC-H 中 Nation 表和 Region 表的大小是固定的,Region 表有 5 个元组,Nation 表有 25 个元组,其他表的元组个数根据 SF 的值伸缩。本书的 TPC-H 测试数据集使用 SF＝1 生成,各个表中元组数如表 1-1 所示。

表 1-1　TPC-H 测试数据集各表的元组数

表　　名	元组数/个	表　　名	元组数/个
Region	5	Partsupp	800 000
Nation	25	Customer	150 000
Supplier	10 000	Orders	1 500 000
Part	200 000	Lineitem	6 000 000

针对决策支持系统,TPC-H 设计了 22 个商业分析查询,每个查询都模拟了特定的商业

决策支持任务，如销售分析、库存管理、客户支持等。这组 SQL 语句会进行多表连接、多表聚合、子查询和复杂条件过滤等操作。

按如下步骤，将 TPC-H 测试数据集导入数据库 test 的模式 tpch 下。

（1）执行下面的 ksql 命令，在数据库 test 中创建模式 tpch：

```
[kingbase@dbsvr ~]$ ksql -d test -U system -c "CREATE SCHEMA tpch"
```

（2）执行下面的命令，创建 KingbaseES 服务名 sntest，下一步将使用它导入 TPC-H 测试数据集：

```
[kingbase@dbsvr ~]$ cat >/home/kingbase/.sys_service.conf<<EOF
[sntest]
host=192.168.100.22
port=54321
dbname=test
EOF
```

（3）导入 TPC-H 测试数据集。

将本书的资源文件 tpch.dmp 上传到/home/kingbase 目录下，使用用户 kingbase 执行下面的命令，将 TPC-H 测试数据集导入数据库 test 的模式 tpch：

```
[kingbase@dbsvr ~]$ imp system/Passw0rd@sntest file=tpch.dmp fromuser=tpch touser=tpch ignore=y
```

执行下面的命令，可以验证是否成功地导入了 TPC-H 测试数据集：

```
[kingbase@dbsvr ~]$ ksql -d test -U system
system@test=# ANALYZE;
system@test=# SELECT nspname AS schema_name, relname AS table_name,
test-#           ROUND(reltuples) AS estimated_row_count
test-#       FROM pg_class c JOIN pg_namespace n ON n.oid = c.relnamespace
test-#       WHERE nspname = 'tpch' AND c.relkind = 'r';
 schema_name | table_name | estimated_row_count
-------------+------------+---------------------
 tpch        | nation     |                  25
 tpch        | customer   |              150000
 tpch        | orders     |             1500000
 tpch        | lineitem   |             6001329
 tpch        | partsupp   |              800000
 tpch        | region     |                   5
 tpch        | part       |              200000
 tpch        | supplier   |               10000
(8 rows)
system@test=# \q
[kingbase@dbsvr ~]$
```

（4）运行 TPC-H 测试语句。

将本书的资源文件 tpch_sql.2024ok.tar 上传到/home/kingbase 目录下，使用用户 kingbase 执行下面的命令，解压缩文件包 tpch_sql.2024ok.tar：

```
[kingbase@dbsvr ~]$ tar xf tpch_sql.2024ok.tar
[kingbase@dbsvr ~]$ cd tpch_sql/
[kingbase@dbsvr tpch_sql]$ ls
stream_0_10.sql   stream_0_14.sql   stream_0_18.sql   stream_0_21.sql   stream_0_4.sql   stream_0_8.sql
stream_0_11.sql   stream_0_15.sql   stream_0_19.sql   stream_0_22.sql   stream_0_5.sql   stream_0_9.sql
stream_0_12.sql   stream_0_16.sql   stream_0_1.sql    stream_0_2.sql    stream_0_6.sql   tpch.sql
stream_0_13.sql   stream_0_17.sql   stream_0_20.sql   stream_0_3.sql    stream_0_7.sql   tpchtest
[kingbase@dbsvr tpch_sql]$
```

解压缩文件包后,可以看到一些 TPC-H 的 SQL 测试语句和 shell 测试脚本。

以用户 kingbase 的身份运行测试脚本 tpchtest,可以仿真一个有一定负载活动的生产数据库:

```
[kingbase@dbsvr tpch_sql]$ sh tpchtest
```

从输出可以看到,执行该脚本会持续不断地运行 TPC-H 测试的 22 条 SQL 语句。

要停止运行 TPC-H 测试,可以持续按 Ctrl+C 组合键,终止 tpchtest 脚本的执行。

(5) 备份虚拟机。

执行下面的命令,关闭虚拟机:

```
[kingbase@dbsvr ~]$ sys_ctl stop
waiting for server to shut down.... done
server stopped
[kingbase@dbsvr ~]$ su -
Password: #输入用户 root 的密码"root123"
Last login: Fri Feb 16 18:56:43 CST 2024 on pts/1
[root@dbsvr ~]# shutdown -h now
```

使用 WinRAR 软件将安装好 KingbaseES 数据库的虚拟机所在目录(本书目录为 G:\dbserver)压缩为 dbserver.rar 文件,压缩完成后,改名为 dbserver-2DB-2data.rar。

1.4.3 命令行提示符约定

本书的命令有两种:①在 shell 提示符下执行操作系统命令;②在 ksql 提示符下执行 SQL 语句或 ksql 的元命令。

本书对命令行提示符约定如下。

(1) 如下的 shell 提示符:

```
[root@dbsvr ~]#
```

表示在服务器 dbsvr 上,以操作系统超级用户 superuser root 的身份,在用户 root 的主目录下执行操作系统命令,其中的符号"~"表示用户 root 的主目录/root。

(2) 如下的 shell 提示符:

```
[kingbase@dbsvr ~]$
```

表示在服务器 dbsvr 上，以操作系统普通用户 kingbase 的身份，在用户 kingbase 的主目录下执行操作系统命令。shell 提示符中的符号"～"表示用户 kingbase 的主目录/home/kingbase。用户 kingbase 是 KingbaseES 数据库的 DBA。

（3）如下的 ksql 提示符：

```
system@test=#
```

表示使用 ksql，以数据库用户 system 的身份连接到数据库 test。ksql 提示符中的符号"=#"表示数据库用户 system 是 KingbaseES 数据库的超级用户。

（4）如下的 ksql 提示符：

```
newuser@newdb=>
```

表示使用 ksql，以数据库用户 newuser 的身份连接到数据库 newdb。ksql 提示符中的符号"=>"表示数据库用户 newuser 是 KingbaseES 数据库的普通用户。

1.4.4　本书提供的资源文件

本书提供了以下资源文件。

（1）KingbaseES V9 数据库安装包 Linux 操作系统发行版（x86-64 平台）。

（2）KingbaseES V9 数据库安装包 Windows 操作系统发行版（x86-64 平台）。

（3）KingbaseES V9 数据库测试用开发版软件许可证。

（4）KingbaseES V9 数据库测试用企业版软件许可证。

（5）资源文件"KingBaseES 数据库实战脚本"是本书各章节的命令和 SQL 语句。

（6）资源文件 tpch.dmp 是 TPC-H 测试数据集。

（7）资源文件 tpch.sql.2024ok.tar 是 TPC-H 测试的 SQL 语句和运行脚本。

（8）资源文件 dbserver-1OS.rar 是安装好 CentOS 7 操作系统的虚拟机备份文件，运行它可用来安装 KingbaseES 数据库（第 1 章），Linux 操作系统的用户 root 的密码是 root123。

（9）资源文件 dbserver-2DB-2data.rar 是安装好 KingbaseES 数据库，并且已经导入 TPC-H 测试数据集的虚拟机备份文件，可以使用它来完成本书第 2～7 章的实战任务。Linux 操作系统的用户 root 的密码是 root123；用户 kingbase 的密码是 kingbase123；KingbaseES 数据库用户 system 的密码是 Passwr0d（密码已经将 Password 中的小写字母 o 替换为数字 0）。

请在第 2～7 章，每章的开始，重新解压缩资源文件 dbserver-2DB-2data.rar，启动运行后，按照 4.4 节的操作方法，将 KingbaseES 数据库的软件许可证更新为企业版软件许可证（除限制使用 90 天外无任何其他限制）。

1.4.5　制作自己的虚拟机备份

为了在学习和实战 KingbaseES 数据库管理的过程中提高效率，建议创建以下虚拟机备份。

（1）制作 KingbaseES 数据库的客户端虚拟机备份文件 dbclient-2OK.rar，制作方法参见 2.1.1 节。

（2）制作 KingbaseES 数据库备份恢复测试虚拟机的备份文件 dbtest-2DB.rar。

① 创建1个虚拟机(4核CPU、16GB内存、1个主机模式(Host-only)类型的网卡、1个网络地址转(network address translation,NAT)类型的网卡、1个900GB大小的磁盘、3个1200GB大小的磁盘)。

② 按照附录A安装CentOS 7操作系统(将主机名设置为dbtest,对外服务的IP地址是192.168.100.19/24)。

③ 按照1.2.3节准备安装KingbaseES数据库的存储空间。

④ 按照1.3节安装KingbaseES数据库。

⑤ 执行下面的命令,关闭虚拟机:

```
[kingbase@dbtest ~]$ sys_ctl stop
waiting for server to shut down.... done
server stopped
[kingbase@dbtest ~]$ su -
Password: #输入用户root的密码"root123"
Last login: Fri Feb 16 18:56:43 CST 2024 on pts/1
[root@dbtest ~]# shutdown -h now
```

使用winrar软件将安装好KingbaseES数据库的虚拟机所在目录(本书目录为G:\dbtest)压缩为dbtest.rar文件,压缩完成后,改名为dbtest-2DB.rar。

(3) 制作KingbaseES数据库的最佳实践虚拟机备份文件dbsvr-3DB-BestPractice.rar。制作的步骤如下。

① 使用本书提供的虚拟机资源文件dbserver-1OS.rar,按照1.2.3节准备安装KingbaseES数据库的存储空间。

② 按照1.3节安装KingbaseES数据库。

③ 按照附录B完成KingbaseES数据库的最佳实践安装。

完成附录B的所有操作之后,将拥有一个名为dbserver-3DB-BestPractice.rar的虚拟机备份文件。

在第8~13章,主要使用虚拟机备份文件dbserver-3DB-BestPractice.rar进行测试。在这些章的开始处,均需重新释放并运行虚拟机备份文件dbserver-3DB-BestPractice.rar。

注意:对于这些安装有KingbaseES数据库的虚拟机备份,在解压缩启动运行之后,需要将KingbaseES数据库的软件许可证更新为企业版软件许可证(参见4.4节)。

第 2 章

KingbaseES 数据库客户端工具

KingbaseES 数据库的客户端主要包括命令行工具(ksql)和图像化开发管理工具(KStudio),可以单独安装在与 KingbaseES 数据库服务器不同的计算机上,可以是 Windows 操作系统,也可以是 Linux 操作系统。

2.1 安装 KingbaseES 数据库客户端

2.1.1 在 Linux 操作系统上安装 KingbaseES 数据库客户端

需要准备一台 x86_64 架构的计算机。如果无法获得,可以使用 VMware Workstation 软件,创建 1 个 CentOS 7 操作系统的虚拟机(配置:4 个以上 CPU 核芯、4GB 内存以上、1 个 Host-only 类型的网络、1 个 NAT 类型的网卡、1 个 900GB 大小的磁盘),作为安装 KingbaseES 数据库客户端的测试环境。

准备好计算机后,可以按照附录 A 为其安装 CentOS 7 操作系统(将主机名设置为 dbclient,网卡 ens33 作为对外服务的网卡,IP 地址设置为 192.168.100.18/24,另一个网卡 ens34 采用动态主机配置协议(dynamic host configuration protocol,DHCP)自动获取 IP 配置)。

接下来,使用用户 root 按照下面步骤完成安装 KingbaseES 数据库客户端的准备工作。

(1) 执行下面的命令,创建用户组 dba 和用户 kingbase,并将用户 kingbase 的密码设置为 kingbase123(在生产环境中需要设置为强度更高的用户密码):

```
[root@dbclient ~]# groupadd dba -g 3000
[root@dbclient ~]# groupadd kingbase  -g 3001
[root@dbclient ~]# useradd  kingbase  -g 3000 -G 3001 -u 3001
[root@dbclient ~]# echo "kingbase123"|passwd --stdin kingbase
[root@dbclient ~]# id kingbase
[root@dbclient ~]# userdel -r test
```

最后一条命令删除了安装操作系统过程中创建的用户 test。

(2) 配置 IP 地址-主机名映射文件/etc/hosts。

执行下面的命令,在文件/etc/hosts 中添加三条记录:

```
[root@dbclient ~]# cat >>/etc/hosts<<EOF
```

```
192.168.100.22    dbsvr
192.168.100.19    dbtest
192.168.100.18    dbclient
EOF
```

（3）创建安装 KingbaseES 数据库客户端需要的目录。

执行下面的命令，创建下面的目录并修改目录的属主和属组，KingbaseES 数据库客户端将使用这些目录：

```
[root@dbclient ~]# mkdir -p /opt/media        #存放 KingbaseES 数据库安装包文件
[root@dbclient ~]# mkdir -p /u00/Kingbase/ES/V9/client
                                              #存放 KingbaseES 数据库客户端二进制程序
[root@dbclient ~]# chown -R kingbase:dba /u00/Kingbase
```

（4）将 KingbaseES 数据库的安装包文件 KingbaseES_V009R001C001B0025_Lin64_install.iso 上传到客户端主机的目录/opt/media 下，执行下面的命令，挂载 KingbaseES 数据库安装包文件：

```
[root@dbclient ~]# mkdir -p /mnt/iso          #创建挂载点目录/mnt/iso
[root@dbclient ~]# cd /opt/media
[root@dbclient media]# mount -t iso9660 -o loop KingbaseES_V009R001C001B0025_Lin64_install.iso /mnt/iso
```

完成以上准备工作后，需要使用用户 kingbase（密码是 kingbase123）登录到 CentOS 7 操作系统的 GUI，打开一个终端，执行下面的命令，启动 KingbaseES 数据库安装程序：

```
[kingbase@dbsvr ~]$ sh /mnt/iso/setup.sh
```

安装 KingbaseES 数据库客户端的方法和安装 KingbaseES 数据库的方法基本相同，区别有 2 个：第 1 个是选择安装 KingbaseES 数据库时选中 Client 单选按钮（图 2-1），只安装 KingbaseES 数据库客户端组件；第 2 个是将客户端的二进制程序安装到目录/u00/Kingbase/ES/V9/client 下（图 2-2）。

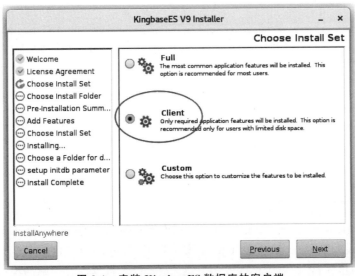

图 2-1 安装 KingbaseES 数据库的客户端

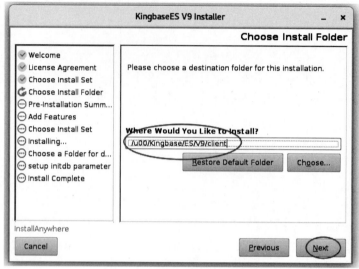

图 2-2 设置客户端的安装目录

安装完成 KingbaseES 数据库客户端后,需要使用 vi 编辑器,为用户 kingbase 在用户初始化环境文件 /home/kingbase/.bashrc 的末尾,添加以下的内容:

```
export KINGBASE_CLIENT_HOME=/u00/Kingbase/ES/V9/client
export PATH=$KINGBASE_CLIENT_HOME/ClientTools/bin:$PATH
export PATH=$KINGBASE_CLIENT_HOME/ClientTools/guitools/KStudio:$PATH
export PATH=$KINGBASE_CLIENT_HOME/ClientTools/guitools/DeployTools:$PATH
export PATH=$KINGBASE_CLIENT_HOME/ClientTools/guitools/KDts:$PATH
```

要让用户初始化环境文件生效,需要用户 kingbase 退出系统并重新登录到 CentOS 7 操作系统的 GUI。

如果 KingbaseES 数据库服务器(IP 地址为 192.168.100.22/24)此时已经处于运行状态,在 KingbaseES 数据库客户端的主机上执行下面的命令,可以测试 KingbaseES 数据库的客户端是否已经正确安装:

```
[kingbase@dbclient ~]$ ksql -U system -d test -h 192.168.100.22 -p 54321
Password for user system:                 --输入数据库用户 system 的密码 Passw0rd
Type "help" for help.
test=# SELECT version();
                    version
--------------------------------------------------------------------
KingbaseES V009R001C001B0025 on x86_64-pc-linux-gnu, compiled by gcc (GCC) 4.8.5 20150623
(Red Hat 4.8.5-28), 64-bit
(1 row)                                   --已经对输出进行了排版美化
test=# \q
[kingbase@dbclient ~]$
```

如果输出如上显示,表明 KingbaseES 数据库客户端可以和 KingbaseES 数据库服务器建立连接会话,说明已经在主机 dbclient 上安装好了 KingbaseES 数据库客户端。

执行下面的步骤,备份这个已经安装好 KingbaseES 数据库客户端的虚拟机:

```
[kingbase@dbclient ~]$ su -
Password: #输入用户 root 的密码 root123
Last login: Fri Feb 16 18:56:43 CST 2024 on pts/1
[root@dbclient ~]# shutdown -h now
```

使用 winrar 软件将 KingbaseES 数据库客户端的虚拟机所在目录(本书的目录为 G:\dbclient),压缩为 dbclient.rar 文件,压缩完成后,改名为 dbclient-2OK.rar。

2.1.2　在 Windows 操作系统上安装 KingbaseES 数据库客户端

在本书的示例中,Windows 操作系统的网卡 IP 地址为 192.168.100.17/24。

使用虚拟光驱软件(如 UltraISO),将下载的 Windows 操作系统的 KingbaseES 数据库安装包文件映射为一个虚拟磁盘(如盘符名 F:)。运行虚拟磁盘根目录下的 KINGBASE.exe 程序,将开始安装 KingbaseES 数据库的客户端。

安装完成 KingbaseES 数据库的客户端后,首先需要为用户(示例为 zqf)设置 Path 环境变量,为 Path 环境变量添加如下两个目录。

(1) C:\Program Files\Kingbase\ES\V9\KESRealPro\V009R001C001B0025\ClientTools\bin。

(2) C:\Program Files\Kingbase\ES\V9\KESRealPro\V009R001C001B0025\ClientTools\guitools\Kstudio。

为用户 zqf 增加一个环境变量 KSQLRC,值为 C:\Users\zqf\.ksqlrc。

接下来在用户 zqf 的主目录 C:\Users\zqf 下,使用 Notepad 文本编辑器,创建文件.ksqlrc,内容如下:

```
\set PROMPT1 '%n@%/%R%#'
```

最后,打开一个 cmd 窗口,执行命令 ksql -U system -d test -h 192.168.100.22,如果显示如图 2-3 所示,则表示已经成功地安装了 KingbaseES 数据库的客户端。

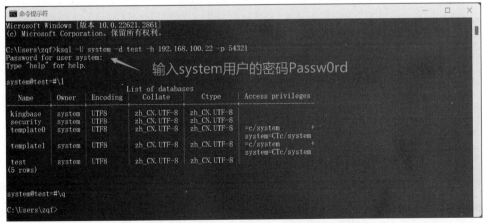

图 2-3　使用 ksql 连接 KingbaseES 数据库服务器

2.2 客户端与服务器的连接方式

KingbaseES 数据库客户端连接服务器时,可以使用如下两种方式。

(1) 客户端(如 ksql)在命令行中直接提供连接所需的所有信息:IP、TCP 端口号、数据库名、数据库用户名(客户端身份认证信息,如数据库用户口令)。

(2) 客户端(如 ksql)使用服务名(service name)文件(包含 IP、端口、数据库名)进行连接。

下面以 ksql 为例,介绍客户端与服务器的连接方式,其他 KingbaseES 数据库客户端工具的使用和 ksql 一样。

2.2.1 使用命令行选项

ksql 使用下面的选项来标识要连接的目标 KingbaseES 数据库服务器。

(1) -h:指定连接的服务器地址。

(2) -p:指定服务器端口号,默认是 54321。

(3) -d:指定数据库名。

(4) -U:指定数据库用户名。

例 2.1 以数据库用户 system 的身份,连接 IP 地址为 192.168.100.22、监听端口号为 54321 的 KingbaseES 数据库服务器,数据库名是 test。

```
[kingbase@dbclient ~]$ ksql -h 192.168.100.22 -p 54321 -d test -U system
Password for user system: --输入数据库用户 system 的密码 Passw0rd
test=# SELECT inet_client_addr() AS client_ip;
   client_ip
----------------
 192.168.100.18
(1 row)
test=# SELECT inet_server_addr() AS server_ip;
   server_ip
----------------
 192.168.100.22
(1 row)
test=# \q
[kingbase@dbclient ~]$
```

2.2.2 使用服务名

可以在 KingbaseES 数据库客户端的主机上创建一个服务名文件来简化客户端与服务器之间的会话连接。

服务名文件的名字一般为.sys_service.conf。如果是 Linux 操作系统,服务名配置文件一般位于用户的主目录下。如果是 Windows 操作系统,可以设置用户环境变量 KINGBASE_SERVICEFILE 指定服务名配置文件的位置(如 C:\Users\zqf\sys_service.conf)。

以 Linux 操作系统为例，以用户 kingbase 的身份执行下面的 cat 命令，创建如下的服务名文件：

```
[kingbase@dbsvr ~]$ cat >/home/kingbase/.sys_service.conf<<EOF
[sntest]
host=192.168.100.22
port=54321
dbname=test
user=system
password=Passw0rd
[snkingbase]
host=192.168.100.22
port=54321
dbname=kingbase
[sntestdb]
host=192.168.100.22
port=54321
dbname=testdb
EOF
```

这里配置了 3 个 KingbaseES 数据库服务名。

（1）服务名 sntest：使用数据库用户 system，连接 IP 地址为 192.168.100.22 的数据库服务器，连接端口号为 54321，连接的数据库的名字为 test，并且服务名 sntest 还提供了用户 system 的密码 Passw0rd（不推荐在服务名中配置数据库用户的密码）。

（2）服务名 snkingbase：连接 IP 地址为 192.168.100.22 的数据库服务器，连接端口号为 54321，连接的数据库的名字为 kingbase。

（3）服务名 sntestdb：连接 IP 地址为 192.168.100.22 的数据库服务器，连接端口号为 54321，连接的数据库的名字为 testdb。

注意：KingbaseES 数据库的一个服务名不支持连接到多个数据库。用户可以配置多个服务名来访问不同的数据库。

如果在数据库服务名中配置了用户密码，则使用该服务名指定的用户名连接指定的数据库时，系统不再提示输入密码。例如，服务名 sntest 中配置了用户密码，服务名 snkingbase 中没有配置用户密码。

例 2.2 使用服务名 sntest，以数据库用户 system 的身份连接 KingbaseES 数据库服务器上的 test 数据库，不需要提供数据库用户 system 的密码：

```
[kingbase@dbclient ~]$ ksql system@sntest
test=# SELECT current_user;
 current_user
--------------
 system
(1 row)
test=# SELECT current_database;
 current_database
------------------
 test
(1 row)
```

```
system@test=#
```

例 2.3 在 KingbaseES 数据库中创建一个新的数据库用户 kingbase，然后使用服务名 snkingbase，在 KingbaseES 数据库的客户端以数据库用户 kingbase 的身份连接到 KingbaseES 数据库服务器上的 kingbase 数据库，需要提供数据库用户 kingbase 的密码：

```
test=# CREATE USER kingbase WITH Superuser PASSWORD 'Passw0rd';
test=# \q
[kingbase@dbclient ~]$ ksql kingbase@snkingbase
Password for user kingbase: --输入数据库用户 kingbase 的密码 Passw0rd
kingbase=# \q
[kingbase@dbclient ~]$
```

2.2.3 数据库用户密码文件 .kbpass

在 KingbaseES 数据库客户端的主机上，可以把 KingbaseES 数据库的用户密码保存在用户主目录下的文件 .kbpass 中，这样可以免去每次使用 ksql 访问 KingbaseES 数据库服务器时输入密码的过程。每个数据库用户都可以在这个文件中有一行，格式如下：

主机名:端口:数据库名:用户名:密码

以 Linux 操作系统为例，使用 vi 编辑器，创建文件 /home/kingbase/.kbpass，添加以下两行内容：

```
192.168.100.22:54321:*:system:Passw0rd
localhost:54321:*:system:Passw0rd
```

第 1 行表示以数据库用户 system 的身份访问 IP 地址是 192.168.100.22、端口号为 54321 的 KingbaseES 数据库时，使用密码 Passw0rd；第 2 行表示以数据库用户 system 的身份访问本地服务器（localhost）上的 KingbaseES 数据库时，使用密码 Passw0rd。

创建完数据库用户密码文件后，还需要执行下面的命令，更改文件访问权限，确保密码安全：

```
[kingbase@dbclient ~]$ chmod 600 /home/kingbase/.kbpass
```

例 2.4 在 KingbaseES 数据库客户端的主机上，配置完数据库用户密码文件后，使用 ksql，以数据库用户 system 的身份访问 IP 地址是 192.168.100.22/24、端口号为 54321 的 KingbaseES 数据库时，不再提示输入数据库用户 system 的密码：

```
[kingbase@dbclient ~]$ ksql -d test -U system -h 192.168.100.22 -p 54321
Type "help" for help.
test=# \q
[kingbase@dbclient ~]$
```

2.3 客户端程序 ksql

ksql 是 KingbaeES 数据库客户端的命令行程序，它允许用户与 KingbaseES 数据库建立连接，并在命令行环境下执行 SQL 语句和管理数据库。ksql 有以下的关键特点和功能。

(1) **交互式 SQL 查询**：ksql 提供一个交互式的命令行界面，允许用户在命令提示符下直接输入 SQL 语句和命令，并立即获取结果。可以在 ksql 中执行各种 SQL 语句，包括 SELECT、INSERT、UPDATE、DELETE 及数据库模式定义语言(data definition language, DDL)语句。

(2) **元命令**：ksql 提供了一系列的元命令，以方便管理和操作数据库。元命令以反斜杠(\)开头，用于执行一些特定的任务，如导入和导出数据、创建和修改数据库对象(database object)、查看服务器状态等。

(3) **历史记录**：ksql 会保存用户在客户端会话中输入的命令和查询历史记录，这样用户可以方便地回顾和重新执行之前执行过的 SQL 语句。

(4) **输出格式控制**：ksql 提供多种输出格式选项，如表格形式、CSV 格式、JSON 格式等，以满足不同的数据显示和导出需求。

ksql 是一个功能强大且灵活的工具，无论是在开发、运维还是数据查询方面，ksql 都是 KingbaseES 数据库用户常用的工具之一，它为用户提供了与 KingbaseES 数据库进行交互和管理的便捷方式。

2.3.1　ksql 的命令行选项

KingbaseES 数据库的客户端程序 ksql 的语法格式如下：

```
ksql [OPTION]… [DBNAME [USERNAME]]
```

其中的[OPTION]用于指定命令行选项，在 2.2.1 节中介绍了-h、-p、-d 和-U 等选项。

例 2.5　使用-d 选项指定连接的数据库名，-U 选项指定连接的数据库用户名，建立与 KingbaseES 数据库的连接。

在 KingbaseES 数据库的服务器上，执行下面的 ksql 命令，建立与 KingbaseES 数据库的连接：

```
[kingbase@dbsvr ~]$ ksql -U system -d test
Type "help" for help.
system@test=# \q
[kingbase@dbsvr ~]$
```

使用 ksql 的元命令\q 可以退出 ksql，也可以按 Ctrl＋D 组合键退出。

例 2.6　使用-c 选项直接在命令行中运行 SQL 语句。

以 Linux 操作系统为例，执行下面的 ksql 命令，运行 SQL 语句：

```
[kingbase@dbclient ~]$ ksql -d test -h 192.168.100.22 -U system -c "SELECT l_orderkey,l_partkey,l_suppkey,l_linenumber FROM tpch.lineitem LIMIT 1;"
 l_orderkey | l_partkey | l_suppkey | l_linenumber
------------+-----------+-----------+--------------
          1 |    155190 |      7706 |            1
(1 row)
[kingbase@dbclient ~]$
```

例 2.7　使用-f 选项直接在命令行中运行 SQL 脚本。

以 Linux 操作系统为例，先生成一个测试用的 SQL 脚本文件 my.sql：

```
[kingbase@dbclient ~]$ cat > my.sql<<EOF
SELECT l_orderkey,l_partkey,l_suppkey,l_linenumber FROM tpch.lineitem LIMIT 1
EOF
```

然后，执行下面的 ksql 命令，运行 SQL 脚本文件 my.sql：

```
[kingbase@dbclient ~]$ ksql -d test -h 192.168.100.22 -U system  -f my.sql
 l_orderkey | l_partkey | l_suppkey | l_linenumber
------------+-----------+-----------+--------------
          1 |    155190 |      7706 |            1
(1 row)
[kingbase@dbclient ~]$
```

2.3.2 ksql 的元命令

数据库用户和 DBA 可以使用客户端程序 ksql 的元命令，管理和查询 KingbaseES 数据库。

以 Linux 操作系统为例，执行 ksql 命令，使用元命令\? 获取 ksql 元命令的帮助信息：

```
 [kingbase@dbclient ~]$ ksql -d test -h 192.168.100.22 -U system
test=# \?
General
  \crosstabview [COLUMNS] execute query and display results in crosstab
--省略了一些输出

Help
  \? [commands]          show help on backslash commands
  \? options             show help on ksql command-line options
--省略了一些输出

Query Buffer
  \e [FILE] [LINE]       edit the query buffer (or file) with external editor
  \ef [FUNCNAME [LINE]]  edit function definition with external editor
--省略了许多输出
test=#
```

1. 查询 KingbaseES 数据库信息

可以在 ksql 中使用如表 2-1 所示的元命令查询数据库的相关信息。

表 2-1 ksql 的元命令

元命令	用途
\l 命令	列出数据库服务器中的数据库信息
\du 命令和\dg 命令	列出用户或角色
\db 命令	列出表空间
\dn 命令	列出数据库模式
\d 命令	显示数据库中的对象

续表

元 命 令	用 途
\dt 命令和\dt＋命令	显示当前数据库中当前模式下有哪些表
\d TableName 命令	显示 TableName 表的详细信息
\di 命令	显示数据库中的索引信息
\dv 命令	显示数据库中的视图信息
\ds 命令	显示数据库中序列的信息
\df 命令	显示数据库中存储函数的信息
\dx 命令	显示数据库中已安装插件的信息

例 2.8　列出数据库服务器中的数据库信息。

```
system@test=# \l
                       List of databases
   Name    | Owner  | Encoding |  Collate   |   Ctype    | Access privileges
-----------+--------+----------+------------+------------+--------------------
 kingbase  | system | UTF8     | zh_CN.UTF-8| zh_CN.UTF-8|
 security  | system | UTF8     | zh_CN.UTF-8| zh_CN.UTF-8|
 template0 | system | UTF8     | zh_CN.UTF-8| zh_CN.UTF-8| =c/system         +
           |        |          |            |            | system=CTc/system
 template1 | system | UTF8     | zh_CN.UTF-8| zh_CN.UTF-8| =c/system         +
           |        |          |            |            | system=CTc/system
 test      | system | UTF8     | zh_CN.UTF-8| zh_CN.UTF-8|
(5 rows)
system@test=#
```

例 2.9　列出当前连接的数据库 test 中的模式。

```
test=# \dn
      List of schemas
      Name       | Owner
-----------------+--------
 anon            | system
--省略了一些输出
 tpch            | system
 xlog_record_read| system
(10 rows)
```

例 2.10　列出当前数据库 test 的 tpch 模式下所有的数据库对象的详细信息。

```
test=# \d tpch.*
                   Table "tpch.customer"
   Column    |          Type           | Collation | Nullable | Default
-------------+-------------------------+-----------+----------+---------
 c_custkey   | integer                 |           | not null |
 c_name      | character varying(25 char) |        | not null |
 c_address   | character varying(40 char) |        | not null |
```

```
 c_nationkey  | integer                    |          | not null |
 c_phone      | character(15 char)         |          | not null |
 c_acctbal    | numeric(15,2)              |          | not null |
 c_mktsegment | character(10 char)         |          | not null |
 c_comment    | character varying(117 char)|          | not null |
Indexes:
    "customer_pkey" PRIMARY KEY, btree (c_custkey)
Foreign-key constraints:
    "customer_c_nationkey_fkey" FOREIGN KEY (c_nationkey) REFERENCES tpch.nation(n_nationkey)
Referenced by:
    TABLE "tpch.orders" CONSTRAINT "orders_o_custkey_fkey" FOREIGN KEY (o_custkey) REFERENCES tpch.customer(c_custkey)

       Index "tpch.customer_pkey"
  Column   |  Type   | Key? | Definition
-----------+---------+------+------------
 c_custkey | integer | yes  | c_custkey
primary key, btree, for table "tpch.customer"
--省略了许多输出
test=#
```

例 2.11 列出当前数据库 test 的 tpch 模式下所有表的信息。

```
test=# \dt tpch.*
         List of relations
 Schema |   Name   | Type  | Owner
--------+----------+-------+--------
 tpch   | customer | table | system
--省略了一些输出
(8 rows)
```

例 2.12 列出当前数据库 test 的 tpch 模式下所有表的详细信息。

```
test=# \dt+ tpch.*
                List of relations
 Schema |   Name   | Type  | Owner  | Size  | Description
--------+----------+-------+--------+-------+-------------
 tpch   | customer | table | system | 28 MB |
--省略了一些输出
(8 rows)
```

例 2.13 查看表 tpch.customer 的详细信息。

```
test=# \d tpch.customer
                  Table "tpch.customer"
   Column    |         Type          | Collation | Nullable | Default
-------------+-----------------------+-----------+----------+---------
 c_custkey   | integer               |           | not null |
 c_name      | character varying(25 char) |      | not null |
 c_address   | character varying(40 char) |      | not null |
 c_nationkey | integer               |           | not null |
```

```
 c_phone        | character(15 char)          | not null |
 c_acctbal      | numeric(15,2)               | not null |
 c_mktsegment   | character(10 char)          | not null |
 c_comment      | character varying(117 char) | not null |
Indexes:
    "customer_pkey" PRIMARY KEY, btree (c_custkey)
Foreign-key constraints:
    "customer_c_nationkey_fkey" FOREIGN KEY (c_nationkey) REFERENCES tpch.nation
(n_nationkey)
Referenced by:
    TABLE "tpch.orders" CONSTRAINT "orders_o_custkey_fkey" FOREIGN KEY (o_
custkey) REFERENCES tpch.customer(c_custkey)
```

例 2.14 查看当前数据库 test 的 tpch 模式下所有索引的信息。

```
test=# \di tpch.*
              List of relations
 Schema | Name          | Type  | Owner  | Table
--------+---------------+-------+--------+----------
 tpch   | customer_pkey | index | system | customer
 tpch   | i_l_orderkey  | index | system | lineitem
--省略了一些输出
(12 rows)
```

例 2.15 列出当前数据库 test 的 tpch 模式下所有索引的详细信息。

```
test=# \di+ tpch.*
                          List of relations
 Schema | Name          | Type  | Owner  | Table    | Size    | Description
--------+---------------+-------+--------+----------+---------+-------------
 tpch   | customer_pkey | index | system | customer | 3336 kB |
 tpch   | i_l_orderkey  | index | system | lineitem | 129 MB  |
--省略了一些输出
(12 rows)
```

例 2.16 显示 tpch 模式下索引 customer_pkey 的详细信息。

```
test=# \di tpch.customer_pkey
              List of relations
 Schema | Name          | Type  | Owner  | Table
--------+---------------+-------+--------+----------
 tpch   | customer_pkey | index | system | customer
(1 row)
```

例 2.17 显示当前数据库 test 的视图信息。

元命令\dv 用于查看数据库中视图的信息。如果不指定模式,则显示当前模式下的视图信息。

```
test=# \dv
          List of relations
 Schema | Name                   | Type | Owner
--------+------------------------+------+--------
```

```
 public | sys_stat_statements      | view | system
 public | sys_stat_statements_all  | view | system
(2 rows)
```

例 2.18 显示当前数据库 test 的序列信息。

样例数据库目前还没有序列。创建一个表,该表两列的数据类型都是 SERIAL。由于 SERIAL 数据类型的列会自动创建序列,因此可以使用元命令\ds 查询在数据库 test 中创建的这两个序列:

```
test=# CREATE TABLE test_table(id SERIAL PRIMARY KEY,testnum SERIAL);
CREATE TABLE
system@test=# \ds
              List of relations
 Schema |        Name           |  Type    | Owner
--------+-----------------------+----------+--------
 public | test_table_id_seq     | sequence | system
 public | test_table_testnum_seq| sequence | system
(2 rows)
```

删除表 test_table,序列也将同时被删除,使用元命令\ds 将查不到任何序列:

```
test=# DROP TABLE IF EXISTS test_table;
DROP TABLE
system@test=# \ds
Did not find any relations.
```

2. 设置查询结果的输出格式

元命令\x 用于设置查询结果的输出模式。默认情况下,查询结果按行显示。如果执行元命令\x on,则查询结果的元组按列显示。

例 2.19 设置查询结果按列显示。

```
test=# SELECT l_orderkey,l_partkey,l_suppkey,l_linenumber FROM tpch.lineitem
LIMIT 2;
 l_orderkey | l_partkey | l_suppkey | l_linenumber
------------+-----------+-----------+--------------
          1 |    155190 |      7706 |            1
          1 |     67310 |      7311 |            2
(2 rows)
test=# \x on
Expanded display is on.          #修改显示方式为列方式
test=# SELECT l_orderkey,l_partkey,l_suppkey,l_linenumber FROM tpch.lineitem
LIMIT 2;
-[ RECORD 1 ]+-------          #以列的方式显示第 1 行结果数据
l_orderkey   | 1
l_partkey    | 155190
l_suppkey    | 7706
l_linenumber | 1
-[ RECORD 2 ]+-------          #以列的方式显示第 2 行结果数据
l_orderkey   | 1
l_partkey    | 67310
```

```
 l_suppkey     | 7311
 l_linenumber  | 2
test=# \x off
Expanded display is off.           #修改显示方式为行方式
```

元命令\t 用于设置查询结果是否输出结果集的字段信息。如果设置为 on,执行 SQL 语句时,将只输出结果行;如果设置为 off,则会显示输出结果行的字段名。系统默认为 off。

例 2.20 设置查询结果集显示结果字段信息。

```
test=# \t on
test=# SELECT l_orderkey,l_partkey,l_suppkey,l_linenumber FROM tpch.lineitem
LIMIT 1;
        1 |   155190 |      7706 |           1
--设置为 on,只显示结果行
test=# \t off
test=# SELECT l_orderkey,l_partkey,l_suppkey,l_linenumber FROM tpch.lineitem
LIMIT 1;
 l_orderkey| l_partkey | l_suppkey | l_linenumber
---------+---------+---------+--------------
        1 |   155190 |      7706 |           1
(1 row)
--设置为 off,将显示字段名、结果行和返回的行数
```

元命令\timing 用于设置是否显示 SQL 语句的执行时间。如果设置为 on,SQL 语句执行完后将显示其执行时间;如果设置为 off,则不显示 SQL 语句的执行时间。系统默认为 off。

例 2.21 设置显示 SQL 语句的执行时间。

```
test=# SELECT l_orderkey,l_partkey,l_suppkey,l_linenumber FROM tpch.lineitem
LIMIT 1;
 l_orderkey| l_partkey | l_suppkey | l_linenumber
--------- +---------+---------+--------------
        1 |   155190 |      7706 |           1
(1 row)                    --默认不显示 SQL 语句的执行时间
test=# \timing on
Timing is on.              --设置显示 SQL 语句的执行时间
test=# SELECT l_orderkey,l_partkey,l_suppkey,l_linenumber FROM tpch.lineitem
LIMIT 1;
 l_orderkey| l_partkey | l_suppkey | l_linenumber
--------- +---------+---------+--------------
        1 |   155190 |      7706 |           1
(1 row)
Time: 0.952 ms             --显示了 SQL 语句的执行时间
test=# \timing off
Timing is off.             --设置不显示 SQL 语句的执行时间
```

3. 切换连接的数据库

元命令\ c[DBNAME]用于切换连接的数据库;元命令\conninfo 用于显示会话的连接

信息。

例 2.22 切换连接的数据库。

```
test=# -- 显示当前 ksql 会话的连接信息
test=# \conninfo
You are connected to database "test" as user "system" on host "192.168.100.22" at port "54321".
test=# -- 切换连接到数据库 kingbase
test=# \c kingbase
You are now connected to database "kingbase" as userName "system".
kingbase=# -- 显示当前 ksql 会话的连接信息
kingbase=# \conninfo
You are connected to database "kingbase" as user "system" on host "192.168.100.22" at port "54321".
kingbase=# SELECT current_database();
 current_database
------------------
 kingbase
(1 row)
kingbase=# -- 切换回数据库 test
kingbase=# \c test
You are now connected to database "test" as userName "system".
test=# \q
[kingbase@dbclient ~]$
```

4. 执行 SQL 脚本

元命令\i file.sql 用于在 ksql 中执行 SQL 脚本文件 file.sql。

例 2.23 在 ksql 中运行 SQL 脚本。

首先创建一个 SQL 脚本文件 my.sql，然后在 ksql 中执行 SQL 脚本文件 my.sql：

```
[kingbase@dbclient ~]$ cat > my.sql<<EOF
SELECT l_orderkey,l_partkey,l_suppkey,l_linenumber FROM tpch.lineitem LIMIT 1;
EOF
[kingbase@dbclient ~]$ ksql -h 192.168.100.22 -d test -U system
test=# \i my.sql
 l_orderkey | l_partkey | l_suppkey | l_linenumber
------------+-----------+-----------+--------------
          1 |    155190 |      7706 |            1
(1 row)
```

5. 把查询结果输出到文件中

元命令\o[fileName]用于重定向输出到文件 fileName。

例 2.24 设置输出重定向到文件。

```
test=# -- 重定向 SQL 语句的输出到文件 myoutputfile
test=# \o myoutputfile
test=# SELECT l_orderkey,l_partkey,l_suppkey,l_linenumber FROM tpch.lineitem LIMIT 1;
--没有输出,因为输出重定向到文件 myoutputfile 中了
```

```
test=# -- 查看文件 myoutputfile 的内容
test=# \! cat myoutputfile
 l_orderkey | l_partkey | l_suppkey | l_linenumber
------------+-----------+-----------+--------------
          1 |    155190 |      7706 |            1
(1 row)
test=# \q
[kingbase@dbclient ~]$
```

6. 显示执行的历史操作

元命令\s 用于显示在 ksql 中曾经执行过的元命令和 SQL 语句，它们被记录在操作系统用户 kingbase 的文件/home/kingbase/.ksql_history 中。

例 2.25 显示执行过的历史操作。

```
[kingbase@dbclient ~]$ ksql -h 192.168.100.22 -d test -U system
test=# --在 ksql 中使用元命令\s 查看执行过的历史操作
test=# \s
SELECT inet_client_addr() AS client_ip;
--省略了许多输出
\q
\s
test=# --在 ksql 中执行操作系统命令 cat,查看执行过的历史操作
test=# \! cat /home/kingbase/.ksql_history
SELECT inet_client_addr() AS client_ip;
--省略了许多输出
\q
\s
\! /home/kingbase/.ksql_history
system@test=#
```

7. 执行操作系统命令

元命令\![os_command]用于在 ksql 中执行操作系统命令。

如果在 KingbaseES 数据库客户端上使用这种方式执行操作系统命令，则该命令是在客户端本地操作系统上执行，而不是在 KingbaseES 数据库服务器的操作系统上执行。

例 2.26 执行操作系统命令，显示系统日期和时间。

```
test=# \! date
Wed Mar  6 20:21:32 CST 2024
test=#
```

在\! 和命令 date 之间，必须有一个空格。如果没有空格会报错：

```
test=# \!date
invalid command \!date
Try \? for help.
test=#
```

2.3.3 ksql 初始化文件.ksqlrc

文件.ksqlrc 是用于设置 ksql 行为的初始化文件。启动 ksql 时，默认会在 Linux 操作

系统用户（示例是用户 kingbase）的主目录/home/kinbase 中查找.ksqlrc 文件,如果该文件存在,就执行其中的命令。

也可以通过设置环境变量 KSQLRC,指定要执行的 ksql 初始文件.ksqlrc 的目录位置：

```
export KSQLRC=PathYourWantItToBe
```

在初始化文件.ksqlrc 中,可以进行如下的配置。

(1) 设置输出格式：如表格、CSV 或扩展模式。

(2) 设置显示的方式：如行数限制、输出截断等。

(3) 设置连接信息：避免在命令行中输入密码。需要注意的是,在.ksqlrc 文件存储密码可能会带来安全风险,因此请谨慎考虑。

(4) 加载外部脚本：在.ksqlrc 文件中加载 SQL 脚本,每次启动 ksql 时自动执行特定操作。

(5) 创建自定义 ksql 命令：在.ksqlrc 文件中创建自定义 ksql 命令的方法。

例 2.27 在 KingbaseES 数据库客户端的主机上创建 ksql 初始化文件.ksqlrc,设置 ksql 的登录提示符。

以 Linux 操作系统为例,以用户 kingbase 的身份,使用 vi 编辑器,编辑文件/home/kingbase/.ksqlrc,在文件尾部添加下面的内容：

```
\set PROMPT1 '%n@%/%R%# '
```

然后执行下面的 ksql 命令：

```
[kingbase@dbclient ~]$ ksql -h 192.168.100.22 -d test -U system
Type "help" for help.
system@test=#
```

从输出可以看到,ksql 的提示符被设置为"登录的数据库用户名@登录的数据库名＝♯"。

2.4 图形客户端程序 KStudio

以 Linux 操作系统为例,使用用户 kingbase 的身份登录 GUI 界面,打开一个终端,执行下面的命令,启动图形客户端程序 KStudio：

```
[kingbase@dbclient ~]$ KStudio
```

在出现启动界面和欢迎界面后,关闭 KStudio 的欢迎界面,显示如图 2-4 所示的 KStudio 第一次启动时的初始界面。

执行下面的操作,配置 KStudio 访问 KingbaseES 数据库。首先如图 2-4 所示,单击 Connect 按钮,打开 New connection 对话框,如图 2-5 所示,单击 Next 按钮,按照图 2-6 填写要连接的目标数据库信息后,单击 Finish 按钮,打开如图 2-7 所示的界面。

使用图形客户端 KStudio,KingbaseES 数据库用户可以完成以下的任务：管理数据库、管理表空间、管理用户和角色、查看系统参数,以及执行 SQL 语句。

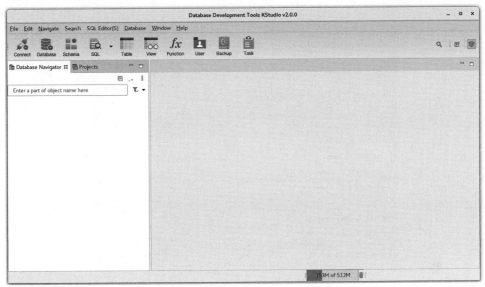

图 2-4 KStudio 第一次启动时的初始界面

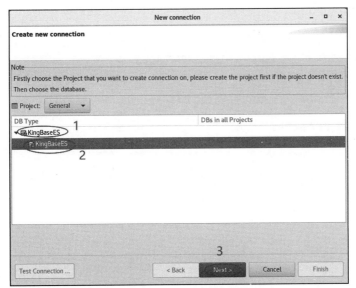

图 2-5 New connection 对话框

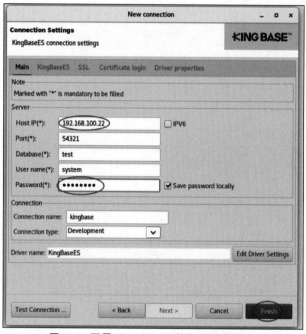

图 2-6　配置 KingbaseES 数据库连接信息

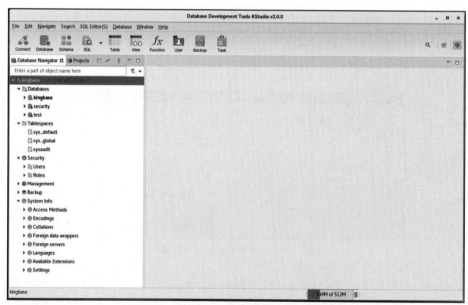

图 2-7　在 KStudio 中配置完成数据库连接

第 3 章

KingbaseES 数据库体系结构

KingbaseES 数据库采用客户端/服务器(Client/Server,C/S)的计算模式,客户端和服务器可以运行在不同的主机上,它们之间通过 TCP/IP 进行通信。

服务器是数据库的核心组件,负责存储和管理数据。服务器通过监听指定的端口,等待客户端的连接请求。一旦连接建立,服务器将响应客户端的请求,执行查询、更新和其他数据库操作。服务器还负责数据的持久化,提供数据完整性和安全性的保障。

客户端包括命令行工具(如 ksql)、图形化开发管理工具(如 KStudio)和用户应用程序等。客户端负责将用户的请求传递给服务器,并将查询结果返回给用户,用户可以通过客户端来连接、查询、更新和管理数据库。

3.1 KingbaseES 数据库服务器

KingbaseES 数据库服务器 指的是 KingbaseES 数据库管理系统服务端的部分,它由 **KingbaseES 数据库实例(database instance)** 和 **KingbaseES 数据库集簇** 组成。如图 3-1 所示为 KingbaseES 数据库服务器的体系结构。

图 3-1 的上半部分是 KingbaseES 数据库实例,主要包括多个数据库进程和一个共享内存区,称为**系统全局区(system global area,SGA)**。系统全局区包括数据缓冲区、重做日志缓冲区和各个进程需要共享的控制结构(如进程控制块、事务控制块、锁表等)。KingbaseES 数据库实例是 KingbaseES 数据库服务器的动态组件。

图 3-1 的下半部分是 KingbaseES 数据库集簇,包括控制文件、数据文件、重做日志文件。KingbaseES 数据库集簇是 KingbaseES 数据库服务器在磁盘存储上的静态组件。数据库集簇除了存储用户数据外,还存储 KingbaseES 数据库服务器自身用于管理用户数据的元数据。

安装 KingbaseES 数据库实际包含以下两个阶段。

(1) 安装软件(二进制程序),包括 KingbaseES 数据库服务器和(或)KingbaseES 数据库客户端的软件组件。

(2) 系统的初始化,使用 initdb 命令初始化磁盘上的数据存储区,创建数据库集簇。

使用 initdb 命令可以创建多个数据库集簇,每个数据库集簇位于一个独立的数据目录(文件系统目录)下。一个 KingbaseES 数据库实例只能访问一个特定数据目录下的数据库集簇。不过,可以在一个操作系统中启动多个 KingbaseES 数据库实例(需要为每个实例配置不同的监听端口),每个实例访问一个不同的数据库集簇。

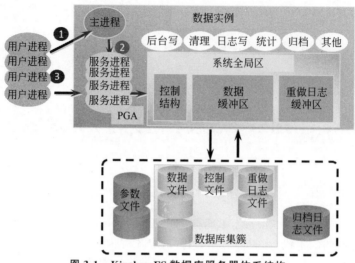

图 3-1　KingbaseES 数据库服务器体系结构

"启动 KingbaseES 数据库"的含义是启动一个 KingbaseES 数据库实例。首先会启动 kingbase 进程,它负责完成如下的任务。

（1）启动 KingbaseES 数据库实例。

① 分配共享内存区,如数据缓冲区、重做日志缓冲区等。

② 初始化 KingbaseES 数据库的内存控制结构。

③ 注册信号处理函数。

④ 启动 KingbaseES 数据库的其他后台进程。

（2）数据库实例启动完成后,kingbase 进程将成为一个监听进程,在接收客户端的连接请求并通过安全认证后,为客户端分配一个后端服务进程。

（3）在后端服务进程出现错误时,执行恢复（REDO）操作。

（4）关闭 KingbaseES 数据库。

数据库用户只能通过数据库实例访问数据库集簇中的数据。

3.2　KingbaseES 数据库的进程结构

KingbaseES 数据库的进程按照功能的不同可以分为以下三类。

（1）**主进程**（监听进程）：kingbase 进程,负责整个系统的启动和关闭,监听并接收客户端的连接请求,为其分配服务进程。

（2）**后台进程**：KingbaseES 数据库服务器在启动或运行过程中自动启动的一些进程,称为后台进程,包括：

- 后台写（background writer）进程；
- 日志写（walwriter）进程；
- 检查点（checkpointer）进程；
- 自动清理启动（autovacuum launcher）进程；
- 统计信息收集（stats collector）进程；

- 归档日志(archiver)进程；
- 系统运行日志(logger)进程；
- 逻辑复制启动(logical replication launcher)进程；
- 会话信息写(ksh writer)进程；
- 会话信息采集(ksh collector)进程；
- 工作负载信息采集(kwr collector)进程等。

（3）**服务进程**：应用程序发起数据库连接请求并通过了安全验证后，KingbaseES 数据库服务器的 kingbase 进程会创建一个后端服务进程，用来处理来自应用程序的数据库服务请求（如执行 SQL 语句）。

3.2.1 KingbaseES 数据库主进程

kingbase 进程是 KingbaseES 数据库服务器的主进程，它是所有其他数据库实例进程的父进程。

3.2.2 KingbaseES 数据库服务进程

KingbaseES 数据库采用 2N 的专用服务器进程结构，即对每个用户的连接请求，创建一个服务进程服务该用户。

客户端用户进程需要访问数据库时，首先要建立与数据库的连接。客户端应用程序（用户进程）调用 KingbaseES 数据库的客户端驱动，如 Java 数据库连接（Java database connectivity，JDBC）、开放式数据库连接（open database connectivity，ODBC）等，向 KingbaseES 数据库服务器发出连接请求。kingbase 进程收到连接请求并完成用户身份认证后，将为发出请求的用户进程创建一个后端服务进程，负责完成这个用户进程的数据库请求任务。当用户进程断开会话连接时，用户进程所对应的后端服务进程也将自动退出。

服务进程主要完成以下任务。

（1）解析并执行用户进程所提交的 SQL 语句。
（2）在数据库缓存中查找用户进程所访问的数据，如果没有，从数据文件中读取。
（3）根据所执行的 SQL 语句类别，更新数据或将数据返回给用户进程。

3.2.3 KingbaseES 数据库后台进程

KingbaseES 数据库的后台进程通常不直接参与 SQL 语句的处理，它们在后台运行，辅助服务进程更高效地处理客户端请求。后台进程具有不同的生命周期，有些后台进程在服务器启动时自动创建，随服务器关闭而退出；有些后台进程根据系统的配置来决定是否启动，一旦工作完成就自动退出。

下面介绍 KingbaseES 数据库后台进程的主要功能及相关配置。

1. 后台写进程

KingbaseES 数据库的后台写进程负责将数据缓冲区中的脏页面（即修改过的页面）写入磁盘数据文件中，使服务进程在需要数据缓存块时减少写操作，从而加快系统的响应时间。后台写进程随数据库服务一起启动，并且在数据库服务的运行过程中一直存在。

后台写进程周期性地把数据缓冲区的脏页面写回到磁盘的数据文件中，写的速度不能太快,也不能太慢。如果写得太快,例如,缓冲区的数据页面更新了多次,每次更新都写到磁盘上,会增加系统的输入/输出(input/output,I/O)负担。如果写得太慢,则服务进程需要缓冲区页面时,可能需要自己将脏页面写到磁盘,从而增加系统的响应时间。

在 KingbaseES 数据库的参数文件 kingbase.conf 中有多个以 bgwriter_ 开头的系统配置参数,配置了后台写进程的行为,例如,参数 bgwriter_delay 设置后台写进程的启动周期,默认值是 200ms(毫秒);参数 bgwriter_lru_maxpages 设置每次从内存写出到数据文件的最大页数,默认值为 100。

2. 日志写进程

KingbaseES 数据库的 WAL 日志写进程负责将共享内存区中日志缓冲区的内容写入磁盘上的 WAL 文件中。

当 SQL 语句对数据进行更新时,更新操作在数据缓冲区中进行,同时会将数据变更记录写到日志缓冲区。当事务提交或日志缓冲区充满到一定程度时,日志缓冲区的日志记录需要写到磁盘上的 WAL 文件中。日志写进程周期性地运行,将大量的随机写改善为 WAL 的顺序写,减少了磁盘 I/O 操作,从而极大地提高了系统的性能。

日志写进程也随数据库服务一起启动,并且在数据库服务的运行过程中一直存在。用户可以通过修改系统配置参数 wal_writer_delay 来设置该进程的写日志间隔时间。

3. 检查点进程

KingbaseES 数据库的检查点进程负责处理系统的检查点操作。数据库的检查点是一个事件,当该事件发生时会触发后台写进程,将数据库缓存中的脏缓存块将全部写入数据文件,同时在 WAL 文件写入一条日志记录,并对数据库控制文件进行更新。可以看出,检查点是一个非常耗 I/O 资源的动作。

如果 KingbaseES 数据库服务器非正常关闭,则下次系统重新启动时需要进行数据库系统恢复。首先从控制文件中读取最后一个检查点位置,然后在 WAL 文件中找到最后的检查点,从这里开始进行系统恢复,直到 WAL 文件结束。如果检查点执行时间间隔太大,则当系统崩溃时,会增加系统恢复的时间;反之,如果检查点执行时间间隔太小,则会增加系统的 I/O 负担,降低系统的吞吐量。

检查点进程也随数据库服务一起启动,并且在数据库服务的运行过程中一直存在。用户可以通过修改系统配置参数 checkpoint_timeout 来设置检查点进程执行检查点操作的间隔时间。

4. 统计信息收集进程

KingbaseES 数据库的统计信息收集进程负责收集 SQL 语句执行过程中的统计信息,例如,在表或索引上进行了多少次增、删、改操作,磁盘的读写次数,元组的读写数量等。这些信息可以辅助 DBA 进行系统性能诊断,找出可能存在的性能瓶颈。

统计信息收集进程也随数据库服务一起启动,并且在数据库服务的运行过程中一直存在。收集统计信息会给系统增加负荷,系统配置文件中有以 track_ 开头的配置参数可以用来配置收集哪些统计信息。

5. 信息采集进程

KingbaseES 数据库提供了类似 Oracle 数据库的性能问题诊断工具 KingbaseES 自动负载信息库(KingbaseES auto workload repertories,KWR)、KingbaseES 会话历史

(KingbaseES session history，KSH)和 KingbaseES 自动数据库诊断监控(KingbaseES auto database diagnostic monitor，KDDM)，它们通过信息采集进程采集相关的信息，给用户提供相关的性能分析报告。安装 KWR 后，系统会自动启动这些进程。

会话信息采集进程负责以每秒采样的方式收集会话的信息，采集的数据主要包括会话、应用、等待事件、命令类型、QueryId 等。

会话信息写进程负责将采集的会话信息写到表中。

工作负载信息采集进程负责周期性地(默认每小时)自动采集数据库实例，运行过程中不断产生一些统计数据，例如，对某个表的访问次数，数据页的内存命中次数，某个等待事件发生的次数和总时间，以及 SQL 语句的解析时间等。

6. 自动清理启动进程

KingbaseES 数据库采用多版本并发控制策略，当事务对表中元组进行删除操作时，并不会立刻从表中直接删除这些数据行，而是在元组上打上删除标记。当事务对表中元组进行更新操作时，采用的是删除旧元组，插入新元组的策略，因此在数据库中会保留元组的多个版本。如果没有其他并发事务需要读取旧元组，则可以将它们清除，否则数据库占用的存储会越来越大。

KingbaseES 数据库的自动清理启动进程负责向 kingbase 进程请求创建自动清理工作进程(autovacuum worker process)。那些过时的、并发事务不再需要的多版本并发控制(multiversion concurrency control，MVCC)旧元组，将由自动清理工作进程负责清除。

7. 归档日志进程

数据库的 WAL 文件记录了数据库中所有对数据页面的更新，一旦系统出现故障，可以使用 WAL 文件进行故障恢复。系统在运行过程中会一直产生日志，当系统执行了一个完整的检查点后，系统故障恢复就不再需要该检查点之前的 WAL 文件了，因此为了防止日志文件的个数不断增长，KingbaseES 数据库会重用早期不再需要的 WAL 文件(以修改日志文件的内容，代替删除旧日志文件)。但是，如果磁盘介质出现了故障，则需要使用数据库的备份和自备份以来的所有 WAL 文件，来恢复一个完整的数据库文件。因此生产系统的 KingbaseES 数据库，通常运行在归档模式下，即在 WAL 文件被重用之前，将其复制到一个特定的地方，这样在将来因介质故障而恢复时，可以使用这些归档的 WAL 文件。

KingbaseES 数据库的归档日志进程负责将 WAL 文件复制到归档日志目录。数据库在归档模式下运行，才会启动该进程，系统配置参数 archive_mode 设置 KingbaseES 数据库是否运行在归档模式下，默认是 off(运行在非归档方式)。

8. 系统运行日志进程

系统运行日志是 DBA 获得系统运行状态的有效工具。当数据库出现故障时，系统运行日志文件可以提供有效信息。KingbaseES 数据库的系统运行日志进程负责将 Kingbase 数据库实例中所有进程的运行日志输出到系统运行日志文件中。系统配置参数 logging_collector 的值，决定是否会启动系统运行日志进程，参数 logging_collector 的默认值是 on。

9. 逻辑复制启动进程

KingbaseES 数据库复制有两种实现方式：物理复制和逻辑复制，它们都是基于 WAL 文件实现的。物理复制基于流复制技术，将主库的 WAL 文件物理复制到备库，然后在备库中重做 WAL 文件，从而实现备库对主库的复制。逻辑复制采用发布者/订阅者模型，在发布端对

WAL 文件进行逻辑解析，在订阅端回放 WAL 文件中的逻辑条目，保持复制表的数据同步。

逻辑复制启动（logical replication launcher）进程是 KingbaseES 数据库逻辑复制体系的一个核心部分，它确保了逻辑复制的稳定运行和管理。

例 3.1 查看 KingbaseES 数据库实例的所有进程。

KingbaseES 数据库服务器启动后，执行 kbps 命令可以查看 KingbaseES 数据库的所有进程：

```
[kingbase@dbsvr ~]$ kbps
kingbase   1889      1  0 Feb29 ?    00:00:00 /u00/Kingbase/ES/V9/kingbase/KESRealPro/
V009R001C001B0025/Server/bin/kingbase -D /u00/Kingbase/ES/V9/data
kingbase   2036   1889  0 Feb29 ?    00:00:00 kingbase: logger
kingbase   2048   1889  0 Feb29 ?    00:00:00 kingbase: checkpointer
kingbase   2049   1889  0 Feb29 ?    00:00:00 kingbase: background writer
kingbase   2050   1889  0 Feb29 ?    00:00:00 kingbase: walwriter
kingbase   2051   1889  0 Feb29 ?    00:00:00 kingbase: autovacuum launcher
kingbase   2052   1889  0 Feb29 ?    00:00:00 kingbase: stats collector
kingbase   2053   1889  0 Feb29 ?    00:00:00 kingbase: kwr collector
kingbase   2054   1889  0 Feb29 ?    00:00:00 kingbase: ksh writer
kingbase   2055   1889  0 Feb29 ?    00:00:00 kingbase: ksh collector
kingbase   2056   1889  0 Feb29 ?    00:00:00 kingbase: logical replication launcher
[kingbase@dbsvr ~]$
```

从输出可以看到，KingbaseES 数据库的 kingbase 进程的进程号（PID）为 1889，该进程也是 KingbaseES 数据库的监听进程，其他进程是后台进程。

此时在 KingbaseES 数据库的客户端执行下面的 ksql 命令，以数据库用户 system 的身份连接到数据库 test：

```
[kingbase@dbclient ~]$ ksql -h 192.168.100.22 -d test -U system
system@test=#
```

回到 KingbaseES 数据库服务器，再次执行 kbps 命令，可以看到增加了一个 KingbaseES 数据库后端服务进程：

```
[kingbase@dbsvr ~]$ kbps
--省略了一些输出
kingbase   3621   1889  0 18:06 ?    00:00:00 kingbase: system test 192.168.100.18
(56848) idle
[kingbase@dbsvr ~]$
```

PID 为 3621 的后端服务进程是远程 KingbaseES 数据库客户端（IP 地址是 192.168.100.18）上的用户进程发起的数据库服务请求，被 PID 为 1889 的 kingbase 进程接收后，在 KingbaseES 数据库服务器上创建的后端服务进程。从后端服务进程的进程名可以看出，客户端上的用户进程以数据库用户 system 的身份连接访问数据库 test。

3.3　KingbaseES 数据库的内存结构

KingbaseES 数据库的使用的内存总体上可以分为两部分。

（1）**系统全局区**：是一组共享内存结构，其中包含一个 KingbaseES 数据库实例的数据和控制信息。系统全局区由所有服务进程和后台进程共享，它使用的是操作系统的共享内存。

（2）程序全局区（program global area，PGA）：KingbaseES 数据库服务器单个进程私有数据和控制信息使用的内存区域。它是在启动服务进程内部使用的内存。每个服务进程和后台进程都有自己的程序全局区用于内部操作，例如，排序和哈希（hash）操作等。

3.3.1 系统全局区

系统全局区是 KingbaseES 数据库实例的主要组成部分，它保存 KingbaseES 数据库系统所有进程的共享信息。KingbaseES 数据库服务器使用系统全局区来实现后台进程间的数据共享。在 KingbaseES 数据库服务器启动时，为系统全局区分配内存，在终止时，系统全局区释放内存。

系统全局区主要由数据缓冲区（Shared buffers）、日志缓冲区（WAL buffers）和进程共享的控制结构组成。

数据缓冲区是系统全局区中最大的一块共享内存，保存最近从数据文件中读取的数据块，以提高数据库的处理性能。理论上来说，数据缓冲区越大越好，但是 KingbaseES 数据库是运行在操作系统之上，操作系统也是有文件缓存的，因此建议该值配置为系统可用内存的 25%～40%。数据缓冲区大小由系统配置参数 shared_buffers 确定。shared_buffers 的默认值是 128MB，通常不能满足生产数据库的需求，DBA 需要根据系统的实际硬件情况和系统负载设置该参数。

日志缓冲区用于缓存在对数据进行修改操作过程中生成的 REDO 记录。该日志缓冲区是一个顺序使用的、循环的结构，当写满时，再从头部开始。由参数 wal_buffers 定义大小，较大的缓冲区可以减少写日志的磁盘 I/O，但是事务提交时日志记录一定要写盘，因此，日志缓冲区不需要太大。wal_buffers 默认值是 −1，将自动配置 WAL 缓冲区的大小范围。

（1）等于 shared_buffers 的 1/32（大约 3%）。

（2）不小于 64KB。

（3）不大于 WAL 段的尺寸。

参数 wal_buffers 的默认配置在大部分情况下能满足生产数据库的需要。

系统全局区中的其他数据结构，如锁表、进程控制块、事务控制块等，都是 KingbaseES 数据库内部使用的控制结构，以保证多个数据库进程正确运行。

3.3.2 程序全局区

每个数据库进程，无论是服务进程还是后台进程，工作时都需要独立的内存空间存放服务器进程工作过程中使用的数据和运行信息，例如，处理 SQL 语句时需要进行内部排序、创建哈希表、对临时表的访问，后台维护进程进行数据清理等。这是每个数据库进程工作过程中使用的内存，由该进程根据需要向操作系统申请和释放，并由该进程独享。

因为数据库是多用户共享的系统，单个服务进程不能耗尽系统的所有资源，因此 KingbaseES 数据库对每个进程使用的内存资源是有控制的。

对于每个进程可以使用的内存，KingbaseES 数据库可以控制其在以下几方面的使用。

（1）**maintenance_work_mem**：执行维护性操作时使用的内存。

（2）**temp_buffer**：访问临时表数据所使用的缓冲区。

（3）**work_mem**：事务执行内部排序或哈希表写入临时文件之前使用的内存缓冲区。

服务进程在处理 SQL 语句时，经常会有排序或创建哈希表这些非常需要内存的操作，对于单个进程来说，这些操作如果能在内存中进行，性能会更好，但是单个进程不能耗尽所有的内存资源，以免影响其他进程的正常运行，所以 KingbaseES 数据库使用配置参数 work_mem 来设定每个进程可以使用的内存空间。work_mem 设置过小，会造成排序或哈希操作变成外存操作，极大地降低系统性能。work_mem 设置过大，则当多个进程同时进行排序或哈希操作时，会导致系统的内存资源耗尽。work_mem 的设置需要同时考虑每个进程处理 SQL 语句的复杂度、系统的并发数及可用的内存资源总量。

数据库中临时表的数据不是所有服务进程共享的，因此服务进程在访问临时表时不使用系统全局区中的数据缓存区，而是把临时表的数据读到自己进程内部的临时缓存区中。配置参数 temp_buffer 设置每个服务进程的临时缓冲区可以使用的最大内存空间，默认值是 8MB。如果服务进程需要访问比较大的临时表时，可以重新设置该值来提升性能。

数据库在做系统维护操作时，如 VACUUM、CREATE INDEX、REINDEX 等时，需要读入大量的数据，在执行这些操作时，把配置参数 maintenance_work_mem 设置为较大的值，可以加速这些操作的执行。

进程对私有内存的使用是根据需要逐渐向系统请求分配的，并不是一开始就分配所有的内存。这些参数都是会话级的，可以为每个数据库会话单独设置这些参数的值。

3.3.3 内存参数初始优化建议

假设服务器拥有 16GB 的物理内存，初始安装 KingbaseES 数据库时，可用采用以下的方法来规划 KingbaseES 数据库使用的物理内存。

（1）为操作系统预留 2GB 物理内存。

（2）为将来 KingbaseES 数据库调优预留 2GB 物理内存。

（3）计算所有会话连接需要的物理内存。假定每个会话进程需要 20MB 物理内存，如果最大的会话连接数为 300，则需要预留 6GB 的物理内存，这部分内存已经包含了数据排序用到的内存。

（4）剩下的物理内存用于 KingbaseES 数据库服务器的系统缓存。

初次安装完 KingbaseES 数据库后，建议 DBA 在系统启动参数文件 kingbase.conf 中，设置以下的内存系统配置参数值：

```
shared_buffers=4096MB          # 不超过内存的 40%
work_mem=8MB                   # 用于排序和哈希操作的内存
maintenance_work_mem=1024MB    # 用于维护操作的内容，最大不超过 1024MB
autovacuum_work_mem = -1       # 设置 VACUUM 操作的内存
                               # 值为-1 表示最大不超过 maintenance_work_mem 参数的值
```

3.4 KingbaseES 数据库逻辑结构

KingbaseES 数据库的逻辑结构是指从用户角度来看的数据库中的内容。

3.4.1 数据库集簇

数据库集簇是多个用户数据库的集合。如图 3-2 所示，数据库集簇具有如下的逻辑层次结构：一个数据库集簇包含多个数据库，一个数据库只能属于一个数据库集簇；一个数据库包含多个模式，一个模式只能属于一个数据库；一个模式里有许多模式对象（schema object），一个模式对象只能属于一个模式。

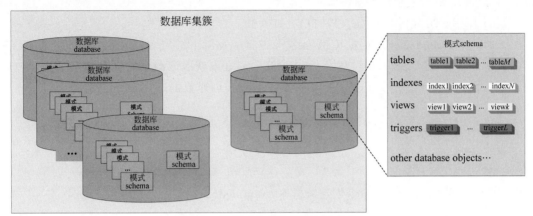

图 3-2　数据库集簇的逻辑层次结构

一个 KingbaseES 数据库实例只能管理一个数据库集簇，换一种说法，一个 KingbaseES 数据库实例可以管理多个用户数据库（数据库集簇包含了一系列用户数据库）。

3.4.2　数据库

如图 3-2 所示，一个数据库集簇包含多个数据库。

在 KingbaseES 数据库系统初始化创建数据库集簇时，会创建两个模板数据库：TEMPLATE0 和 TEMPLATE1。TEMPLATE0 是最原始的模板库，不允许建立数据库连接。TEMPLATE1 根据 TEMPLATE0 创建，是创建其他数据库使用的模板数据库，可以进行定制，如自定义的表、函数、触发器等，以及一些常用的配置设置，软件开发商可以用它部署初始的产品数据库。

系统默认还创建了用户数据库 kingbase 和 test，以及存放系统安全信息的数据库 security。本书使用数据库 test 作为测试数据的驻留数据库。

在 KingbaseES 数据库服务器上，执行下面的 ksql 元命令\l，查看数据库集簇中有哪些数据库：

```
[kingbase@dbsvr ~]$ ksql -d test -U system
system@test=# \l
                                List of databases
   Name    | Owner  | Encoding | Collate    | Ctype      | Access privileges
-----------+--------+----------+------------+------------+--------------------
 kingbase  | system | UTF8     | zh_CN.UTF-8| zh_CN.UTF-8|
 security  | system | UTF8     | zh_CN.UTF-8| zh_CN.UTF-8|
 template0 | system | UTF8     | zh_CN.UTF-8| zh_CN.UTF-8| =c/system         +
           |        |          |            |            | system=CTc/system
 template1 | system | UTF8     | zh_CN.UTF-8| zh_CN.UTF-8| =c/system         +
           |        |          |            |            | system=CTc/system
 test      | system | UTF8     | zh_CN.UTF-8| zh_CN.UTF-8|
(5 rows)
```

```
system@test=#
```

1. 创建数据库

可以使用如下两种方法创建新的数据库。

（1）使用 SQL 语句：CREATE DATABASE。

（2）使用 kingbaseES 数据库的命令行工具 createdb，它是基于 SQL 语句 CREATE DATABASE 封装实现的。可以执行 createdb --help 来获取该命令的使用方法。

例 3.2 使用 SQL 语句 CREATE DATABASE 创建数据库 exampledb1：

```
system@test=# CREATE DATABASE exampledb1;
CREATE DATABASE
system@test=# \q
[kingbase@dbsvr ~]$
```

例 3.3 使用 createdb 命令创建数据库 exampledb2：

```
[kingbase@dbsvr ~]$ createdb  -U system  --owner=system exampledb2
[kingbase@dbsvr ~]$ ksql -d test -U system
system@test=# \l
                    List of databases
    Name    | Owner  | Encoding |  Collate   |   Ctype    | Access privileges
----------+--------+----------+------------+------------+-----------------
 exampledb1| system | UTF8     | zh_CN.UTF-8| zh_CN.UTF-8|
 exampledb2| system | UTF8     | zh_CN.UTF-8| zh_CN.UTF-8|
--省略了一些输出
(7 rows)
system@test=# \q
[kingbase@dbsvr ~]$
```

从输出可以看到，已经成功地创建了两个数据库：exampledb1 和 exampledb2。

2. 模板数据库

软件开发商可以定制公司的产品模板数据库。工程师可在用户现场基于公司的产品模板数据库来部署公司的产品数据库。一般不允许用户连接产品模板数据库。

可以创建一个新的数据库，并在该数据库中创建软件产品相关的模式和模式对象，然后将这个数据库修改为产品模板数据库。

例 3.4 生成软件产品的模板数据库。

基于 TPC-H 数据集，制作一个没有数据、只有空表的模板数据库 ptemplate：

```
[kingbase@dbsvr ~]$ sys_dump -U system -d test -n tpch -F p -f ptemplate.sql \
                 --no-owner  --no-tablespaces  --no-privileges  --schema-only
[kingbase@dbsvr ~]$ ksql -d test -U system -c "CREATE DATABASE ptemplate;"
[kingbase@dbsvr ~]$ ksql -d ptemplate -U system  -f ptemplate.sql -q
[kingbase@dbsvr ~]$ ksql -d test -U system
system@ test = # ALTER DATABASE ptemplate  WITH ALLOW_CONNECTIONS false  IS_
TEMPLATE true;
```

基于模板数据库 ptemplate 创建产品数据库 newtpch：

```
system@test=# CREATE DATABASE newtpch TEMPLATE ptemplate ;
system@test=# \q
[kingbase@dbsvr ~]$ ksql -d newtpch -U system
system@newtpch=# \dt tpch.*
        List of relations
 Schema |   Name   | Type  | Owner
--------+----------+-------+--------
 tpch   | customer | table | system
--省略了一些输出
(8 rows)
system@newtpch=# SELECT count(*) FROM tpch.customer ;
 count
-------
     0
(1 row)
system@newtpch=# \q
[kingbase@dbsvr ~]$
```

从输出可以看到，新创建的数据库 newtpch，拥有与模板数据库 ptemplate 一样的 8 张空表。

3. 删除数据库

使用 DROP DATABASE 语句可以删除一个数据库，但是不能删除一个正在使用的数据库，也不能删除模板数据库。

例 3.5 无法删除一个正在使用的数据库。

（1）打开第 1 个终端，使用数据库用户 system，登录到数据库 newtpch：

```
[kingbase@dbsvr ~]$ ksql -d newtpch -U system
system@newtpch=#
```

（2）打开第 2 个终端，使用数据库用户 system，删除数据库 newtpch：

```
[kingbase@dbsvr ~]$ ksql -d test -U system
system@test=# DROP DATABASE newtpch;
ERROR:  database "newtpch" is being accessed by other users
DETAIL:  There is 1 other session using the database.
system@test=#
```

从输出可以看到，无法删除一个正在使用的数据库。

（3）为了删除数据库 newtpch，需要返回到第 1 个终端，执行元命令\q 退出 ksql：

```
system@newtpch=# \q
[kingbase@dbsvr ~]$
```

（4）返回第 2 个终端，再次删除数据库 newtpch：

```
system@test=# DROP DATABASE newtpch;
DROP DATABASE
system@test=#
```

从输出可以看到，这一次数据库 newtpch 被删除掉了。

例 3.6 无法直接删除一个模板数据库。

例 3.4 中已经把数据库 ptemplate 设置为模板数据库，因此无法删除模板数据库 ptemplate：

```
system@test=# DROP DATABASE ptemplate;
ERROR:  cannot drop a template database
system@test=#
```

要删除模板数据库 ptemplate，首先要将它修改为非模板数据库：

```
system@test=# ALTER DATABASE ptemplate IS_TEMPLATE false;
ALTER DATABASE
system@test=# DROP DATABASE ptemplate;
DROP DATABASE
system@test=#
```

3.4.3 模式

数据库中可以包含许多模式对象，可以将这些模式对象进行逻辑分组，每个分组称为一个模式。如图 3-2 所示，一个数据库可以有多个模式，一个模式只能属于一个数据库。模式又称命名空间，一个数据库的某个模式下的模式对象不能重名，但是一个数据库的不同模式下的模式对象可以重名。每个数据库都有一个默认的模式 public。

在 Oracle 数据库中，每个用户有一个同名的数据库模式。KingbaseES 数据库与此不同，数据库模式与用户没有关系。如果用户拥有适当的权限，则该用户可以在任意的模式中创建模式对象。

可以使用 ksql 元命令 \dn 来查看当前数据库中有哪些模式：

```
system@test=# \dn
     List of schemas
    Name      |  Owner
--------------+--------
 anon         | system
--省略了一些输出
(10 rows)
```

模式对象通常属于某个模式，可以使用下面的方法标识一个模式对象：

数据库名.模式名.对象名

由于 KingbaseES 数据库的每个用户连接只能访问一个数据库，因此数据库名可以省略，即可以使用下面的方法标识一个模式对象：

模式名.对象名

可以使用 SQL 语句 CREATE SCHEMA 来创建新的模式，在不同的模式下可以创建同名的模式对象。

例 3.7 在不同的模式下可以创建同名的模式对象。

（1）在数据库 exampledb1 中，创建模式 newschema1 和 newschema2：

```
[kingbase@dbsvr ~]$ ksql -d exampledb1 -U system
system@exampledb1=# CREATE SCHEMA newschema1;
system@exampledb1=# CREATE SCHEMA newschema2;
```

此时,这些模式下没有任何表:

```
system@exampledb1=# \dt public.*
Did not find any relation named "public.*"
system@exampledb1=# \dt newschema1.*
Did not find any relation named "newschema1.*"
system@exampledb1=# \dt newschema2.*
Did not find any relation named "newschema2.*"
system@exampledb1=#
```

(2) 分别在模式 public、newschema1 和 newschema2 中创建表 mytable,并插入 1 条数据:

```
system@exampledb1=# CREATE TABLE public.mytable(col varchar(20) PRIMARY KEY);
system@exampledb1=# INSERT INTO  public.mytable VALUES('000000');
system@exampledb1=# CREATE TABLE newschema1.mytable(col varchar(20) PRIMARY KEY);
system@exampledb1=# INSERT INTO  newschema1.mytable VALUES('111111');
system@exampledb1=# CREATE TABLE newschema2.mytable(col varchar(20) PRIMARY KEY);
system@exampledb1=# INSERT INTO  newschema2.mytable VALUES('222222');
```

(3) 执行下面的 SQL 语句,可以看出数据库 exampledb1 的这些模式下的同名表是不同的表:

```
system@exampledb1=# SELECT * FROM public.mytable;
  col                       --使用模式名.表名指定一个表
--------
 000000
(1 row)
system@exampledb1=# SELECT * FROM newschema1.mytable;
  col                       --使用模式名.表名指定一个表
--------
 111111
(1 row)
system@exampledb1=# SELECT * FROM exampledb1.newschema2.mytable;
  col                       --使用数据库名.模式名.表名指定一个表
--------
 222222
(1 row)
system@exampledb1=#
```

一个数据库可以有多个模式,如果 SQL 语句中的模式对象名前面没有指定模式名,那么将按照系统配置参数 search_path 指定的搜索顺序,定位诸如表、索引、序列等模式对象。

```
system@exampledb1=# SHOW search_path;
  search_path
-----------------
```

```
"$user", public
(1 row)
system@exampledb1=#
```

配置参数 search_path 的默认值是"$user",public,其含义是：查找一个模式对象时，首先在与数据库用户同名的模式下查找；如果不存与数据库用户同名的模式，或者在这个同名模式下没有找到这个指定名字的模式对象，则转到模式 public 下去查找，如果在模式 public 下也找不到，则向用户返回错误信息。

例 3.8 根据配置参数 search_path 的值查找到正确的模式对象。

```
system@exampledb1=# CREATE SCHEMA system;
                                --创建与登录用户 system 同名的模式 system
CREATE SCHEMA
system@exampledb1=# CREATE TABLE system.mytable(col varchar(20) PRIMARY KEY);
system@exampledb1=# INSERT INTO  system.mytable VALUES('333333');
system@exampledb1=# SELECT * FROM mytable;
  col
--------
 333333
(1 row)          --由于存在与连接的数据库用户 system 同名的模式 system,因此访问的表是
                 --system.mytable
system@exampledb1=# DROP SCHEMA system CASCADE;
                                --删除与连接的数据库用户 system 同名的模式 system
NOTICE:  drop cascades to table mytable
DROP SCHEMA
system@exampledb1=# SELECT * FROM mytable;
  col
--------
 000000
(1 row)   --由于删除了与数据库用户同名的模式,此时访问的表是 public.mytable
system@exampledb1=# DROP TABLE public.mytable;
                                --删除 public 模式下的表 mytable
DROP TABLE
system@exampledb1=# SELECT * FROM mytable;
ERROR:  relation "mytable" does not exist
LINE 1: SELECT * FROM mytable;
                      --无法在模式搜索路径下找到任何名字为 mytable 的表,因此报错
                      ^
system@exampledb1=#
```

例 3.9 设置配置参数 search_path 的值,设置默认访问的模式。

```
system@exampledb1=# SET search_path TO newschema2;
                                -- 设置参数 search_path 的值为模式 newschema2
system@exampledb1=# SHOW search_path;
 search_path
-------------
 newschema2
(1 row)
system@exampledb1=# SELECT * FROM mytable;
```

```
  col
--------
 222222
(1 row) --由于当前的搜索模式为newschema2,因此此时访问的表是newschema2.mytable
system@exampledb1=# \q
[kingbase@dbsvr ~]$
```

3.4.4 数据库对象

数据库对象是KingbaseES数据库中用于存储和引用数据的数据结构。

KingbaseES数据库中的一切皆称为数据库对象。有些数据库对象必须在模式下创建,如表、索引、视图、系列、函数和过程等,这类数据库对象称为**模式对象**。如数据库、表空间和用户等数据库对象与模式无关,称为**非模式对象**(Non-Schema Object)。

就像用身份证号码唯一标识一个人一样,每个数据库对象也有一个唯一标识自身的数据库对象标识(object identifier,OID)。

例 3.10 查询数据库对象的OID。

查看数据库test中用户表tpch.orders的OID。

```
 [kingbase@dbsvr ~]$ ksql -d test -U system
system@test=# SELECT c.oid, c.relname
test-#          FROM sys_class c JOIN sys_namespace n ON n.oid = c.relnamespace
test-#          WHERE n.nspname = 'tpch' AND c.relname='orders';
  oid  | relname
-------+-------------
 16394 | orders
(1 row)
```

查看数据库test的OID。

```
system@test=# SELECT datname,oid FROM sys_database WHERE datname='test';
 datname |  oid
---------+-------
 test    | 14386
(1 row)
```

查看KingbaseES数据库用户system的OID。

```
system@test=# SELECT rolname,oid FROM  sys_authid WHERE rolname='system';
 rolname | oid
---------+-----
 system  |  10
(1 row)
```

3.5 KingbaseES数据库物理结构

KingbaseES数据库的物理结构是指数据库集簇在操作系统上的物理文件,包括数据文件、控制文件和WAL文件等。

KingbaseES 数据库在初始化数据库集簇时，会将数据库集簇（包括数据文件、控制文件和 WAL 文件）存储在 KingbaseES 数据库的数据目录下。执行下面的 ksql 命令，可以查看 KingbaseES 数据库的数据目录位置：

```
system@test=# SHOW data_directory;
     data_directory
-------------------------
 /u00/Kingbase/ES/V9/data
(1 row)
```

3.5.1 数据文件

KingbaseES 数据库的数据文件用来存储 KingbaseES 数据库的元数据和用户数据，像表、索引等数据库对象的内容都存储在数据文件中。

KingbaseES 数据库使用**表空间**如图 3-3 所示，来组织用户**数据库**的数据（表或者索引）。**表空间**是数据库对象的物理容器。（TABLESPACE）表空间和数据库之间有多对多的联系：在一个表空间中可以存储多个数据库的表或者索引；一个数据库的表或者索引，可以存储在多个表空间中。

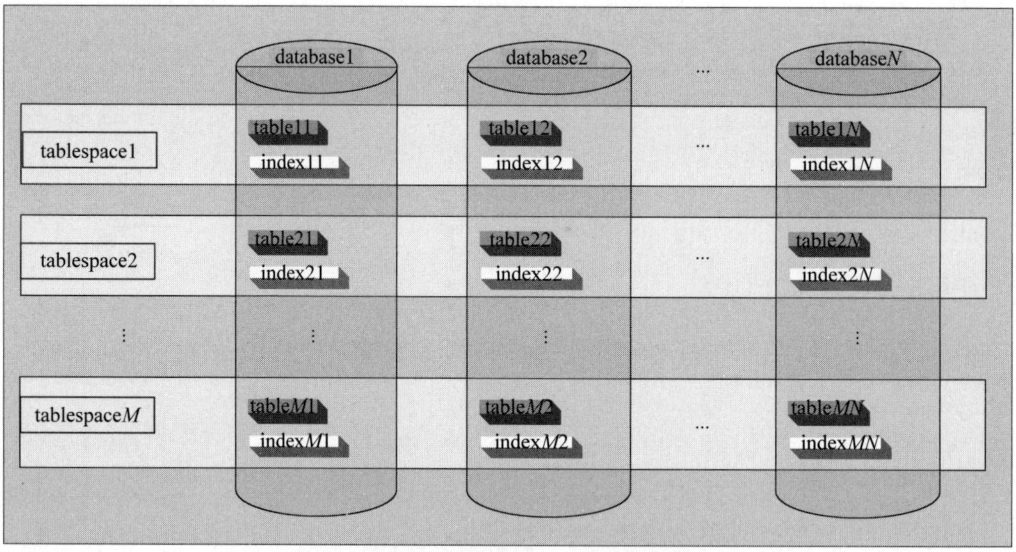

图 3-3　表空间和数据库之间的多对多联系

KingbaseES 数据库的**表空间对应操作系统的一个目录**，在这个操作系统目录下，表空间所存储的数据库有一个以数据库 OID 为名字的子目录。当在数据库中创建诸如表或者索引等数据库对象时，会在这个子目录下存储数据库对象。每个数据库对象对应一个或多个数据文件，数据文件名的前缀是该数据库对象的 OID 值。

系统初始化时，会在 KingbaseES 数据库的数据目录下，默认创建下面的表空间。

(1) 表空间 sys_default：存放数据库中数据，对应的操作系统目录是 data/base。

(2) 表空间 sys_global：存放全局系统表的数据，对应的操作系统目录是 data/global。

(3) 表空间 sysaudit：存放数据库的审计数据信息，对应的操作系统目录是 data/

sysaud。

执行下面的命令和 SQL 语句，查看表空间 sys_default 的情况：

```
system@test=# \! ls -l /u00/Kingbase/ES/V9/data/base/
total 112
drwx------. 2 kingbase dba 12288 Feb 29 21:27 1
drwx------. 2 kingbase dba 12288 Feb 29 21:27 14385
drwx------. 2 kingbase dba 12288 Mar  7 19:04 14386
drwx------. 2 kingbase dba 12288 Mar  7 19:06 14387
drwx------. 2 kingbase dba 12288 Feb 29 21:28 14388
drwx------. 2 kingbase dba 12288 Mar  7 19:11 16628
drwx------. 2 kingbase dba 12288 Mar  7 19:06 16629
drwx------. 2 kingbase dba     6 Mar  7 11:20 syssql_tmp
system@test=# SELECT datname, oid FROM sys_database;
  datname   |  oid
------------+-------
 test       | 14386
 kingbase   | 14387
 template1  |     1
 template0  | 14385
 security   | 14388
 exampledb1 | 16628
 exampledb2 | 16629
(7 rows)
```

从输出可以看到，在表空间 sys_default 对应的操作系统目录 /u00/Kingbase/ES/V9/data/base/ 下，当前存储了 7 个数据库的数据。

在数据库中创建表或索引时，如果不指定表空间，将存放在默认表空间 sys_default 中。

例 3.11 查询数据库对象所对应的数据文件。

在系统默认的表空间中，为数据库 test 创建表 t1：

```
system@test=# CREATE TABLE t1(col1 int, col2 int);
CREATE TABLE                       -- 在默认的表空间下创建表 t1
system@test=#                      -- 查看数据库 test 的 OID
system@test=# SELECT oid, datname, dattablespace FROM sys_database WHERE datname='test';
  oid  | datname | dattablespace
-------+---------+---------------
 14386 | test    |          1663
(1 row)
system@test=#                      -- 查看 OID=1663 的表空间的名字
system@test=# SELECT spcname FROM sys_tablespace WHERE oid = 1663;
   spcname
-------------
 sys_default
(1 row)
system@test=# -- 查询表 t1 的 OID 和 relfilenode
system@test=# SELECT oid,relname,relfilenode FROM sys_class WHERE relname='t1';
  oid  | relname | relfilenode
-------+---------+-------------         -- 默认情况下 OID 和 relfilenode 是相同的，
```

```
       16744 | t1         |       16744          -- 如果在表 t1 上执行过 truncate、alter table 等
                                                 -- 需要重新整理数据的 SQL 语句后,
(1 row)                                          -- relfilenode 的值会发生变化
system@test=#                                    -- 查询表 t1 所对应的操作系统文件
system@test=# SELECT sys_relation_filepath('t1');
 sys_relation_filepath
-----------------------
 base/14386/16744
(1 row)
system@test=# \! ls -l /u00/Kingbase/ES/V9/data/base/14386/16744
-rw-------. 1 kingbase dba 0 Mar.  7 19:14 /u00/Kingbase/ES/V9/data/base/
14386/16744
```

从输出可以看到,表 t1 对应的操作系统文件为 /u00/Kingbase/ES/V9/data/base/14386/16744。

执行下面的 SQL 语句,删除表 t1,则表 t1 对应的操作系统文件也被删除了:

```
system@test=# DROP TABLE t1;
DROP TABLE
system@test=#
--需要稍等一会儿(可能是几分钟),直到执行下面的命令显式表 t1 对应的文件被删除
system@test=# \! ls -l /u00/Kingbase/ES/V9/data/base/14386/16744
ls: cannot access /u00/Kingbase/ES/V9/data/base/14386/16744: No such file
or directory
```

3.5.2 控制文件

控制文件是记录数据库内部状态的重要文件。KingbaseES 数据库的控制文件中包括初始化静态信息、数据库日志和检查点的动态信息,以及一些配置信息。当 KingbaseES 数据库启动时,控制文件必须可供数据库正确读取和写入。如果没有控制文件或控制文件不可读,KingbaseES 数据库将无法启动。

KingbaseES 数据库的控制文件是在初始化数据库集簇时创建的,位于目录 data/global 下,控制文件名为 sys_control。控制文件的静态信息在初始化时自动生成,运行过程中不允许修改,如系统标识符;控制文件的配置信息允许用户在初始化时一次性定制,不再允许修改,如 REDO 日志段的大小;控制文件的 WAL 及检查点的动态信息,在 KingbaseES 数据库运行中动态修改。

1. 控制文件的内容

KingbaseES 数据库提供了命令行工具 sys_controldata 来读取控制文件的信息:

```
[kingbase@dbsvr ~]$ sys_controldata /u00/Kingbase/ES/V9/data
sys_control version number:            1201
Catalog version number:                202305151
Database system identifier:            7341015102399125796
Database cluster state:                in production
sys_control last modified:             Fri 01 Mar 2024 08:25:32 AM CST
Latest checkpoint location:            0/62D2DE48
Latest checkpoint's REDO location:     0/62D2DE18
```

Latest checkpoint's REDO WAL file:	000000010000000000000062
Latest checkpoint's WalTimeLineID:	1
Latest checkpoint's PrevTimeLineID:	1
Latest checkpoint's full_page_writes:	on
Latest checkpoint's NextXID:	0:1424
Latest checkpoint's NextOID:	24663
Latest checkpoint's NextMultiXactId:	1
Latest checkpoint's NextMultiOffset:	0
Latest checkpoint's oldestXID:	1012
Latest checkpoint's oldestXID's DB:	1
Latest checkpoint's oldestActiveXID:	1424
Latest checkpoint's oldestMultiXid:	1
Latest checkpoint's oldestMulti's DB:	1
Latest checkpoint's oldestCommitTsXid:	0
Latest checkpoint's newestCommitTsXid:	0
Time of latest checkpoint:	Fri 01 Mar 2024 08:25:30 AM CST
Fake LSN counter for unlogged rels:	0/3E8
Minimum recovery ending location:	0/0
Min recovery ending loc's timeline:	0
Backup start location:	0/0
Backup end location:	0/0
End-of-backup record required:	no
arg setting:	replica
wal_log_hints setting:	off
max_connections setting:	10
max_worker_processes setting:	8
max_wal_senders setting:	10
max_prepared_xacts setting:	0
MaxLocksPerXact setting:	64
track_commit_timestamp setting:	off
Maximum data alignment:	8
Database block size:	8192
Blocks per segment of large relation:	131072
WAL block size:	8192
Bytes per WAL segment:	16777216
Maximum length of identifiers:	64
Maximum columns in an index:	32
Maximum size of a TOAST chunk:	1988
Size of a large-object chunk:	2048
Date/time type storage:	64-bit integers
Float4 argument passing:	by value
Float8 argument passing:	by value
Data page checksum version:	0
Data page checksum device:	0
Mock authentication nonce:	5f3504ad52ca02b256bda1b8b6a874b9c8a8503957fa0b32f65b27af5e5a8c66
database mode:	1
auth method mode:	0

控制文件中包含以下类型的信息。

(1) 系统信息,这些信息一旦初始化数据库后将不再修改。例如,sys_control version

number(控制文件版本)、Catalog version number(系统表版本号)、Database system identifier(数据库系统的唯一标识符)。数据库系统的唯一标识符是一个 64b 的整数,其中包含了创建数据库的时间戳和使用 initdb 命令时初始化的进程号。创建时间可以通过 to_timestamp 转换查看。

```
system@test=# SELECT to_timestamp(((7341015102399125796>>32) & (2^32 - 1)::bigint));
    to_timestamp
---------------------
 2024-02-29 21:27:51
```

(2) 数据库状态信息,包括 Database cluster state(数据库系统的当前状态)、sys_control last modified(控制文件最后修改时间)、database mode(数据库模式)。其中数据库系统的当前状态有以下几种。

① starting up:表示数据库正在启动状态。

② shut down:数据库实例(非备节点(standby))正常关闭后控制文件中就是此状态。

③ shut down in recovery:standby 实例正常关闭后控制文件中就是此状态。

④ shutting down:正常停库时,先做 checkpoint,开始做 checkpoint 时,会把状态设置为此状态,做完后把状态设置为 shut down。

⑤ in crash recovery:数据库实例非异常停止后,重新启动,会先进行实例的恢复,在实例恢复时的状态就是此状态。

⑥ in archive recovery:standby 实例正常启动后,就是此状态。

⑦ in production:数据库实例正常启动后就是此状态。Standby 数据库正常启动后不是此状态。

(3) 检查点信息、事务和 OID 信息、WAL 和恢复信息。

(4) 系统配置信息,包括数据库的配置设置、数据页、WAL 块大小等,如 max_connections、max_worker_processes、Database block size、WAL block size 等。

(5) 数据类型和存储信息、安全和认证的备份信息,以及表示标识符最大长度、索引中最大列数等其他技术信息,如 Backup start location、Backup end location、End-of-backup record required 以及其他技术信息,如 Maximum length of identifiers、Maximum columns in an index 等。

2. 控制文件的多元化

在进行系统初始化时,默认只有一个控制文件,位于数据库目录下的 global/sys_control。控制文件对于 KingbaseES 数据库的启动过程、恢复、备份和事务一致性至关重要。如果控制文件损坏,系统则不能启动,有可能造成数据的丢失。为了控制文件的安全性,KingbaseES 数据库支持同时写多个控制文件的功能,这个功能称为控制文件多元化(multiplex)。KingbaseES 数据库支持最多 8 个控制文件副本。

例 3.12 设置 KingbaseES 数据库控制文件多元化。

(1) 以用户 kingbase 的身份执行 sys_ctl stop 命令,关闭 KingbaseES 数据库。

(2) 以用户 kingbase 的身份,使用 vi 编辑器编辑 KingbaseES 数据库的启动参数文件/u00/Kingbase/ES/V9/data/kingbase.conf,修改参数 control_file_copy(注意下面是 1 行):

```
control_file_copy = '/u00/Kingbase/ES/V9/data/global/sys_control1;/u00/
Kingbase/ES/V9/data/global/sys_control2;/u00/Kingbase/ES/V9/data/global/sys_
control3'
```

(3) 以用户 kingbase 的身份执行 sys_ctl start 命令,重新启动 KingbaseES 数据库。

(4) 执行下面的命令,查看控制文件的副本:

```
[kingbase@dbsvr ~]$ ksql -d test -U system
system@test=# SHOW control_file_copy;
                             control_file_copy
------------------------------------------------------------------
/u00/Kingbase/ES/V9/data/global/sys_control1;
/u01/Kingbase/ES/V9/data/global/sys_control2;
/u02/Kingbase/ES/V9/data/global/sys_control3            --注意已经美化了输出
 (1 row)
```

从输出可以看到,KingbaseES 数据库现在有 3 个控制文件。

3.5.3 WAL 文件

KingbaseES 数据库的日志文件记录了操作对数据库中数据的所有修改。当系统出现故障时,使用日志文件可以将数据库恢复到一致的状态。

KingbaseES 数据库日志系统遵守 WAL 协议,KingbaseES 数据库日志文件又称 WAL 文件。

(1) 数据页面的每个修改都应该生成日志记录,并且在数据页面刷到磁盘之前,其对应的日志记录必须先刷到持久化的日志设备上。

(2) 数据库的日志记录必须顺序刷出到设备上。

(3) 事务的提交是以该事务的日志记录刷到磁盘上来界定。

在 KingbaseES 数据库中,当 SQL 语句对数据进行修改时,不仅在共享缓冲区中修改相应的数据,还会同时将数据变更记录写到文件缓冲区。共享缓冲区中的数据修改后不会立刻写入磁盘文件中,而是会定期地由 KingbaseES 数据库的后台写进程刷写到数据文件中。在事务提交时,该提交点之前的所有日志记录都会刷写到磁盘上的 WAL 文件中。

为了提高系统的性能,事务成功提交时并不确保该事务的所有数据已经被刷写到数据文件中。因此当系统出现故障时,数据库中的数据就会不一致。在下次启动 KingbaseES 数据库服务器时,就会首先根据控制文件中的信息决定数据库是否需要恢复,如果需要,则根据日志文件把系统恢复到一致的状态,然后系统才正常启动。

因此,利用 WAL 文件保证了数据库事务的原子性和持久性,还可以实现数据库复制。

(1) KingbaseES 数据库实例失败后,重新启动 KingbaseES 数据库时,根据日志文件需要进行数据库的数据恢复。

(2) KingbaseES 数据库执行联机备份(online dackup)时,需要复制数据库集簇的数据文件。由于复制数据库集簇的数据文件需要一定的时间,在此期间 KingbaseES 数据库会继续修改这些数据文件。在恢复时,首先还原数据库备份,然后将 WAL 文件(也许还有归档日志文件)应用到还原的数据库备份上,确保数据库集簇的数据文件能够恢复到最新的一致状态。

（3）使用 WAL 文件还可以实现 KingbaseES 数据库复制。对于物理复制，将 WAL 文件从源服务器传输到目标服务器，并重做这些 WAL 文件。对于逻辑复制，在源服务器上根据 WAL 文件提取 SQL 语句，并在目标服务器上执行这些 SQL 语句。

例 3.13 查看 KingbaseES 数据库当前的 WAL 文件信息。

WAL 文件位于目录 /u00/Kingbase/ES/V9/data/sys_wal 下。每个 WAL 文件的大小由参数 wal_segment_size 控制，默认为 16MB。可以使用 SQL 语句查看 WAL 文件的信息，也可以使用操作系统命令查看日志文件。

```
system@test=# SELECT pg_ls_waldir();
                  pg_ls_waldir
--------------------------------------------------------------
 (000000010000000000000007E,16777216,"2024-03-07 11:14:01+08")
 (0000000100000000000000095,16777216,"2024-03-07 11:14:18+08")
--省略了一些输出
(64 rows)
system@test=# \q
[kingbase@dbsvr ~]$ ls -lh /u00/Kingbase/ES/V9/data/sys_wal
total 1.0G
-rw-------. 1 kingbase dba 16M Mar  7 19:20 00000001000000000000006B
-rw-------. 1 kingbase dba 16M Mar  7 11:14 00000001000000000000006C
--省略了一些输出
drwx------. 2 kingbase dba   6 Feb 29 21:27 archive_status
[kingbase@dbsvr ~]$
```

在 7.7 节，将更为深入地介绍 WAL 文件的相关内容。

3.5.4 配置文件

KingbaseES 数据库的配置文件定制了 KingbaseES 数据库的运行状态，对 KingbaseES 数据库的高效正确地运行至关重要。主要有以下三种配置文件：

（1）系统配置参数文件 kingbase.conf；

（2）客户端访问控制文件 sys_hba.conf；

（3）本地操作系统用户与数据库用户的身份映射文件 sys_ident.conf。

例 3.14 查看 KingbaseES 数据库配置文件的位置。

```
[kingbase@dbsvr ~]$ ksql -d test -U system
system@test=# SELECT name, setting
test-#          FROM pg_settings
test-#          WHERE name IN ('config_file','hba_file','ident_file');
   name     |                setting
------------+----------------------------------------------
 config_file| /u00/Kingbase/ES/V9/data/kingbase.conf   --启动参数文件
 hba_file   | /u00/Kingbase/ES/V9/data/sys_hba.conf    --客户端访问控制文件
 ident_file | /u00/Kingbase/ES/V9/data/sys_ident.conf
                                      --本地操作系统用户与数据库用户的身份映射文件
(3 rows)
```

在4.2节，将深入地讨论配置文件 kingbase.conf；在5.3节，将深入地讨论配置文件 sys_hba.conf；在5.4节，将深入地讨论配置文件 sys_ident.conf。

3.6 系统表与系统视图

系统表与**系统视图**又称**数据字典**，存储数据库中数据的元信息，如表定义、函数定义、系统的统计信息、用户权限信息等。KingbaseES 数据库会使用系统表中的信息对 SQL 语句进行分析、重写和优化，DBA 和数据库用户也可以通过它们查询数据库中的数据情况。

从存储结构上看，KingbaseES 数据库的系统表与普通的用户表相同，只是用户不能（不应该）通过 INSERT、UPDATE、DELETE 语句对其进行修改，只能查询其中的数据信息。

在执行 DDL 语句时，KingbaseES 数据库会更新系统表来记录数据库对象的定义；在执行 vacuum 或 analyze 命令时，系统会更新系统表中的统计信息。

3.6.1 系统表与系统视图的类别

KingbaseES 数据库的系统表和系统视图主要有以下几个模式。

（1）sys_catalog：KingbaseES 数据库的系统表和系统视图。

（2）pg_catalog：兼容 PostgreSQL 数据库的系统视图。

（3）sys：兼容 Oracle 数据库的系统视图。

（4）information_schema：SQL 标准定义的系统视图。

KingbaseES 数据库的系统表和系统视图以 sys_ 开头命名，DBA 经常查询的系统表如下。

（1）sys_database：存储 KingbaseES 数据库中数据库的信息。

（2）sys_namespace：存储 KingbaseES 数据库中数据库模式的信息。

（3）sys_class：存储 KingbaseES 数据库中表或索引等数据库对象的信息。

（4）sys_attribute：存储 KingbaseES 数据库中表或索引的属性信息。

（5）sys_proc：存储 KingbaseES 数据库中函数或存储过程信息。

（6）sys_tablespace：存储 KingbaseES 数据库中表空间信息。

KingbaseES 数据库在 sys 模式下提供的兼容 Oracle 数据库的视图，分为以下几类。

（1）具有 DBA_ 前缀的视图显示在整个数据库中的所有相关信息。

（2）具有 ALL_ 前缀的视图，是站在用户角度，从整体上看待数据库。

（3）具有 USER_ 前缀的视图指的是用户在数据库中的私有环境，包括用户所创建的数据库对象的元数据，对该用户的授权等。

（4）DUAL 是数据字典中的一个很小的表，数据库和用户编写的程序可以引用它，以保证一个已知的结果。

KingbaseES 数据库还提供一些动态视图来记录数据库操作过程中的当前活动状态。例如，系统和会话参数、内存使用和分配、CPU 和 I/O 使用情况、文件状态、会话和事务进度等，这些视图信息是动态的，在数据库运行过程中会不断更新。

动态视图基于从数据库内存结构生成的虚拟表，不是存储在数据库中的常规表，其数据取决于数据库和实例的状态。由于数据是动态更新的，所以不需要保证视图的一致性。

数据库管理工具可以使用这些视图获取有关数据库的信息。DBA 也可以使用这些视图，用于性能监控和调试。例如，查询 sys_stat_activity 视图可以看到数据库所有实时的会话信息。

3.6.2 查询数据库对象信息

下面的示例，通过系统表和系统视图来查询数据库对象信息。

例 3.15 查看数据库集簇有哪些数据库及数据库的属主。

```
system@test=# SELECT datname, sys_catalog.sys_get_userbyid(datdba) AS owner
FROM sys_database;
  datname  | owner
-----------+--------
 test      | system
 kingbase  | system
--省略了一些输出
(7 rows)
```

例 3.16 查看当前数据库中有哪些模式。

```
system@test=# SELECT nsp.oid,nspname, rolname
test-#          FROM sys_namespace nsp,sys_authid au
test-#          WHERE nsp.nspowner= au.oid;
  oid  |     nspname      | rolname
-------+------------------+---------
  7012 | pg_bitmapindex   | system
--省略了一些输出
(15 rows)
```

例 3.17 查看当前数据库中不同数据库对象的数量。

```
system@test=# --查询系统中表的数量,其中:
system@test=# --'r'表示普通表,'t'表示 TOAST 表,'f'表示外部表,'p'表示分区表
system@test=# SELECT count(*)
test-#          FROM sys_class
test-#          WHERE relkind='r' or relkind='t' or relkind='f' or relkind='p';
 count
-------
   175
(1 row)
system@test=# --查询系统中视图的数量
system@test=# SELECT count(*) FROM sys_class WHERE relkind='v';
 count
-------
   499
(1 row)
system@test=# --查询系统中序列的数量
system@test=# SELECT count(*) FROM sys_class WHERE relkind='S';
 count
-------
     2
```

```
(1 row)
system@test=# --查询系统中函数的数量,其中'f'表示普通函数,'a'表示聚集函数,'w'表示窗
              --口函数
system@test=# SELECT count(*) FROM sys_proc WHERE prokind='f' or prokind='a' or
prokind='w';
 count
-------
  4461
(1 row)
system@test=# --查询系统中存储过程的数量
system@test=# SELECT count(*) FROM sys_proc WHERE prokind='p';
 count
-------
    41
(1 row)
system@test=# \q
[kingbase@dbsvr ~]$
```

另外,从 sys_tables、sys_indexes、sys_views、sys_triggers 等这些系统视图中可以很方便查询出数据库对象的信息,如对象名、对象所在的模式、对象的属主等信息。

在第 2 章中介绍了使用 ksql 的元命令查询数据库对象的信息。实际上 ksql 的这些元命令就是通过执行对系统表和系统视图的查询来实现的。使用 ksql 命令的-E 选项,可以在执行元命令时,显示实际执行的 SQL 语句。

例 3.18 查看当前 KingbaseES 数据库中有哪些数据库。

```
[kingbase@dbsvr ~]$ ksql -d test -U system -E
system@test=# \l
********* QUERY **********
 SELECT d.datname as "Name",
   pg_catalog.pg_get_userbyid(d.datdba) as "Owner",
       pg_catalog.pg_encoding_to_char(d.encoding) as "Encoding",
       d.datcollate as "Collate",
   d.datctype as "Ctype",
       pg_catalog.array_to_string(d.datacl, E'\n') AS "Access privileges"
FROM pg_catalog.pg_database d
ORDER BY 1;
**************************
                List of databases
    Name     | Owner  | Encoding |  Collate   |   Ctype    | Access privileges
-------------+--------+----------+------------+------------+------------------
 exampledb1  | system | UTF8     | zh_CN.UTF-8| zh_CN.UTF-8|
--省略了一些输出
(7 rows)
```

这个例子中执行 ksql 的元命令\l,同时显示了其对应的 SQL 语句,阅读这些 SQL 语句,可以更好地了解 KingbaseES 数据库的系统表和系统视图。

3.7 连接和会话

KingbaseES 数据库支持众多的客户端驱动程序：JDBC、ODBC、Python、PHP-DBI、PHP-PDO、ADO.NET、Node.js、DCI、GOKb 等。KingbaseES 数据库客户端程序使用这些驱动程序，创建与 KingbaseES 数据库服务器的连接和会话。

连接是客户端用户进程和 KingbaseES 数据库服务器之间的一个通信信道。在连接的基础上，数据库用户通过了 KingbaseES 数据库的身份认证后，可以创建一个**会话**。

客户端用户进程与 KingbaseES 数据库服务器建立连接会话的过程如图 3-4 所示。

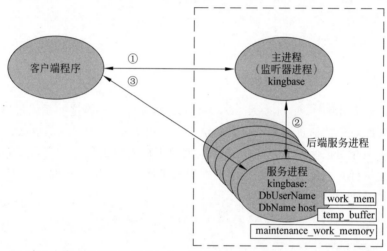

图 3-4　客户端用户进程与 KingbaseES 数据库服务器建立连接会话的过程

客户端用户进程首先发出一个连接请求，kingbase 进程收到这个请求后，向客户端用户进程索要身份认证信息。如果客户端用户进程没有提供正确的身份认证信息（如密码），这个连接将被销毁。如果客户端用户进程向 kingbase 进程发送了正确密码，kingbase 进程将为用户创建一个新的后端服务进程，名字为 kingbase: DbUserName DbName host，并将与这个新创建的后端服务进程建立的会话信息，传回给客户端用户进程，客户端用户进程使用这些信息与这个新创建的后端服务进程建立直接的联系，此时客户端用户进程与 KingbaseES 数据库的后端服务进程之间，建立了一个会话。

客户端用户进程将要执行的 SQL 语句，通过这个会话发送给这个特定的后端服务进程，由这个后端服务进程执行这些 SQL 语句，并向客户端用户进程返回执行结果。

从上述的过程看，连接建立在客户端用户进程和 kingbase 进程之间，会话建立在客户端用户进程和后端服务进程之间。在不引起混淆的情况下，客户端不区分连接和会话，称为一个会话连接。

KingbaseES 数据库的用户、会话连接与数据库之间的关系如下。

（1）一个用户在一个会话连接上的只能访问一个数据库。

（2）一个用户可以使用多个会话连接同时访问同一个数据库。

（3）一个用户可以使用多个会话连接同时访问多个不同的数据库。

3.8 SQL 语句的执行过程

SQL 语句的正确执行需要 KingbaseES 数据库服务器的多个组件互相配合,如数据缓存区、日志缓冲区、后台进程等。用户执行 SQL 语句首先需要与数据库实例建立会话连接,这时 KingbaseES 数据库服务器就会创建一个专用的后端服务进程为该用户服务。根据不同类型的 SQL 语句,后端服务进程使用不同的组件。

(1) SELECT 语句需要把结果集返回给客户端用户进程。

(2) DML 语句需要日志系统。

(3) COMMIT 语句必须保证一个事务可以恢复。

有些数据库后台进程并不参与 SQL 语句的处理,只是为了提高数据库性能。

以 DML 语句为例,说明 SQL 语句执行的关键步骤。

(1) 后端服务进程接收到 SQL 语句后,如果是 DML 语句,看需要的数据是否在数据缓冲区,如果在,则直接执行步骤(3),否则执行步骤(2)。

(2) 从数据库文件中将数据调入数据缓冲区。

(3) 完成数据缓冲区中数据的修改。

(4) 在 REDO 日志中生成该操作的日志,记录数据的修改,写到日志缓冲区。

(5) 给客户端用户进程返回 SQL 语句的执行情况,成功还是失败,修改的数据元组数。

如果后端服务进程接收到 COMMIT 语句,则生成一条事务提交日志记录写到日志缓存区,并把该日志之前所有的日志都写到日志文件,给客户端用户进程返回 SQL 语句的执行情况。

第 4 章

管理 KingbaseES 数据库实例

KingbaseES 数据库服务器由两个部分组成：KingbaseES 数据库实例和 KingbaseES 数据库集簇。本章将聚焦于 KingbaseES 数据库实例的管理，包括数据库实例的启动与关闭、系统配置参数、管理数据库扩展插件和数据库软件许可证等。

4.1 数据库实例的启动与关闭

4.1.1 启动数据库

KingbaseES 数据库在给用户提供服务前，必须启动 KingbaseES 数据库实例。

启动 KingbaseES 数据库实例时，首先检测并处理系统环境变量；然后运行 kingbase 进程，运行前会先读取 KingbaseES 数据库的系统配置文件 kingbase.conf，根据其中的参数值来分配 KingbaseES 数据库实例内存、处理信号、打开监听和 socket 通道，接下来会继续启动 KingbaseES 数据库的各个功能后台进程，最后监听端口等待来自 KingbaseES 数据库客户端的连接请求。

在启动过程中，如果发现 KingbaseES 数据库实例的最后一次关闭是非正常关闭，例如，

（1）以 immediate 模式关闭 KingbaseES 数据库实例。

（2）由于发生实例故障而关闭。

① KingbaseES 数据库服务器硬件故障导致的宕机。

② 操作系统或者 KingbaseES 数据库软件故障导致的宕机。

③ 系统掉电导致的宕机。

此时，kingbase 进程在启动的过程中将派生一个实例恢复进程执行实例恢复，执行完恢复操作后该进程退出，kingbase 进程继续 KingbaseES 数据库服务器的启动过程。

启动 KingbaseES 数据库实例有以下两种方式。

（1）直接运行 kingbase 程序启动。

（2）使用 sys_ctl 命令行工具启动。

kingbase 程序是 KingbaseES 数据库服务器的主程序，它位于：

/u00/Kingbase/ES/V9/KESRealPro/V009R001C001B0025/Server/bin/kingbase

DBA 可以在 KingbaseES 数据库服务器上直接运行 kingbase 程序启动 KingbaseES 数

据库实例：

```
[kingbase@dbsvr ~]$ sys_ctl stop   # 关闭正在运行的 KingbaseES 数据库实例
waiting for server to shut down.... done
server stopped
[kingbase@dbsvr ~]$ # 直接运行 kingbase 程序启动 KingbaseES 数据库实例
[kingbase@dbsvr ~]$ kingbase -D /u00/Kingbase/ES/V9/data >logfile 2>&1 &
[1] 4454
[kingbase@dbsvr ~]$ kbps
kingbase  74664  73558  0 21:25 pts/3    00:00:00 kingbase - D /u00/Kingbase/ES/V9/data
kingbase  74671  74664  0 21:25 ?        00:00:00 kingbase: logger
#省略了一些输出
[kingbase@dbsvr ~]$
```

kingbase 程序的-D 选项用于指定 KingbaseES 数据库的数据目录位置。

sys_ctl 用于启动、停止、重启，以及管理 KingbaseES 数据库实例，它位于：

/u00/Kingbase/ES/V9/kingbase/KESRealPro/V009R001C001B0025/Server/bin/sys_ctl

DBA 可以使用 sys_ctl 来停止和启动 KingbaseES 数据库实例：

```
[kingbase@dbsvr ~]$ sys_ctl stop   # 关闭正在运行的 KingbaseES 数据库实例
waiting for server to shut down.... done
server stopped
[1]+  Done                    kingbase -D /u00/Kingbase/ES/V9/data > logfile 2>&1
[kingbase@dbsvr ~]$ sys_ctl -D /u00/Kingbase/ES/V9/data start
                         # 启动 KingbaseES 数据库
 waiting for server to start.... 2024 - 01 - 24 10: 54: 28.726 CST [4453] LOG:
sepapower extension initialized
#省略了一些输出
 done
server started
[kingbase@dbsvr ~]$
```

建议 KingbaseES 数据库的 DBA，为用户 kingbase 配置以下几个环境变量。

(1) KINGBASE_HOME。

(2) KINGBASE_DATA。

(3) KINGBASE_PORT。

设置好这几个环境变量之后，使用 sys_ctl 启动 KingbaseES 数据库实例时，就不再需要使用-D 选项指定 KingbaseES 数据库的数据目录了：

```
[kingbase@dbsvr ~]$ echo $KINGBASE_HOME
/u00/Kingbase/ES/V9/kingbase
[kingbase@dbsvr ~]$ echo $KINGBASE_DATA
/u00/Kingbase/ES/V9/data
[kingbase@dbsvr ~]$ echo $KINGBASE_PORT
54321
[kingbase@dbsvr ~]$ sys_ctl stop
                  # 关闭 KingbaseES 数据库实例,不再需要使用-D 选项指定数据目录
```

```
waiting for server to shut down.... done
server stopped
[kingbase@dbsvr ~]$ sys_ctl start
                            # 启动 KingbaseES 数据库实例,不再需要使用-D 选项指定数据目录
waiting for server to start....2024-01-24 11:04:22.173 CST [4640] LOG:  sepapower
extension initialized
#省略了一些输出
server started
[kingbase@dbsvr ~]$
```

建议将这几个环境变量设置在初始化文件/homg/kingbase/.bashrc 中。

4.1.2 关闭数据库

为了维护数据的安全性和完整性,减少数据损坏的风险,以及保证数据库系统的稳定性和可靠性,应当在关闭运行 KingbaseES 数据库的服务器硬件之前,先关闭 KingbaseES 数据库实例,以规避直接关机可能会导致数据库中数据处于不一致的状态,甚至损坏。

关闭 KingbaseES 数据库实例有三种模式。

(1) smart 模式:不再允许新的连接,但是允许所有活跃的会话正常完成工作,必须等待所有客户端断开连接后才关闭 KingbaseES 数据库实例。

(2) fast 模式:默认的关闭模式,开始关闭后,不允许建立新的数据库连接,已有的连接将回滚未提交的事务,回滚结束后将关闭 KingbaseES 数据库实例。

(3) immediate 模式:不等待现有连接回滚事务,立即关闭 KingbaseES 数据库实例,属于危险操作,相当于发生了一次 KingbaseES 数据库实例故障,会导致在重启 KingbaseES 数据库时,需要进行数据库恢复(Recovery)操作。建议只有在紧急的情况下,才使用这个模式关闭数据库实例。

KingbaseES 数据库的 DBA 可以使用以下两种方法来关闭 KingbaseES 数据库实例。

(1) 使用 Linux 操作系统命令 kill,通过给 kingbase 进程发以下的三种信号来关闭 KingbaseES 数据库实例。

① 发送信号 SIGTERM,以 smart 模式关闭数据库实例。

② 发送信号 SIGINT,以 fast 模式关闭数据库实例。

③ 发送信号 SIGOUIT,以 immediate 模式关闭数据库实例。

(2) 使用 sys_ctl 的-m 选项来关闭 KingbaseES 数据库实例。

① -m smart:以 smart 模式关闭数据库实例。

② -m fast:以 fast 模式关闭数据库实例。

③ -m immediate:以 immediate 模式关闭数据库实例。

例 4.1 使用命令 kill 关闭 KingbaseES 数据库实例。

```
 [kingbase@dbsvr ~]$ sys_ctl status         # 查看 KingbaseES 数据库的运行状态
sys_ctl: server is running (PID: 4920)
/u00/Kingbase/ES/V9/KESRealPro/V009R001C001B0025/Server/bin/kingbase
[kingbase@dbsvr ~]$ # 以 smart 模式关闭正在运行的 KingbaseES 数据库实例
[kingbase@dbsvr ~]$ kill -TERM `head -1 /u00/Kingbase/ES/V9/data/kingbase.pid`
[kingbase@dbsvr ~]$ sys_ctl status         # 查看 KingbaseES 数据库的运行状态
```

```
sys_ctl: no server running
[kingbase@dbsvr ~]$ sys_ctl start            # 启动 KingbaseES 数据库实例
waiting for server to start....2024-01-24 11:21:07.438 CST [5059] LOG:  sepapower
extension initialized
#省略了一些输出
server started
[kingbase@dbsvr ~]$ # 以 fast 模式关闭正在运行的 KingbaseES 数据库实例
[kingbase@dbsvr ~]$ kill -QUIT `head -1 /u00/Kingbase/ES/V9/data/kingbase.pid`
[kingbase@dbsvr ~]$ sys_ctl status           # 查看 KingbaseES 数据库的运行状态
sys_ctl: no server running
[kingbase@dbsvr ~]$ sys_ctl start            # 启动 KingbaseES 数据库实例
waiting for server to start....2024-01-24 11:23:15.244 CST [5115] LOG:  sepapower
extension initialized
#省略了一些输出
server started
[kingbase@dbsvr ~]$ # 以 immediate 模式关闭正在运行的 KingbaseES 数据库实例
[kingbase@dbsvr ~]$ kill -INT `head -1 /u00/Kingbase/ES/V9/data/kingbase.pid`
[kingbase@dbsvr ~]$ sys_ctl status           # 查看 KingbaseES 数据库的运行状态
sys_ctl: no server running
[kingbase@dbsvr ~]$ sys_ctl start            # 启动 KingbaseES 数据库实例
waiting for server to start....2024-01-24 11:26:25.185 CST [5185] LOG:  sepapower
extension initialized
#省略了一些输出
server started
[kingbase@dbsvr ~]$
```

例 4.2 使用 sys_ctl 关闭 KingbaseES 数据库实例。

```
[kingbase@dbsvr ~]$ sys_ctl status           # 查看 KingbaseES 数据库的运行状态
sys_ctl: server is running (PID: 5185)
/u00/Kingbase/ES/V9/KESRealPro/V009R001C001B0025/Server/bin/kingbase
[kingbase@dbsvr ~]$ sys_ctl stop -m smart    #以 smart 模式停止 KingbaseES 数据库实例
waiting for server to shut down.... done
server stopped
[kingbase@dbsvr ~]$ sys_ctl status           # 查看 KingbaseES 数据库的运行状态
sys_ctl: no server running
[kingbase@dbsvr ~]$ sys_ctl start            # 启动 KingbaseES 数据库实例
waiting for server to start....2024-01-24 11:29:50.776 CST [5279] LOG:  sepapower
extension initialized
#省略了一些输出
server started
[kingbase@dbsvr ~]$ sys_ctl status           # 查看 KingbaseES 数据库的运行状态
sys_ctl: server is running (PID: 5279)
/u00/Kingbase/ES/V9/KESRealPro/V009R001C001B0025/Server/bin/kingbase
[kingbase@dbsvr ~]$ sys_ctl stop -m fast     #以 fast 模式停止 KingbaseES 数据库实例
waiting for server to shut down.... done
server stopped
[kingbase@dbsvr ~]$ sys_ctl status           # 查看 KingbaseES 数据库的运行状态
sys_ctl: no server running
[kingbase@dbsvr ~]$ sys_ctl start            # 启动 KingbaseES 数据库实例
waiting for server to start....2024-01-24 11:32:44.377 CST [5384] LOG:  sepapower
```

```
extension initialized
#省略了一些输出
server started
[kingbase@dbsvr ~]$ sys_ctl status            # 查看 KingbaseES 数据库的运行状态
sys_ctl: server is running (PID: 5384)
/u00/Kingbase/ES/V9/KESRealPro/V009R001C001B0025/Server/bin/kingbase
[kingbase@dbsvr ~]$ sys_ctl stop -m immediate
                                              #以 immmediate 模式停止 KingbaseES 数据库实例
waiting for server to shut down.... done
server stopped
[kingbase@dbsvr ~]$ sys_ctl status            # 查看 KingbaseES 数据库的运行状态
sys_ctl: no server running
[kingbase@dbsvr ~]$ sys_ctl start             # 启动 KingbaseES 数据库实例
[kingbase@dbsvr ~]$
```

4.1.3 kingbase 服务

安装 KingbaseES 数据库时，使用用户 root 执行 root.sh 脚本将会为 KingbaseES 数据库生成 Linux 操作系统的服务 kingbase。因此可以使用 Linux 操作系统的服务管理命令 systemctl 来管理 kingbaseES 数据库：

```
[root@dbsvr ~]# systemctl status   kingbase       # 查看 KingbaseES 数据库的状态
● kingbased.service - LSB: Start and stop the kingbase server
   Loaded: loaded (/etc/rc.d/init.d/kingbased; bad; vendor preset: disabled)
   Active: active (exited) since Wed 2024-01-24 02:33:44 CST; 9h ago
     Docs: man:systemd-sysv-generator(8)
  Process: 1480 ExecStart=/etc/rc.d/init.d/kingbased start (code=exited, status
=0/SUCCESS)
    Tasks: 0
#省略了一些输出
[root@dbsvr ~]# systemctl stop     kingbase       # 停止 KingbaseES 数据库实例
[root@dbsvr ~]# systemctl status   kingbase       # 查看 KingbaseES 数据库的状态
● kingbased.service - LSB: Start and stop the kingbase server
   Loaded: loaded (/etc/rc.d/init.d/kingbased; bad; vendor preset: disabled)
   Active: failed (Result: exit-code) since Wed 2024-01-24 11:40:50 CST; 52s ago
     Docs: man:systemd-sysv-generator(8)
  Process: 69199 ExecStop=/etc/rc.d/init.d/kingbased stop (code=exited, status
=1/FAILURE)
  Process: 1480 ExecStart=/etc/rc.d/init.d/kingbased start (code=exited, status
=0/SUCCESS)
#省略了一些输出
[root@dbsvr ~]# systemctl start    kingbase       # 启动 KingbaseES 数据库实例
[root@dbsvr ~]# systemctl restart  kingbase       # 重启 KingbaseES 数据库
[root@dbsvr ~]#
```

4.2 系统配置参数

KingbaseES 数据库的系统配置参数在数据库管理和优化中扮演着关键角色。系统配置参数允许 DBA 调整和优化 KingbaseES 数据库实例的行为，以满足不同的应用需求、性能目标和操作环境。

4.2.1 系统配置参数概述

每次启动 KingbaseES 数据库服务器时，kingbase 进程会读取 KingbaseES 数据库的系统配置文件 kingbase.conf，并根据读到的系统配置参数值，为数据库用户提供指定的数据库功能特性。

系统视图 sys_settings 中包含系统配置参数的各种信息，例如，参数描述、取值范围、当前取值等。查询系统视图 sys_settings 可以获取系统配置参数的相关信息。

1. 系统配置参数的基本信息

在系统配置文件 kingbase.conf 中的系统配置参数，从功能的角度可以分为连接相关的参数、资源使用相关的参数、WAL 文件相关的参数、查询优化相关的参数等。

系统视图 sys_settings 的 category 字段，指出了系统配置参数所属的分类。下面的查询可以获得系统配置参数的功能分类：

```
[kingbase@dbsvr ~]$ ksql -d test -U system
system@test=# SELECT DISTINCT category FROM sys_settings ORDER BY category;
                  category
----------------------------------------------------------
 Autovacuum
#省略了一些输出
(45 rows)
```

从输出可以看到，KingbaseES 数据库的系统配置参数有 45 个类别。

从系统视图 sys_settings 中查询可以获得参数的描述信息，包括参数的类别、名称、单位、参数说明、默认值等信息。

例 4.3　查询系统最大连接数的参数信息。

```
system@test=# SELECT name,setting,reset_val,unit,short_desc
test-#        FROM sys_settings
test-#        WHERE name like 'max_connections';
      name       | setting | reset_val | unit |            short_desc
-----------------+---------+-----------+------+----------------------------------
 max_connections | 100     | 100       |      |
                                                Sets the maximum number of concurrent connections.
(1 row)
test=#
```

其中：

(1) unit 字段是系统配置参数的单位；

(2) setting 字段是系统配置参数的当前值；

(3) reset_val 字段是重新启动后或者重新加载后系统配置参数的新值；

(4) short_desc 字段是关于该系统配置参数的简短说明。

使用 show 命令可以显示系统配置参数的当前取值，也可以从系统视图 sys_settings 中查询系统配置参数的当前取值。

例 4.4 查询当前数据库服务器数据缓冲区的大小。

```
system@test=# SHOW  shared_buffers;
 shared_buffers
----------------
 128MB
(1 row)
system@test=# SELECT name, setting, unit FROM sys_settings WHERE name='shared_
buffers';
      name       | setting | unit
-----------------+---------+------
 shared_buffers  | 16384   | 8kB
(1 row)
```

这两种方式都查出了当前数据缓冲区大小是 128MB,只是显示方式不同,从系统视图 sys_settings 查询出来的值是 16 384×8KB,也就是 128MB。

2. 系统配置参数的取值

KingbaseES 数据库的系统配置参数的取值有以下五种数据类型。

(1) 布尔值:布尔值是大小写无关的,可以是 on/off、1/0、yes/no、true/false。

(2) 字符串:字符串的值包含在单引号内。如果字符串内部包含单引号,则需要双写。如果字符串是数字,可以不使用单引号。

(3) 整数:可以在整数后面指定单位,例如,一些内存配置参数可以指定 KB、MB 或 GB 等。如果不指定,系统会有一个默认的单位。

(4) 浮点数:不能使用引号和逗号(千位分隔符),只能有一个小数点。

(5) 枚举值:枚举值需要包含括号内,只能取有限个值。

可以在系统视图 sys_settings 中查询系统配置参数的数据类型、当前值、取值范围等信息。如果参数的数据类型是枚举值,还可以查出其可能的取值。

例 4.5 查询各种类型系统配置参数的取值信息。

```
system@test=# SELECT  name, setting, vartype, min_val, max_val,enumvals
test-#         FROM sys_settings
test-#         WHERE name IN ('deadlock_timeout','log_directory',
test(#                'wal_level','seq_page_cost', 'max_connections','autovacuum')
test-#         ORDER BY  name;
      name        | setting | vartype | min_val |  max_val    |    enumvals
------------------+---------+---------+---------+-------------+------------------
 autovacuum       | on      | bool    |         |             |
 deadlock_timeout | 1000    | integer | 1       | 2147483647  |
 log_directory    | sys_log | string  |         |             |
 max_connections  | 100     | integer | 1       | 262143      |
 seq_page_cost    | 1       | real    | 0       | 1.79769e+308|
 wal_level        | replica | enum    |         |             |
                                                               {minimal,replica,logical}
(6 rows)
```

其中:

(1) setting 字段显示了系统配置参数的当前值;

（2）vartype 字段显示了系统配置参数的数据类型；

（3）min_val 字段显示了系统配置参数（数据类型为整数或浮点数时）的最小值；

（4）max_val 字段显示了系统配置参数（数据类型为整数或浮点数时）的最大值；

（5）emunvals 字段显示了系统配置参数（数据类型为枚举时）的所有枚举值。

3. 系统配置参数的级别

在 KingbaseES 数据库实例启动的时候读取系统配置文件可以设置系统配置参数，在 KingbaseES 数据库正常运行的时候，还可以使用 SQL 语句设置系统配置参数的值。但是，在 KingbaseES 数据库运行期间，并不是任何参数在任意情况下都可以修改其值。有些参数不能修改，有些参数重新设置后，必须重新启动 KingbaseES 数据库实例才能生效，有些参数只能超级用户才能修改。

在 KingbaseES 数据库中，系统配置参数分为以下几种级别。

（1）internal：该级别参数是只读参数，这些参数在 KingbaseES 数据库安装或初始化时就已经确定且不能再进行修改。例如，数据块的大小 block_size，在系统初始化指定后，就不能再修改。

（2）kingbase：该级别参数是实例级参数，如果修改了这类参数后，必须重新启动 KingbaseES 数据库实例才能生效，这类参数只能在系统配置文件中修改。例如，数据库服务器的数据缓冲区配置参数 shared_buffers。

（3）sighup：该级别参数也是实例级参数，但是修改该类参数后，不需要重新启动 KingbaseES 数据库实例，只需要向 kingbase 进程发送 SIGHUP 信号就可以在所有会话连接中生效。例如，设置系统自动清理空间的参数 autovacuum。

（4）backend/superuser-backend：该级别参数也是实例级参数，修改该系统配置参数后，不需要重新启动 KingbaseES 数据库实例，只需要向 kingbase 进程发送 SIGHUP 信号就可以重新加载该参数，但是只影响未来新创建的连接会话，对已有的连接会话没有影响。例如，设置在系统运行日志中记录用户登录信息的参数 log_connection。

（5）superuser/user：该级别参数是会话级参数，可以在会话中使用 SQL 语句修改该类参数并立即生效。superuser 表示只能由超级用户修改，user 表示可以由所有用户修改。

系统视图 sys_settings 中的 context 列说明该系统配置参数的级别。

例 4.6 查询系统配置参数的级别。

```
system@test=# SELECT name,context,setting,unit FROM sys_settings
test-#          WHERE name IN ('deadlock_timeout','shared_buffers','block_size',
test(#                         'work_mem','log_connections','autovacuum')
test-#          ORDER BY  name,context;
      name       |       context       | setting | unit
-----------------+---------------------+---------+------
 autovacuum      | sighup              | on      |
 block_size      | internal            | 8192    |
 deadlock_timeout| superuser           | 1000    | ms
 log_connections | superuser-backend   | off     |
 shared_buffers  | kingbase            | 16384   | 8kB
 work_mem        | user                | 4096    | kB
(6 rows)
```

4.2.2 设置系统配置参数

设置 KingbaseES 数据库系统配置参数的方法有如下几种。
(1) 在系统配置文件 kingbase.conf 中,设置系统配置参数的值。
(2) 使用 SQL 语句设置系统配置参数。
① 使用 ALTER SYSTEM 语句设置系统配置参数。
② 使用 ALTER DATABASE 语句为指定的数据库设置系统配置参数。
③ 使用 ALTER USER 语句为指定的数据库用户设置系统配置参数。
④ 使用 ALTER SESSION 语句为当前会话设置系统配置参数。
⑤ 使用 SET 语句为当前会话或事务设置系统配置参数。
(3) 在数据库启动的命令行中设置系统配置参数。

1. 在系统配置文件中设置系统配置参数

系统配置文件 kingbase.conf 位于数据目录下(/u00/Kingbase/ES/V9/data)。在启动 KingbaseES 数据库实例时,kingbase 进程会读取系统配置文件来获得系统配置参数的值。如果在系统配置文件 kingbase.conf 中没有设置某个系统配置参数,则将使用默认值。

系统配置文件 kingbase.conf 中可以设置除了 internal 级别的参数外的所有系统配置参数,每个系统配置参数的设置格式如下:

参数名 = 参数值

在系统配置文件 kingbase.conf 中,注释的语法采用与 shell 脚本语法一样,即采用符号"#"进行注释。

2. 使用 ALTER SYSTEM 语句设置系统配置参数

可以在 KingbaseES 数据库系统正常运行的时候,使用 ALTER SYSTEM 语句设置系统配置参数的值。语法如下:

```
ALTER SYSTEM SET    parameterName TO [value|default]
ALTER SYSTEM SET    parameterName=value
ALTER SYSTEM RESET parameterName
```

前两条语句,将系统配置参数 parameterName 的值设置为指定的值 value 或默认值 default。最后一条语句,将系统配置参数 parameterName 的值,恢复为之前的值。

ALTER SYSTEM 语句可以设置除了 internal 级别的参数外的所有系统配置参数值。通过 ALTER SYSTEM 语句设置系统配置参数值,实际执行的操作是把对系统配置参数的设置写到文件 kingbase.auto.conf(位于数据目录下)中。

如果希望系统配置参数生效,则需要执行 sys_reload_conf()重新加载系统配置文件。首先读取文件 kingbase.conf 中的配置,然后读取文件 kingbase.auto.conf 中的配置,如果某个系统配置参数在这两个文件中的值不一样,则文件 kingbase.auto.conf 中的值将会覆盖文件 kingbase.conf 中的值,即优先使用文件 kingbase.auto.conf 中的系统配置参数值。

注意:如果修改系统配置参数的级别是 kingbase,即使执行 sys_reload_conf()重新加载系统配置文件,系统配置参数也不会生效,因为 kingbase 级别的系统配置参数需要

KingbaseES 数据库实例重新启动后才会生效。

例 4.7 使用 ALTER SYSTEM 语句设置系统配置参数。

```
system@test=# SHOW block_size;
 block_size
------------
 8192
(1 row)
system@test=# SHOW shared_buffers;
 shared_buffers
----------------
 128MB
(1 row)

system@test=# SHOW autovacuum;
 autovacuum
------------
 on
(1 row)
system@test=# ALTER SYSTEM SET block_size = 16384;
ERROR:  parameter "block_size" cannot be changed    --不能修改 internal 级别的参数
system@test=#
system@test=# ALTER SYSTEM SET shared_buffers = 25600;
ALTER SYSTEM
system@test=# ALTER SYSTEM SET autovacuum = off;
ALTER SYSTEM
system@test=# SELECT sys_reload_conf();
 sys_reload_conf
-----------------
 t
(1 row)
system@test=# SHOW shared_buffers;
 shared_buffers
----------------
 128MB           --修改 kingbase 级别的参数 shared_buffers 的值,执行 sys_reload_conf()后不
                   会生效
(1 row)
system@test=# SHOW autovacuum;
 autovacuum
------------
 Off             --修改 sighup 级别的参数 autovacuum 的值,执行 sys_reload_conf()后会生效
(1 row)
system@test=# \q
[kingbase@dbsvr ~]$
```

重新启动 KingbaseES 数据库实例后,参数 shared_buffers 值生效:

```
[kingbase@dbsvr ~]$ sys_ctl stop
[kingbase@dbsvr ~]$ sys_ctl start
[kingbase@dbsvr ~]$ ksql -d test -U system
system@test=# SHOW shared_buffers;
```

```
 shared_buffers
----------------
 200MB
(1 row)
```

注意：千万不要手工编辑文件 kingbase.auto.conf 中的系统配置参数值。

想要删除文件 kingbase.auto.conf 中的系统配置参数，可以执行下面的命令和 SQL 语句：

```
system@test=# \! grep ^shared_buffers /u00/Kingbase/ES/V9/data/kingbase.auto.conf
shared_buffers = '25600'
system@test=# ALTER SYSTEM RESET shared_buffers;
ALTER SYSTEM
system@test=# \! grep ^shared_buffers /u00/Kingbase/ES/V9/data/kingbase.auto.conf
system@test=#
--没有输出,说明已经在文件/u00/Kingbase/ES/V9/data/kingbase.auto.conf 中删除了参数 share_buffers
system@test=# \q
[kingbase@dbsvr ~]$
```

从输出可以看到，无法再使用 grep 命令在文件 kingbase.auto.conf 中搜索到参数 shared_buffers，也就是说，已经删除了文件 kingbase.auto.conf 中的参数 shared_buffers。

3. 使用 ALTER DATABASE 语句设置数据库的系统配置参数

使用 ALTER DATABASE 语句可以为 KingbaseES 数据库集簇中的某个数据库设置该数据库专属的系统配置参数值。ALTER DATABASE 语句只能设置 user/superuser 级别的系统配置参数。

通过 ALTER DATABASE 语句为数据库设置的系统配置参数值保存在数据字典中，作用范围是这个指定名字的数据库，不会影响其他的数据库，并且只有当用户新创建连接到该数据库的会话时，才会起作用。

例 4.8 使用 ALTER DATABASE 语句设置连接到该数据库的系统配置参数。

（1）打开第 1 个终端，设置数据库 test 的系统配置参数：

```
[kingbase@dbsvr ~]$ ksql -d test -U system
system@test=# ALTER DATABASE test SET autovacuum = on;
ERROR:  parameter "autovacuum" cannot be changed now
--不能为数据库 test 修改 sighup 级别的参数
system@test=# ALTER DATABASE test SET log_connections=on;
ERROR:  parameter "log_connections" cannot be set after connection start
--不能为数据库 test 修改 backend 级别的参数
system@test=# SHOW work_mem;
 work_mem
----------
 4MB
(1 row)
system@test=# ALTER DATABASE test SET work_mem=8192;
ALTER DATABASE
```

```
system@test=# SHOW work_mem;
 work_mem
----------
 4MB                    --为数据库 test 修改的系统配置参数值,不会在当前连接中立即生效
(1 row)
```

(2) 打开第 2 个终端,使用用户 system 重新创建一个连接到数据库 test 的会话:

```
[kingbase@dbsvr ~]$ ksql -d test -U system
system@test=# SHOW work_mem;
 work_mem
----------
 8MB                    --为数据库 test 修改的系统配置参数值,在新创建的连接中会生效
(1 row)
```

(3) 打开第 3 个终端,使用用户 system 创建一个连接到数据库 kingbase 的会话:

```
[kingbase@dbsvr ~]$  ksql -d kingbase -U system
system@kingbase=# SHOW work_mem;
 work_mem
----------              --为数据库 test 修改的系统配置参数值
 4MB         --在除数据库 test 以外的其他数据库(如数据库 kingbase)上新建的连接中不会生效
(1 row)
```

4. 使用 ALTER USER 语句设置数据库用户的系统配置参数

使用 ALTER USER 语句,可以为 KingbaseES 数据库的某个用户,设置该用户专属的系统配置参数值。ALTER USER 语句也只能设置 user/superuser 级别的系统配置参数。

通过 ALTER USER 语句为用户设置的系统配置参数值保存在数据字典中,作用范围是这个指定名字的用户,不会影响其他用户,并且只有当用户新创建连接会话时,才会起作用。

例 4.9   使用 ALTER USER 语句设置连接到该数据库用户专用的系统配置参数。

(1) 打开第 1 个终端,设置数据库用户 system 的系统配置参数:

```
[kingbase@dbsvr ~]$ ksql -d test -U system
system@test=#   ALTER USER system SET autovacuum = on;
ERROR:  parameter "autovacuum" cannot be changed now
--不能为用户 system 修改 sighup 级别的参数
system@test=# ALTER USER system SET log_connections=on;
ERROR:  parameter "log_connections" cannot be set after connection start
--不能为用户 system 修改 backend 级别的参数
system@test=# SHOW work_mem;
 work_mem
----------
 8MB
(1 row)
system@test=# ALTER USER system SET work_mem = 16384;
ALTER ROLE
system@test=# SHOW work_mem;
 work_mem
----------
```

```
    8MB          --为数据库用户system修改的系统配置参数值,不会在当前连接中立即生效
(1 row)
```

(2) 打开第2个终端,使用用户system重新创建一个连接到数据库test的会话：

```
[kingbase@dbsvr ~]$ ksql -d test -U system
system@test=# SHOW work_mem;
 work_mem
----------
    16MB         --为数据库用户system修改的系统配置参数值,在新创建的连接中才会生效
(1 row)
```

可以看到,使用数据库用户system,新创建的连接到数据库test的会话,系统配置参数work_mem使用的值是为数据库用户system设置的值16MB,也就是说为数据库用户设置的参数值的优先级要高于为数据库设置的参数值。

(3) 打开第3个终端,创建数据库用户kingbase,然后使用数据库用户kingbase创建一个连接数据库test的会话：

```
[kingbase@dbsvr ~]$ ksql -d test -U system
system@test=# CREATE USER kingbase WITH SUPERUSER PASSWORD 'Passw0rd';
system@test=# \q
[kingbase@dbsvr ~]$ ksql -d test -U kingbase
Password for user kingbase:#输入数据库用户kingbase的密码Passw0rd
kingbase@test=# SHOW work_mem;
 work_mem
----------
    8MB          --为数据库用户system修改的系统配置参数值
(1 row)              --不会影响其他数据库用户(如数据库用户kingbase)创建的新连接
```

5. 临时修改当前会话的系统配置参数

使用ALTER SESSION语句可以临时修改当前会话的系统配置参数值。该修改只针对当前会话生效,其他会话及以后新创建的会话连接的系统配置参数值都没有影响。ALTER SESSION语句也只能设置user/superuser级别的系统配置参数。

例4.10 使用ALTER SESSION语句设置当前会话的系统配置参数。

```
[kingbase@dbsvr ~]$ ksql -d test -U system
system@test=# SHOW work_mem;
 work_mem
----------
 16MB
(1 row)
system@test=# ALTER SESSION SET work_mem=32768;
system@test=# SHOW work_mem;
 work_mem
----------
    32MB         --为当前会话修改的系统配置参数值,在当前会话中立即生效
(1 row)
system@test=# \q
[kingbase@dbsvr ~]$ ksql -d test -U system
```

```
system@test=# SHOW work_mem;
 work_mem
----------
 16MB                --在其他会话中修改的系统配置参数值,不会影响其他会话
(1 row)
```

使用 SET 语句也可以修改当前会话或当前事务的系统配置参数值。SET 语句修改系统配置参数值的语法如下:

```
SET [ SESSION | LOCAL ] configuration_parameter { TO | = } { value | 'value' | DEFAULT }
```

直接使用 SET 语句是修改当前会话的系统配置参数值。

使用 SET LOCAL 语句是修改当前事务的系统配置参数值,只能在事务块中使用。在事务块中设置的系统配置参数值优先于当前会话上设置的系统配置参数值。

例 4.11 使用 SET 语句设置当前会话的系统配置参数。

```
[kingbase@dbsvr ~]$ ksql -d test -U system
system@test=# SHOW work_mem;
 work_mem
----------
 16MB
(1 row)
system@test=# SET work_mem=32768;
SET
system@test=# SHOW work_mem;
 work_mem
----------
 32MB                --为当前会话修改的系统配置参数值,在当前会话中立即生效
(1 row)
system@test=# SET LOCAL work_mem=40960;
WARNING:  SET LOCAL can only be used in transaction blocks
SET                  --只能在事务块中执行 SET LOCAL 语句
system@test=# BEGIN;
BEGIN
system@test=# SET LOCAL work_mem=40960;
SET                  --设置当前事务的系统配置参数值,在当前事务中会立即生效
system@test=# SHOW work_mem;
 work_mem
----------
 40MB
(1 row)
system@test=# END;
COMMIT
system@test=# SHOW work_mem;
 work_mem
----------
 32MB                --事务结束,系统配置参数值还原成当前会话的参数值
(1 row)
```

6. 临时修改数据库实例的系统配置参数

使用 kingbase 命令启动 KingbaseES 数据库时，可以在命令行中设置系统配置参数值。kingbase 命令中设置的系统配置参数值优先于在系统配置参数文件中设置的系统参数值。但是，在数据库或用户上设置的系统配置参数值优先于 kingbase 命令中设置的系统配置参数值。

例 4.12 在命令行启动时设置系统配置参数值。

(1) 查看系统配置参数 deadlock_timeout 和 work_mem 当前的值：

```
[kingbase@dbsvr ~]$ ksql -d test -U system
system@test=# SHOW deadlock_timeout;
 deadlock_timeout
------------------
 1s
(1 row)           --显示的是系统配置文件中设置的值
system@test=# SHOW work_mem;
 work_mem
----------
 16MB
(1 row)           --显示的是给用户 system 设置的值
system@test=# \q
[kingbase@dbsvr ~]$
```

(2) 继续执行下面的命令，停止 KingbaseES 数据库。

```
[kingbase@dbsvr ~]$ sys_ctl stop
```

(3) 使用 kingbase 命令启动 KingbaseES 数据库，启动时将系统配置参数 deadlock_timeout 设置成 5000 秒：

```
[kingbase@dbsvr ~]$ kingbase -c deadlock_timeout=5000  -c work_mem=32768 \
                   -D /u00/Kingbase/ES/V9/data >logfile 2>&1 &
[kingbase@dbsvr ~]$ ksql -d test -U system
system@test=# SHOW deadlock_timeout;
 deadlock_timeout
------------------
 5s
(1 row)           --显示的是命令行中配置的值 5000ms=5s
system@test=# SHOW work_mem;
 work_mem
----------
 16MB
(1 row)           --显示的是给用户 system 设置的值
system@test=# \q
[kingbase@dbsvr ~]$
```

执行下面的命令重新启动数据库：

```
[kingbase@dbsvr ~]$ sys_ctl stop
[kingbase@dbsvr ~]$ sys_ctl start
```

4.2.3 查看系统配置参数值的设置来源

在 4.2.2 中,介绍了可以使用多种方式设置系统配置参数值,可以看出数据库实例设置其系统配置参数值的顺序如下。

(1) 在数据库实例启动时,首先加载系统配置文件 kingbase.conf 中的参数值,然后加载 kingbase.auto.conf 中的参数值,如果在启动的命令行中设置参数值,则以该值为最后的参数值。

(2) 建立数据库连接时,服务进程继承 kingbase 进程中的系统配置参数值,加载在数据库上设置的参数值,然后再加载在数据库用户上设置的参数值。

(3) 在会话上设置该会话的系统配置参数值,在事务上可以设置本事务的系统配置参数值。

因此在执行事务中的 SQL 语句时,使用的系统配置参数值是按照上述顺序最后设置的值。

在系统视图 sys_settings 中可以查看某个系统配置参数的当前使用的值是如何设置的。系统视图 sys_settings 中的 source 字段值说明了参数的设置来源,取值如下。

(1) default:默认级别,表示参数采用全局默认值。

(2) file:表示参数是通过系统配置文件(kingbase.conf 或 kingbase.auto.conf)进行设置的。

(3) override:表示参数是通过特定的系统配置文件(kingbase.override.conf)进行设置的。

(4) database:表示参数是通过特定数据库的参数进行设置的。

(5) user:表示参数是由特定用户通过 ALTER ROLE 语句进行设置的。

(6) session:表示参数是由特定会话通过 ALTER SESSION 或 SET 语句进行设置的。

例 4.13 查看系统配置参数的设置来源。

```
[kingbase@dbsvr ~]$ ksql -d test -U system
system@test=# SELECT name, setting, source FROM pg_settings
test-#          WHERE name IN ('deadlock_timeout','shared_buffers','block_size',
test(#                 'work_mem','log_connections', 'autovacuum');
       name       | setting  |       source
------------------+----------+--------------------
 autovacuum       | off      | configuration file
 block_size       | 8192     | override
 deadlock_timeout | 1000     | default
 log_connections  | off      | default
 shared_buffers   | 16384    | configuration file
 work_mem         | 16384    | user
(6 rows)
```

从输出可以看到,初始时,参数 shared_buffers 和 autovacuum 的值来自系统配置文件,参数 deadlock_timeout 和 log_connections 使用的是系统默认值,参数 work_mem 的值来自为用户 system 设置的参数值。

使用 SET 命令为当前会话设置参数 deadlock_timeout 的值:

```
system@test=# SET  deadlock_timeout=5000;
SET
system@test=# SELECT name, setting, source FROM pg_settings
test-#         WHERE name IN ('deadlock_timeout','shared_buffers','block_size',
test(#                        'work_mem','log_connections', 'autovacuum');
       name       | setting |       source
------------------+---------+--------------------
 autovacuum       | off     | configuration file
 block_size       | 8192    | override
 deadlock_timeout | 5000    | session
 log_connections  | off     | default
 shared_buffers   | 16384   | configuration file
 work_mem         | 16384   | user
(6 rows)
```

从输出可以看到，使用 SET 语句修改参数 deadlock_timeout 的值后，其来源已经修改为 session。

例 4.14 查看为当前连接访问的数据库设置的所有系统配置参数。

```
system@test=# SELECT name, setting FROM sys_settings WHERE source != 'default';
         name          |                 setting
-----------------------+----------------------------------------
--省略了一些输出
 block_size            | 8192
 client_encoding       | UTF8
 config_file           | /u00/Kingbase/ES/V9/data/kingbase.conf
 control_file_copy     | /u00/Kingbase/ES/V9/data/global/sys_control1;
                       | /u00/Kingbase/ES/V9/data/global/sys_control2;
                       | /u00/Kingbase/ES/V9/data/global/sys_control3
--省略了一些输出
(46 rows)
```

4.3 管理数据库的扩展插件

KingbaseES 数据库的一个关键特性是其扩展插件系统，这使得用户可以添加自定义功能或第三方开发的扩展插件，以进一步增强数据库的能力。

有效管理 KingbaseES 数据库扩展插件对于确保数据库性能和安全至关重要。扩展插件可以提供额外的功能，如新的数据类型、索引类型、函数或更复杂的分析工具。然而，扩展插件如果管理不当可能会引入性能瓶颈或安全漏洞。因此，合理地安装、配置和维护插件对于 DBA 而言是一个必要的技能。

基于安全方面的考虑，建议 KingbaseES 数据库的 DBA 只使用电科金仓官方提供的扩展插件。在某些情况下需要使用第三方扩展插件时，一定要在测试机上做好充分的测试。

执行下面的 SQL 语句，显示 KingbaseES 数据库官方提供的可用扩展插件：

```
[kingbase@dbsvr ~]$ ksql -d test -U system
system@test=# SELECT * FROM sys_available_extensions;
         name    | default_version | installed_version |                comment
```

```
-------------+-------------+--------------+------------------------------------------
--省略了一些输出
 oracle_fdw  | 1.2         |              | foreign data wrapper for Oracle access
 pageinspect | 1.7         |              | inspect the contents of database pages at a low level
 sys_kwr     | 1.7         |              | KingbaseES auto workload repository and report builder
--省略了一些输出,仅列出了几个扩展插件
(105 rows)
```

执行下面的 SQL 语句,查看 KingbaseES 数据库实例当前已经安装的扩展插件:

```
system@test=# SELECT * FROM sys_extension;
  oid  |   extname   | extowner | extnamespace | extrelocatable | extversion | extconfig | extcondition
-------+-------------+----------+--------------+----------------+------------+-----------+--------------
 13362 | plpgsql     |       10 |           11 | f              | 1.0        |           |
 13364 | kdb_license |       10 |           11 | f              | 1.0        |           |
 13366 | sysaudit    |       10 |        13365 | f              | 1.0        |           |
--省略了一些输出
(19 rows)
```

执行下面的 SQL 语句,安装官方的扩展插件 pageinspect:

```
system@test=# CREATE EXTENSION pageinspect;
CREATE EXTENSION
system@test=#
```

执行下面的 SQL 语句,再次查看 KingbaseES 数据库实例当前已经安装的扩展插件:

```
system@test=# SELECT * FROM sys_extension;
  oid  |      extname       | extowner | extnamespace | extrelocatable | extversion | extconfig | extcondition
-------+--------------------+----------+--------------+----------------+------------+-----------+--------------
--省略了一些输出
 16610 | pageinspect        |       10 |         2200 | t              | 1.7        |           |
 14092 | sys_stat_statements|       10 |         2200 | t              | 1.10       |           |
 14101 | kdb_tinyint        |       10 |           11 | f              | 1.0        |           |
(20 rows)
```

从输出可以看到,KingbaseES 数据库实例已经安装了扩展插件 pageinspect。

要查看扩展插件 pageinspect 的详细信息,可用执行下面的元命令:

```
system@test=# \dx+ pageinspect
          Objects in extension "pageinspect"
                  Object description
------------------------------------------------------------
 function bm_bitmap_page_header(text,integer)
 function bm_bitmap_page_items(bytea)
 function bm_bitmap_page_items(text,integer)
--省略了一些输出
 function tuple_data_split(oid,bytea,integer,integer,text,boolean)
(31 rows)
```

要卸载已经安装的扩展插件 pageinspect,可用执行下面的 SQL 语句:

```
system@test=# DROP EXTENSION pageinspect;
DROP EXTENSION
system@test=#
```

4.4 管理数据库的软件许可证

KingbaseES 数据库的软件许可证有以下几个大类：标准版、专业版、企业版和开发版。

KingbaseES 数据库软件许可证标准版是为了满足政府部门、中小型企业客户的简单应用而推出的一款精简数据库产品，该产品提供除集群、地理信息系统（geographic information system, GIS）以外的全部应用开发与系统管理功能，支持太字节（terabyte, TB）级海量数据，支持多用户并发访问，支持所有国内外主流的 CPU、操作系统与云平台部署。

KingbaseES 数据库软件许可证专业版是一款入选双名录的产品，可满足党政客户、中小型企业客户的中等规模、非核心业务应用场景的要求，提供全部应用开发及系统管理功能，支持主备集群、读写分离集群等集群架构，支持所有国内外主流 CPU、操作系统与云平台部署。

KingbaseES 数据库软件许可证企业版是面向全行业、全客户关键应用的大型通用数据库管理系统，适用于联机事务处理、查询密集型数据仓库、要求苛刻的互联网应用等场景，提供全部应用开发及系统管理功能，提供性能增强特性，可支持主备集群、读写分离集群、多活共享存储集群等全集群架构，具有高性能、高安全、高可用、易使用、易管理、易维护的特点，支持所有国内外主流 CPU、操作系统与云平台部署。

KingbaseES 数据库软件许可证开发版是为了满足用户的产品试用、日常开发需求而推出的一个精简版本，提供除集群、GIS 以外的全部应用开发及系统管理功能。该版本可用 10 个数据库连接，试用 1 年，不能用于商业用途。

下面是更新 KingbaseES 数据库软件许可证的示例。**重新解压缩虚拟机备份文件 dbserver-2DB-2data.rar 并运行虚拟机**。KingbaseES 数据库的软件许可证文件 license.dat 位于安装目录/u00/Kingbase/ES/V9/kingbase 的子目录 KESRealPro/V009R001C001B0025 下：

```
[kingbase@dbsvr ~]$ cd /u00/Kingbase/ES/V9/kingbase
[kingbase@dbsvr kingbase]$ ls -l license.dat
lrwxrwxrwx. 1 kingbase dba 69 Feb 29 21:26 license.dat -> /u00/Kingbase/ES/V9/kingbase/KESRealPro/V009R001C001B0025/license.dat
[kingbase@dbsvr kingbase]$ cat /u00/Kingbase/ES/V9/kingbase/KESRealPro/V009R001C001B0025/license.dat
-------------------BEGIN KINGASE.LICENSE.3.0 PRIVATE KEY--------------
C2jEV4qxkPiL106gQ3qlnxRssF4qjtepwpH/utnuj5vJvRvAiqne6LimVFm7/xDKsMFybE
#省略了一些输出
qg5CM3/xV8ACphcDmVhDa26PdA9Oz5jGcw8N5lAFFb1oNlduHOnB4sdVREpuzpUA==
--------------------------MD5SUM----------------------
b88fdb4be938a6ff40b9f9efcb7244dd
-------------------END KINGASE.LICENSE.3.0 PRIVATE KEY---------------
License 序列号 --- 启用 --- DF3316AC-69B2-11EE-A0E1-000C29CBE49F
生产日期 --- 启用 --- 2023-10-13
产品名称 --- 启用 --- KingbaseES V8
细分版本模板名 --- 启用 --- SALES-开发版 V9R1
#省略了一些输出
[kingbase@dbsvr V9]$
```

从输出可以看到,当前软件许可证的类型为开发版,其中列出了支持的 KingbaseES 数据库特性。

要查看软件许可证的有效时间,可以在 KingbaseES 数据库服务器上使用用户 kingbase 执行下面的命令:

```
[kingbase@dbsvr ~]$ ksql -d kingbase -U system -c " select get_license_validdays ()"
get_license_validdays
-----------------------
                   359
(1 row)
[kingbase@dbsvr ~]$
```

从输出可以看到,当前的软件许可证还可以使用 359 天。

要变更 KingbaseES 数据库的软件许可证,可以执行下面的步骤。

(1) 从电科金仓的官方网站,下载 KingbaseES 数据库的企业版软件许可证,并上传到目录 /home/kingbase。

(2) 使用用户 kingbase,执行下面的命令,解压软件许可证:

```
[kingbase@dbsvr ~]$ cd
[kingbase@dbsvr ~]$ unzip license_企业版.zip
Archive:  license_企业版.zip
   creating: license_34148/
  inflating: license_34148/license_34148_0.dat
[kingbase@dbsvr ~]$
```

(3) 停止 KingbaseES 数据库:

```
[kingbase@dbsvr ~]$ sys_ctl stop
```

(4) 执行下面的命令,替换软件许可证文件:

```
[kingbase@dbsvr ~]$ cd license_34148
[kingbase@dbsvr license_34148]$ cp license_34148_0.dat \
         /u00/Kingbase/ES/V9/kingbase/KESRealPro/V009R001C001B0025/license.dat
```

(5) 重新启动 KingbaseES 数据库:

```
[kingbase@dbsvr license_34148]$ sys_ctl start
```

(6) 重新查看 KingbaseES 数据库的软件许可证文件:

```
[kingbase@dbsvr license_34148]$ cd
[kingbase@dbsvr ~]$ cat     /u00/Kingbase/ES/V9/kingbase/KESRealPro/V009R001C001B0025/license.dat
#省略了一些输出
License 序列号 --- 启用 --- 375DDEEA-69B1-11EE-A0E1-000C29CBE49F
生产日期 --- 启用 --- 2023-10-13
产品名称 --- 启用 --- KingbaseES V8
细分版本模板名 --- 启用 --- SALES-企业版 V9R1
#省略了一些输出
```

```
[kingbase@dbsvr ~]$ ksql -d kingbase -U system -c " select get_license_validdays()"
 get_license_validdays
-----------------------
                     90
(1 row)
[kingbase@dbsvr ~]$
```

从输出可以看到，KingbaseES 数据库的软件许可证已经变更为企业版，目前还可以使用 90 天。

4.5 数据库实例与数据库集簇

数据库服务器由数据库实例和数据库集簇组成。一个 KingbaseES 数据库实例只能访问一个特定数据目录下的数据库集簇，不过，可以在操作系统中启动多个 KingbaseES 数据库实例（需要为每个实例配置不同的监听端口），每个实例访问一个不同的数据库集簇。

当前，KingbaseES 数据库服务器上已经在数据目录/u00/Kingbase/ES/V9/data 下有一个数据库集簇：

```
[kingbase@dbsvr ~]$ ksql -d test -U system
system@test=# SELECT name, setting FROM sys_settings WHERE name = 'data_directory';
      name      |         setting
----------------+--------------------------
 data_directory | /u00/Kingbase/ES/V9/data
(1 row)
```

使用 initdb 命令可以创建多个数据库集簇，每个数据库集簇位于一个独立的数据目录（文件系统目录）下。

例 4.15 初始化一个新的数据库集簇。

使用 initdb 命令，在数据目录/opt/Kingbase/ES/V9/data 下创建一个新数据库集簇：

```
[kingbase@dbsvr ~]$ su -
Password: #输入 Linux 超级用户 root 的密码 root123
[root@dbsvr ~]# mkdir -p /opt/Kingbase/ES/V9/
[root@dbsvr ~]# chown -R kingbase:dba /opt/Kingbase/ES/V9/
[root@dbsvr ~]# exit
[kingbase@dbsvr ~]$ initdb -D/opt/Kingbase/ES/V9/data -EUTF8 -Usystem -W
#省略了一些输出
Enter new superuser password: #输入 system 用户的密码 Passw0rd
Enter it again: #输入 system 用户的密码 Passw0rd
#省略了一些输出
Success. You can now start the database server using:
    sys_ctl -D /opt/Kingbase/ES/V9/data -l logfile start
[kingbase@dbsvr ~]$
```

启动实例，打开这个新创建的数据库集簇之前，需要以用户 kingbase 的身份，使用 vi 编辑器，在系统参数配置文件 kingbase.conf 中配置参数 port，为打开这个数据库集簇的数据库实例指定一个监听端口。在 kingbase.conf 文件中，找到下面的内容：

```
#port = 54321                          # (change requires restart)
```

将监听端口 port 的值设置为：

```
port = 50001
```

启动一个 KingbaseES 数据库实例访问这个新创建的数据库集簇：

```
[kingbase@dbsvr ~]$ sys_ctl -D /opt/Kingbase/ES/V9/data  start
```

执行下面的命令，可以查看这个新启动的 KingbaseES 数据库实例的所有进程：

```
[kingbase@dbsvr ~]$ ps -ef |grep $(head -1 /opt/Kingbase/ES/V9/data/kingbase.pid) |grep -v grep
kingbase   4408      1  0 17:57 ?        00:00:00 /u00/Kingbase/ES/V9/KESRealPro/V009R001C001B0025/Server/bin/kingbase -D /opt/Kingbase/ES/V9/data
kingbase   4409   4408  0 17:57 ?        00:00:00 kingbase: logger
#省略了一些输出
[kingbase@dbsvr ~]$
```

要访问这个新创建的数据库集簇（实例），可以执行下面的 ksql 命令：

```
[kingbase@dbsvr ~]$ ksql -d test -U system -p 50001
system@test=# SELECT version();
                        version
----------------------------------------------------------------
 KingbaseES V009R001C001B0025 on x86_64-pc-linux-gnu,
 compiled by gcc (GCC) 4.8.5 20150623 (Red Hat 4.8.5-28), 64-bit     --已经美化了输出
(1 row)
system@test=# SHOW data_directory;
     data_directory
------------------------
 /opt/Kingbase/ES/V9/data
(1 row)
system@test=# \q
[kingbase@dbsvr ~]$
```

执行下面的命令，关闭这个新创建的 KingbaseES 数据库实例：

```
[kingbase@dbsvr ~]$ sys_ctl -D /opt/Kingbase/ES/V9/data  stop
```

执行下面的命令，查看当前这个新创建的 KingbaseES 数据库实例的状态：

```
[kingbase@dbsvr ~]$ sys_ctl -D /opt/Kingbase/ES/V9/data  status
sys_ctl: no server running
[kingbase@dbsvr ~]$
```

第 5 章

管理用户与会话连接

本章主要讨论关于 KingbaseES 数据库的用户管理和会话连接管理，包括用户和角色的管理、会话连接的管理以及与用户相关的系统配置文件的使用方式。

 ## 5.1 用户管理

5.1.1 创建数据库用户

在 KingbaseES 数据库中，可以使用 CREATE USER 或 CREATE ROLE 语句创建一个 KingbaseES 数据库的新用户或新角色。实际上，CREATE USER 是 CREATE ROLE 的别名，它们可以互换使用。用户和角色的区别就是能否登录数据库。无论是使用 CREATE USER 语句还是使用 CREATE ROLE 语句来创建用户（角色），如果不授予它连接到数据库的权限，那么就可以将这个新创建的用户（角色），仅当作一个权限集合的角色。

创建数据库集簇的时候，系统会默认创建以下用户：

```
[kingbase@dbsvr ~]$ ksql -d test -U system
system@test=# \du
                                    List of roles
 Role name |                         Attributes                          | Member of
-----------+-------------------------------------------------------------+-----------
 kcluster  | Cannot login                                                | {}
 sao       | No inheritance                                              | {}
 sso       | No inheritance                                              | {}
 system    | Superuser, Create role, Create DB, Replication, Bypass RLS  | {}
```

从输出可以看到，系统默认创建的用户有 4 个：kcluster、sao、sso 和 system。
KingbaseES 数据库定义了多个系统权限，在创建用户时可以指定这些权限。

（1）超级用户（superuser）权限：具有该权限的超级用户操作数据库不需要数据库的权限检查。建议非必须最好不要以超级用户的身份执行数据库的管理操作，避免误操作。

（2）创建用户/角色（createrole）权限：具有该权限的用户可以创建新的用户/角色。

（3）登录（login）权限：具有该权限的用户才可以建立数据库连接。CREATE USER 语句和 CREATE ROLE 语句的区别就是在创建用户/角色时，是否具有 login 权限。使用 CREATE USER 创建的用户默认具有 login 权限。

(4) 创建数据库(createdb)权限：具有该权限的用户才可以创建一个新的数据库。

(5) 发起流复制(replication)权限：具有该权限的用户能够发起流复制。

在进行系统初始化时，创建了超级用户 system，该用户可以创建其他新用户并执行其系统权限。新创建的用户在任何数据库上都有创建自己的数据库对象（如表）的权限，但不具有其他数据库对象的权限。使用 ALTER USER/ALTER ROLE 语句可以给用户增加或减少系统权限。

例 5.1 创建用户 user1，并给其相应的权限。

打开第 1 个终端，创建用户 newuser，创建数据库 newdb：

```
[kingbase@dbsvr ~]$ ksql -d test -U system
system@test=# CREATE USER newuser WITH PASSWORD 'newuser123';
CREATE ROLE
system@test=# CREATE DATABASE newdb;
CREATE DATABASE
```

打开第 2 个终端，使用用户 newuser 连接到数据库 newdb：

```
[kingbase@dbsvr ~]$ ksql -d newdb -U newuser
Password for user newuser:          #输入用户 newuser 的密码 newuser123
newuser@newdb=> CREATE TABLE ttt1(col int);
newuser@newdb=> CREATE TABLE ttt2(col int);
newuser@newdb=> CREATE DATABASE newdb1;
ERROR:  permission denied to create database
                                    #目前数据库用户 newuser 没有创建数据库的权限
newuser@newdb=> CREATE USER tpchuser;
ERROR:  permission denied to create role
                                    #目前数据库用户 newuser 没有创建数据库用户的权限
newuser@newdb=> \du
          List of roles
 Role name | Attributes | Member of
-----------+------------+-----------
 newuser   |            | {}
```

从输出可以看到，用户 newuser 在数据库 newdb 上成功地创建了两个表，但是没有创建数据库和用户的系统权限。

回到第 1 个终端，给用户 newuser 授予 CREATEROLE 权限，并把数据库 newdb 的属主给 newuser：

```
system@test=# ALTER USER newuser CREATEROLE ;
ALTER ROLE
system@test=# ALTER DATABASE newdb OWNER TO newuser ;
ALTER DATABASE
```

回到第 2 个终端，查看用户 newuser 的属性，并创建新用户：

```
newuser@test=> \du
          List of roles
 Role name | Attributes | Member of
-----------+------------+-----------
```

```
 newuser    | Create role | {}
newuser@test=> CREATE USER tpchuser;
CREATE ROLE
```

从输出可以看到,用户 newuser 有 Create role 的权限,并且现在可以成功地创建用户 tpchuser 了。

回到第 1 个终端,回收用户 newuser 的 CREATEROLE 权限:

```
system@test=# ALTER USER newuser NOCREATEROLE ;
ALTER ROLE
```

例 5.2 设置用户的连接属性。

可以设置用户的属性,例如,该用户可以建立的数据库连接的个数、用户口令的有效期等。

在第 1 个终端,设置用户 newuser 只能建立 1 个数据库连接:

```
system@test=# ALTER USER newuser CONNECTION LIMIT 1;
ALTER ROLE
```

打开第 3 个终端,尝试使用用户 newuser 连接到数据库 test:

```
[kingbase@dbsvr ~]$  ksql -d test -U newuser
Password for user newuser: #输入用户 newuser 的密码 newuser123
ksql: error: could not connect to server: FATAL:  too many connections for role "newuser"
[kingbase@dbsvr ~]$
```

因为设置了用户 newuser 只能建立 1 个数据库连接,而在第 2 个终端上 newuser 已经建立了一个数据库连接,所以这里会报错。

在第 1 个终端,还可以设置用户 newuser 的口令有效期:

```
system@test=# ALTER USER newuser VALID UNTIL '2030-12-31';
ALTER ROLE
system@test=#
```

为了后面示例方便,不再限制用户 newuser 建立数据库连接的个数:

```
system@test=# ALTER USER newuser CONNECTION LIMIT -1;
ALTER ROLE
```

例 5.3 锁定和解锁数据库用户。

锁定一个用户意味着保留该数据库用户账号,并且不会修改账号的密码,不仅仅是不允许使用该数据库用户账号登录到 KingbaseES 数据库,对于已经建立的数据库连接不受影响。

在第 1 个终端,锁定用户 newuser:

```
system@test=# ALTER USER newuser ACCOUNT LOCK;
```

打开第 3 个终端,尝试使用用户 newuser 连接到数据库 test:

```
[kingbase@dbsvr ~]$ ksql -d test -U newuser
```

```
Password for user newuser: #输入用户 newuser 的密码 newuser123
ksql: error: could not connect to server: FATAL:  role "newuser" is not permitted
to log in
[kingbase@dbsvr ~]$
```

从输出可以看到,由于用户账号 newuser 被锁定,因此无法创建与 KingbaseES 数据库的连接。

回到第 1 个终端,解锁用户 newuser:

```
system@test=# ALTER USER newuser ACCOUNT UNLOCK;
ALTER ROLE
system@test=#
```

回到第 3 个终端,再次尝试使用用户 newuser 连接到数据库 test:

```
[kingbase@dbsvr ~]$ ksql -d test -U newuser
Password for user newuser: #输入用户 newuser 的密码 newuser123
newuser@test=> \q
[kingbase@dbsvr ~]$
```

从输出可以看到,解锁用户账号 newuser 后,可以创建与 KingbaseES 数据库的连接。

5.1.2 删除数据库用户

如果用户正在连接某个数据库,或用户是数据库对象的属主,则不能删除该用户。

例 5.4 删除数据库的用户。

在第 1 个终端上,执行下面的 SQL 语句,删除用户 newuser:

```
system@test=# DROP USER newuser;
ERROR:  current logined user cannot be dropped
system@test=#
```

从输出可以看到,不能删除一个已经连接到 KingbaseES 数据库的用户。

回到第 2 个终端,执行下面的元命令,退出 ksql:

```
newuser@newdb=> \q
[kingbase@dbsvr ~]$
```

再次回到第 1 个终端,继续执行删除用户 newuser 的 SQL 语句:

```
system@test=# DROP USER newuser;
ERROR:  role "newuser" cannot be dropped because some objects depend on it
DETAIL: owner of database newdb
2 objects in database newdb
```

从输出可以看到,现在还是不能删除用户 newuser,原因是一些数据库对象依赖于用户 newuser:

(1) 用户 newuser 是数据库 newdb 的属主;
(2) 数据库 newdb 中有 2 个属于用户 newuser 的数据库对象。

要删除用户 newuser,必须先将数据库的属主修改为其他的用户,或者删除。这里选择将数据库 newdb 的属主修改为用户 system。

在第 1 个终端上运行下面的 SQL 语句：

system@test=# ALTER DATABASE newdb OWNER TO system;

对数据库 newdb 中属于用户 newuser 的数据库对象，有以下两种处理方法：
(1) 假如该对象(如表 ttt1)没用了，可以直接删除掉该对象；
(2) 假如该对象(如表 ttt2)仍然有用，可以将该对象转移给其他用户(如 system)。
转到第 2 个终端，执行下面的 ksql 命令和 SQL 语句：

```
[kingbase@dbsvr ~]$ ksql -d newdb -U system
system@newdb=# \dt
        List of relations
 Schema | Name | Type  | Owner
--------+------+-------+---------
 public | ttt1 | table | newuser
 public | ttt2 | table | newuser
(2 rows)
system@newdb=# DROP TABLE ttt1;
system@newdb=# REASSIGN OWNED BY newuser to system;
system@newdb=# \dt
        List of relations
 Schema | Name | Type  | Owner
--------+------+-------+--------
 public | ttt2 | table | system
(1 row)
system@newdb=# \q
[kingbase@dbsvr ~]$
```

转回到第 1 个终端，执行下面的 SQL 语句，删除用户 newuser：

```
system@test=#   DROP USER newuser;
DROP ROLE
system@test=# \du newuser
          List of roles
 Role name | Attributes | Member of
-----------+------------+-----------

system@test=#
```

从输出可以看到，数据库用户 newuser 已经被删除。

5.1.3 查询用户信息

通过系统视图 sys_shadow、sys_user、sys_roles 可以查询用户的相关信息。使用 ksql 的元命令\du 或\dg，也可以查看到 Kingbase 数据库上所有的用户和角色。

例 5.5 查看系统上的用户/角色信息。

查看不可以登录到 KingbaseES 数据库的用户/角色：

```
system@test=# SELECT rolname FROM sys_roles WHERE rolcanlogin='f';
        rolname
----------------------------
```

```
  pg_monitor
--省略了许多输出
(9 rows)
```

例 5.6 查看为用户设置的系统配置参数信息。

```
system@test=# ALTER USER tpchuser SET work_mem='10MB';
system@test=# SELECT   *   FROM sys_user;
 usename  | usesysid | usecreatedb | usesuper | userepl | usebypassrls | passwd   | valuntil | useconfig
---------+----------+-------------+----------+---------+--------------+----------+----------+----------------
 system  |    10    | t           | t        | t       | t            | ******** |          |
 sao     |     9    | f           | f        | f       | f            | ******** |          |
 sso     |     8    | f           | f        | f       | f            | ******** |          |
 tpchuser|  16512   | f           | f        | f       | f            | ******** |          | {work_mem=10MB}
(4 rows)
```

从输出可以看到,为用户 tpchuser 设置了参数 work_mem。

例 5.7 查看用户是否设置了密码。

```
system@test=# CREATE USER newrole;
system@test=# SELECT usename, passwd FROM sys_shadow;
 usename  |                                                      passwd
---------+------------------------------------------------------------------------------------------------------------------------
 system  | SCRAM-SHA-256$4096:R7zhQyJxrwhc8sCMDxSCoQ==$D7oLvisZNWnN3zKM4OBgG72VkWAd1Ba2qZus275HLUE=:5Uo+fSQwzuQc9tSClgtdvy+e9GcwVc9076MO85mUzds=
 sao     | SCRAM-SHA-256$4096:6pOpAwuNdkwNHyJ1W6MvZA==$WsUmBS7H11GZt1yFoLyXL3B0jBBOdo8nZrn8vYq4xH8=:5GQ1sf1MMGvb27e3yAhs9rgQo0Av0wpunTHBlzMSURo=
 sso     | SCRAM-SHA-256$4096:oW48mmQKo4yxibNDiSD9Aw==$NgpPSuOK1iSQ+CppdhUGUgNkW9ZnQ8rU3w0c3ImlqPg=:2ezBhkZTp48ThPqoak1WQSOpBwqYUBaA1waXLCp54B0=
 tpchuser|
 newrole |
(5 rows)
```

从输出可以看到,目前还没有为 newrole 和 tpchuser 设置过密码。在 KingbaseES 数据库巡检时,可以使用这条 SQL 语句,检查是否存在系统安全漏洞。

例 5.8 查看指定名字的用户有哪些系统权限。

```
system@test=# SELECT rolname AS username,
test-#       CASE WHEN rolsuper THEN 'superuser' ELSE '' END AS superuser,
test-#       CASE WHEN rolcreaterole THEN 'create role' ELSE '' END AS create_role,
test-#       CASE WHEN rolcreatedb THEN 'create db' ELSE '' END AS create_db,
test-#       CASE WHEN rolreplication THEN 'replication' ELSE '' END AS replication
test-#     FROM sys_roles
test-#     WHERE rolname = 'newrole';
 username | superuser | create_role | create_db | replication
---------+-----------+-------------+-----------+-------------
 newrole  |           |             |           |
(1 row)
```

从输出可以看到,当前用户 newrole 还没有任何系统权限。

5.2 管理连接会话

用户访问数据库,首先要建立与数据库服务器的连接,称为连接会话。

5.2.1 设置与会话连接相关的系统配置参数

在系统配置文件 kingbase.conf 中必须进行正确的配置,用户才能建立与数据库服务器的连接,相关的配置如下。

1. 监听的 IP 地址和端口号

KingbaseES 数据库使用参数 listen_addresses 和 port 来配置监听的 IP 地址和端口号。默认的配置如下:

```
listen_addresses = '*'
port=54321
```

默认配置的含义是 KingbaseES 数据库服务器的 kingbase 进程,将监听服务器上所有的 IP 地址,监听 TCP 端口号 54321。

可以修改系统配置文件中的这两个参数的值,让 KingbaseES 数据库服务器只监听指定的 IP 地址和 TCP 端口号,例如,可将这两个参数的值按如下设置:

```
listen_addresses = '192.168.100.22'  # 设置为KingbaseES数据库服务器对外服务的网卡地址
port=54321                            # 设置监听端口
```

2. 最大连接数

在 kingbase.conf 文件中设置参数 max_connections,指定 KingbaseES 数据库服务器允许的最大连接数,参数 max_connections 的默认值为 100,可以在 kingbase.conf 文件中根据应用的实际需求修改该值,例如,将其设置为 300:

```
max_connections=300              # 设置最大连接会话数
```

该值需要在数据库重新启动后生效。

3. 为超级用户保留的最大连接数

参数 superuser_reserved_connections 设置了为超级用户保留的连接数,默认值是 3。为超级用户保留连接数的目的是如果应用程序把系统所有的用户连接数使用完后,超级用户还有机会连接数据库进行数据库管理。

如果设置了参数 superuser_reserved_connections,则普通用户可以使用的最大连接数为 max_connections — superuser_reserved_connections。

注意:参数 superuser_reserved_connections 设置的是为超级用户保留的连接数,并不是超级用户可以建立的最大连接数。

5.2.2 查看会话连接信息

KingbaseES 数据库的动态系统视图 sys_stat_activity 提供了关于当前数据库服务器所有进程的信息。每一行对应一个数据库服务器进程,主要包括以下类型的信息。

1. 数据库服务器进程的基本信息

(1) backend_type:数据库服务器进程的类型,例如,client backend 表示服务进程,autovacuum launcher、logical replication launcher 等,表示后台进程。

(2) pid:数据库服务器进程 ID,是操作系统的进程号。

2. 会话连接的信息

如果是服务进程,则包含该连接会话的用户信息、客户端的地址等信息、连接数据库信息以及应用程序的名称。如果是其他后台进程，则这些字段为空。

(1) usesysid：用户的 OID。

(2) usename：执行当前进程的用户的名称。

(3) client_addr：发起数据库连接的客户端的 IP 地址。

(4) client_hostname：发起连接的客户端的主机名。

(5) client_port：与数据库通信的客户端的端口号。

(6) datid：数据库的 OID。

(7) datname：当前进程所在数据库的名称。

(8) application_name：连接到数据库的应用程序的名称。

3. 会话连接的状态信息

会话连接的状态信息包括进程的当前状态、进程启动时间、事务的启动时间、当前正在执行的 SQL 语句以及查询的启动时间等。

(1) state：进程的当前状态，有以下取值。

① active：正在执行一个查询。

② idle：正在等待客户端发送新的命令。

③ idle in transaction：在事务中，但是没有执行命令。

④ idle in transaction (aborted)：在事务中，没有执行命令，事务中有语句出现了错误。

⑤ fastpath function call：在执行一个 fastpath 函数。

(2) backend_start：当前数据库后台进程启动的时间。

(3) xact_start：当前进程开始最新事务的时间。

(4) query_start：当前正在执行的查询开始的时间。

(5) state_change：进程最后一次改变状态的时间。

(6) backend_xid：当前事务的事务 ID(如果进程在一个事务中)。

(7) backend_xmin：当前进程可见的最早未提交的事务 ID。

(8) query：当前进程正在执行的 SQL 语句。

4. 会话连接的等待时间

(1) wait_event_type：进程当前正在等待的事件类型(如果有)，事件类别包括轻量级锁((wlock)、I/O、客户端、IPC)等。

(2) wait_event：进程当前正在等待的具体事件。

例 5.9 查看数据库服务器进程的状态信息。

系统视图 sys_stat_activity 反映的是数据库进程的实时信息。

首先，在 **KingbaseES 数据库客户端**上，执行一条会持续运行很长时间的 SQL 语句：

```
[kingbase@dbclient ~]$ ksql -h 192.168.100.22 -d test -U system
system@test=# DROP INDEX tpch.lineitem_l_partkey_idx;
--说明:删除该索引将导致接下来的 SQL 语句需要运行很长的时间
system@test=# select sum(l_extendedprice) / 7.0 as avg_yearly
test-#    from tpch.lineitem,tpch.part
test-#    where p_partkey = l_partkey and p_brand = 'Brand#45' and p_container = 'LG PKG'
test-#      and l_quantity < ( select 0.2 * avg(l_quantity)
```

```
test(#                     from tpch.lineitem
test(#                     where l_partkey = p_partkey );
--SQL 语句运行中……
```

然后，在 KingbaseES 数据库服务器上使用系统视图 sys_stat_activity 查看信息：

```
[kingbase@dbsvr ~]$ ksql -d test -U system
system@test=# \! date
Sat Mar  2 02:51:02 CST 2024
system@test=# SELECT pid,state,query_start,query  FROM sys_stat_activity;
  pid  | state  |        query_start         |                          query
-------+--------+----------------------------+---------------------------------------------------------
  2040 |        |                            |
  2048 | idle   |                            |
  2049 |        |                            |
 69028 | active | 2024-03-02 02:51:09.967257+08 | SELECT pid,state,query_start, query  FROM sys_stat_activity;
 68855 | active | 2024-03-02 02:41:11.076161+08 | select sum(l_extendedprice) / 7.0 as avg_yearly         +
       |        |                            | from tpch.lineitem,tpch.part                            +
       |        |                            | where p_partkey = l_partkey and p_brand = 'Brand#45' and p_container = 'LG PKG' +
       |        |                            |      and l_quantity < ( select 0.2 * avg(l_quantity)    +
       |        |                            |                        from tpch.lineitem               +
       |        |                            |                        where l_partkey = p_partkey );
  2038 |        |                            |
  2037 |        |                            |
  2039 |        |                            |
(8 rows)
```

从输出可以看出，查询的开始时间为 02:41:11.07，查询会话信息的时间为 02:51:02，语句运行了约 10min，目前还在运行中。

例 5.10 查找长时间运行 SQL 查询的会话连接。

查找长时间（大于 5min）运行 SQL 查询的会话：

```
system@test=# \x
Expanded display is on.
system@test=# SELECT * FROM sys_stat_activity
test-#         WHERE state = 'active' AND now() - query_start > INTERVAL '5 minutes';
-[ RECORD 1 ]----+-----------------------------------------------------------------
datid            | 14386
datname          | test
pid              | 68855
usesysid         | 10
username         | system
application_name | ksql
client_addr      | 192.168.100.18
client_hostname  |
client_port      | 58822
backend_start    | 2024-03-02 02:41:02.905953+08
xact_start       | 2024-03-02 02:41:11.076161+08
query_start      | 2024-03-02 02:41:11.076161+08
state_change     | 2024-03-02 02:41:11.076163+08
wait_event_type  |
wait_event       |
state            | active
backend_xid      |
backend_xmin     | 1119
query            | select sum(l_extendedprice) / 7.0 as avg_yearly                +
                 | from tpch.lineitem,tpch.part                                   +
                 | where p_partkey = l_partkey and p_brand = 'Brand#45' and p_container = 'LG PKG' +
                 |      and l_quantity < ( select 0.2 * avg(l_quantity)           +
                 |                        from tpch.lineitem                      +
```

```
                |                 where l_partkey = p_partkey );
 backend_type   | client backend

system@test=# \x
Expanded display is off.
```

从输出可以看到,这个长时间执行 SQL 查询的会话进程的 PID 是 68855。

例 5.11 查看系统等待事件的情况。

可以使用下面的查询,查看系统等待事件的情况:

```
system@test=# SELECT pid, wait_event_type, wait_event
test-#          FROM pg_stat_activity
test-#          WHERE wait_event is NOT NULL;
  pid  | wait_event_type |    wait_event
-------+-----------------+---------------------
--省略了一些输出
 68855 | IO              | DataFileRead
--省略了一些输出
(7 rows)
```

5.2.3 处理有问题的连接会话

当一个会话长时间不响应时,KingbaseES 数据库的 DBA 可能需要通过取消这个会话中正在运行的 SQL 语句(会话继续保持),或者删除这个会话来释放占用的数据库资源。

例 5.12 取消会话中运行的 SQL 语句。

在 5.2.2 节中执行时间超过 5min 的会话进程的 PID 是 68855。执行下面的 SQL 语句,取消这个正在运行的 SQL 语句:

```
system@test=# SELECT pg_cancel_backend(68855);
 pg_cancel_backend
-------------------
 t
(1 row)
```

运行 SQL 语句的会话终端,将显示如下信息:

```
ERROR:  canceling statement due to user request
system@test=#
```

在这个会话上重新运行这条 SQL 语句。

例 5.13 终止执行时间比较长的会话连接。

如果想终止这个会话(进程 PID 是 68855),执行以下步骤:

在 KingbaseES 数据库服务器上,执行 kbps 脚本,再次确认系统有 PID 是 68855 的会话:

```
[kingbase@dbsvr ~]$ kbps
--省略了许多输出
kingbase   68855   1887 99 02:41 ?        00:21:14 kingbase: system test 192.168.100.18(58822) SELECT
--省略了输出
```

[kingbase@dbsvr ~]$

在 KingbaseES 数据库服务器上,终止 PID 是 68855 的会话进程:

[kingbase@dbsvr ~]$ ksql -d test -U system -c "SELECT sys_terminate_backend(68855);"

在 KingbaseES 数据库服务器上,执行下面的 kbps 脚本命令:

```
[kingbase@dbsvr ~]$ kbps
kingbase   1887      1  0 01:47 ?        00:00:00 /u00/Kingbase/ES/V9/kingbase/KESRealPro/V009R001C001B0025/Server/bin/kingbase -D /u00/Kingbase/ES/V9/data
kingbase   2016   1887  0 01:47 ?        00:00:00 kingbase: logger
kingbase   2037   1887  0 01:47 ?        00:00:00 kingbase: checkpointer
kingbase   2038   1887  0 01:47 ?        00:00:00 kingbase: background writer
kingbase   2039   1887  0 01:47 ?        00:00:00 kingbase: walwriter
kingbase   2040   1887  0 01:47 ?        00:00:00 kingbase: autovacuum launcher
kingbase   2041   1887  0 01:47 ?        00:00:00 kingbase: stats collector
kingbase   2044   1887  0 01:47 ?        00:00:00 kingbase: kwr collector
kingbase   2046   1887  0 01:47 ?        00:00:00 kingbase: ksh writer
kingbase   2048   1887  0 01:47 ?        00:00:00 kingbase: ksh collector
kingbase   2049   1887  0 01:47 ?        00:00:00 kingbase: logical replication launcher
kingbase  69028   1887  0 02:50 ?        00:00:00 kingbase: system test [local] idle
[kingbase@dbsvr ~]$
```

从输出可以看到,PID 是 68855 的后端服务进程已经消失了。

回到 KingbaseES 数据库客户端上执行 SQL 语句的终端,随便执行一条 SQL 语句(示例中执行的是 SELECT now()语句),显示如下:

```
system@test=# SELECT now();
FATAL:  terminating connection due to administrator command
server closed the connection unexpectedly
    This probably means the server terminated abnormally
    before or while processing the request.
    [kingbase@dbclient ~]$
```

从输出可以看到,KingbaseES 数据库客户端上的 ksql 会话连接已经被终止。

5.3 配置文件 sys_hba.conf

KingbaseES 数据库服务器的客户端访问控制配置文件 sys_hba.conf 在 KingbaseES 数据库服务器的数据目录/u00/Kingbase/ES/V9/data/下,文件名中的 hba 是 Host-Based Authentication 的缩写,意思是基于主机的身份认证。

sys_hba.conf 文件的每一行内容都是一条客户端访问控制配置的记录,格式如下:

```
TYPE    DATABASE    USER    ADDRESS    METHOD    [OPTIONS]
```

其中,

(1) TYPE：指定连接的类型。

① 值为 local，表示客户端使用 UNIX 本地套接字连接访问 KingbaseES 数据库服务器，客户端和 KingbaseES 数据库服务器在同一台计算机上。TYPE 值为 local 的客户端访问控制配置记录的 ADDRESS 字段为空。

② 值为 host，表示客户端使用 TCP/IP 连接访问 KingbaseES 数据库服务器。TYPE 值为 host 的客户端访问控制记录的 ADDRESS 字段，表示可以访问的 KingbaseES 数据库的服务器地址（IP 地址/子网掩码前缀）。

③ 值为 hostssl，表示客户端必须使用 TCP/IP 的安全套接层（secure socket layer，SSL）连接访问 KingbaseES 数据库服务器。

④ 值为 hostnossl，表示客户端不使用 TCP/IP 的 SSL 连接访问 KingbaseES 数据库服务器。

(2) DATABASE：被连接访问的数据库名，如果 DATABASE 字段的值为 all，表示可以访问 KingbaseES 数据库服务器的任何数据库。

(3) USER：访问数据库的用户名，如果 USER 字段的值为 all，表示 KingbaseES 数据库服务器上的任何数据库用户，都可以访问。

(4) ADDRESS：允许连接 KingbaseES 数据库服务器的 IP 地址，格式为"IP 地址/子网掩码前缀"。

(5) METHOD：身份验证方法。

① 值为 trust，表示无须密码就允许连接。
② 值为 scram-sha-256，表示认证使用经过 scram-sha-256 加密的密码。
③ 值为 password，表示认证使用明文密码。
④ 值为 peer，表示使用操作系统用户进行身份验证。
⑤ 值为 ident，表示使用 ident 协议进行身份验证。
⑥ 值为 reject，表示明确拒绝。

(6) OPTIONS：提供额外的选项信息。

DBA 向客户端访问控制文件 sys_hba.conf 中，添加的记录可以是以下几种格式之一：

```
Local       Database  User           Method  [Options]
Host        Database  User  Address  Method  [Options]
Hostssl     Database  User  Address  Method  [Options]
Hostnossl   Database  User  Address  Method  [Options]
```

例 5.14 对于本机使用 trust 配置客户端认证。

trust 表示信任认证，数据库用户可以直接建立数据库连接，不需要口令或其他任何认证。trust 认证非常方便，但是可能会带来安全隐患。在确认安全性的情况下可以使用 trust 方式认证客户端，例如，对于单用户工作的本地连接。

在配置之前，需要在 KingbaseES 数据库服务器上创建一个新的 Linux 操作系统的用户 testdba，它是 DBA 组的成员，使用 Linux 用户 testdba 访问 KingbaseES 数据库：

```
[kingbase@dbsvr ~]$ su -
Password: #输入 Linux 超级用户 root 的密码 root123
[root@dbsvr ~]# useradd -G dba testdba
```

```
[root@dbsvr ~]# rm -rf ~testdba/.bashrc
[root@dbsvr ~]# cp ~kingbase/.bashrc ~testdba/
[root@dbsvr ~]# chown testdba.dba ~testdba/.bashrc
[root@dbsvr ~]# su - testdba
[testdba@dbsvr ~]$ ksql -d test -U system
Password for user system: #输入 system 用户的密码 Passw0rd
test=# \q
[testdba@dbsvr ~]$
```

从输出可以看到，即使是 DBA 组的成员，访问 KingbaseES 数据库也要提供密码。

接下来开始配置本机操作系统 DBA 组的用户，以 trust 的方式，无密码访问 KingbaseES 数据库。配置之前，在 KingbaseES 数据库的服务器上，先以 Linux 操作系统的用户 kingbase 的身份执行下面的命令，备份服务器上的客户端访问控制文件/u00/Kingbase/ES/V9/data/sys_hba.conf：

```
[kingbase@dbsvr ~]$ cd /u00/Kingbase/ES/V9/data
[kingbase@dbsvr data]$ cp sys_hba.conf sys_hba.conf.bak
```

然后在 KingbaseES 数据库服务器上，以用户 kingbase 的身份，使用 vi 编辑器编辑配置文件/u00/Kingbase/ES/V9/data/sys_hba.conf，找到如下两行内容：

```
# "local" 只能用于 UNIX 域套接字
local   all             all                                     scram-sha-256
```

将第 2 行修改为：

```
local   all             all                                     trust
```

配置后，在 KingbaseES 数据库服务器上，任何操作系统用户，使用任何数据库用户，连接任何数据库时，均不再需要提供数据库用户的密码。

接着，在 KingbaseES 数据库服务器上，以用户 kingbase 的身份执行下面的命令，让 KingbaseES 数据库服务器重新装载(读取)，配置文件 sys_hba.conf 才能生效：

```
[kingbase@dbsvr ~]$ ksql -d test -U system -c "SELECT sys_reload_conf();"
```

接下来，在 KingbaseES 数据库服务器上，以用户 testdba 的身份，执行下面的 ksql 命令：

```
[kingbase@dbsvr ~]$ ksql -d test -U system
system@test=# \q
[kingbase@dbsvr ~]$
```

从输出可以看到，现在只要是 DBA 组的成员(原因是只有 DBA 组的用户，才拥有权限访问和执行 KingbaseES 数据库管理系统的二进制程序)，可以无密码访问 KingbaseES 数据库了。

为了继续后面的实战，在 KingbaseES 数据库服务器上，以 kingbase 用户的身份，执行下面的命令，将文件 sys_hba.conf 恢复为原始的配置：

```
[kingbase@dbsvr ~]$ cd /u00/Kingbase/ES/V9/data
[kingbase@dbsvr data]$ cp sys_hba.conf.bak sys_hba.conf
```

```
[kingbase@dbsvr data]$ ksql -d test -U system -c "SELECT sys_reload_conf();"
```

例 5.15 对于所有的客户端使用 scram-sha-256 认证。

scram-sha-256 认证是 KingbaseES 数据库默认的客户端认证方法。

```
host    all    all    0.0.0.0/0              scram-sha-256
```

此处 IP 地址 0.0.0.0/0,代表所有的 IP 地址集合。这行配置的含义是,可以使用任何用户(第 2 个 all),从任意 IP 地址的客户端主机,通过 TCP/IP 访问 KingbaseES 数据库服务器上的任何数据库(第 1 个 all),但需要进行用户密码认证,用户密码的加密算法是 scram-sha-256。

例 5.16 设置只允许 IP 地址为 192.168.100.22/32 的客户端可以建立数据库连接。

```
host    all    all    192.168.100.22/32      scram-sha-256
```

表示可以使用任何用户(第 2 个 all),只允许 IP 地址为 192.168.100.22/32 的客户端,通过 TCP/IP 访问 KingbaseES 数据库服务器上的任何数据库(第 1 个 all),并且需要进行用户密码认证,用户密码的加密算法是 scram-sha-256。

需要特别说明的是,KingbaseES 数据库服务器采用首次匹配和首次拒绝的原则,来处理客户端访问控制文件中的访问规则:**不管是匹配还是被拒绝,都不再继续尝试后面的配置条目了**。因此在配置访问规则时,需要注意访问规则的先后顺序。

5.4 配置文件 sys_ident.conf

数据目录/u00/Kingbase/ES/V9/data/下的 sys_ident.conf 文件是 KingbaseES 数据库用于设置用户映射规则的配置文件,它与客户端认证文件 sys_hba.conf 配合使用。

文件 sys_ident.conf 中定义操作系统用户和 KingbaseES 数据库用户之间的映射关系,这在使用外部认证机制(如 LDAP、GSSAPI 或 SSL 证书)时特别有用。

要映射操作系统用户和数据库用户,可以在文件 sys_ident.conf 中添加一行内容,格式如下。

```
MAPNAME     SYSTEM-USERNAME     KDB-USERNAME
```

其中,

(1) MAPNAME:映射的名字。

(2) SYSTEM-USERNAME:操作系统用户的名字。

(3) KDB-USERNAME:KingbaseES 数据库用户的名字。

修改了文件 sys_ident.conf 之后,需要为 KingbaseES 数据库服务器重新装载该配置文件,文件 sys_hba.conf 中会引用文件 sys_ident.conf 中定义的映射。

例 5.17 定义操作系统用户与数据库用户之间的映射。

现在计划将 DBA 使用的 Linux 操作系统用户 kingbase 映射为数据库用户 kingbase,这样使用用户 kingbase 管理 KingbaseES 数据库将会更方便。

(1) 执行下面的 ksql 命令和 SQL 语句,创建数据库用户 kingbase:

```
[kingbase@dbsvr ~]$ ksql -d test -U system
system@test=# CREATE USER kingbase WITH SUPERUSER PASSWORD 'Passw0rd';
system@test=# \q
[kingbase@dbsvr ~]$ ksql -d test -U kingbase
Password for user kingbase: #输入数据库用户 kingbase 的密码 Passw0rd
kingbase@test=# \q
[kingbase@dbsvr ~]$
```

从输出可以看到，此时使用数据库用户 kingbase 访问 KingbaseES 数据库，需要提供密码。

（2）使用 vi 编辑器，编辑文件/u00/Kingbase/ES/V9/data/sys_ident.conf，在文件末尾添加如下的内容：

```
kingbase_map    kingbase    kingbase
```

（3）使用 vi 编辑器，编辑文件/u00/Kingbase/ES/V9/data/sys_hba.conf，找到如下的内容：

```
# "local" 只能用于 UNIX 域套接字
local   all         all                                     scram-sha-256
```

在这两行之间，添加一行，内容如下：

```
# "local" 只能用于 UNIX 域套接字
local   all         kingbase                                peer map=kingbase_map
local   all         all                                     scram-sha-256
```

修改后的配置如下。

① 使用数据库用户 kingbase，通过 UNIX 本地套接字访问所有数据库时，将使用操作系统用户进行身份认证，映射名为 kingbase_map。

② 所有数据库用户，通过 UNIX 本地套接字访问任何数据库时，都需要提供经过 scram-sha-256 加密后的用户密码进行身份认证。

（4）执行下面的 ksql 命令，重新装载数据库，让修改的配置文件生效：

```
[kingbase@dbsvr ~]$ ksql -d test -U system -c "SELECT sys_reload_conf();"
```

（5）使用 Linux 操作系统的用户 kingbase，执行下面的命令：

```
[kingbase@dbsvr ~]$ ksql -d test -U kingbase
kingbase@test=# \q
[kingbase@dbsvr ~]$
```

从输出可以看到，使用 Linux 操作系统的用户 kingbase 登录操作系统后，以数据库用户 kingbase 的身份访问 test 数据库（任意一个数据库），不再需要提供数据库用户 kingbase 的密码了。

第 6 章

管理用户数据

 ## 6.1 管理表空间

在 KingbaseES 数据库中,表空间(tablespace)是用于组织和管理数据库物理存储的一种机制。它允许将数据库对象(如表、索引等)存储在不同的文件系统路径或磁盘上。也就是说,表空间是数据的物理容器。

6.1.1 创建表空间

KingbaseES 数据库的表空间对应 KingbaseES 数据库服务器上的一个操作系统目录。一个表空间可以存储多个数据库的模式对象,一个数据库可以存储在多个表空间中,在表空间对应的操作系统目录下,KingbaseES 数据库为每个在该表空间中存有模式对象的数据库,创建以数据库的 OID 为名字的子目录。

在系统初始化时,默认创建存放用户数据的表空间 sys_default,对应的操作系统目录是 data/base。使用 SQL 语句 CREATE TABLESPACE 可以创建用户自己的表空间。在创建表或索引时,指定其使用的表空间,从而指定其数据的存放位置。

KingbaseES 数据库表空间对应着操作系统的目录,因此创建表空间之前,需要在操作系统中创建存储数据文件的文件目录。

例 6.1 创建用户表空间,并在该表空间中创建表。

(1) 创建表空间 data_ts 之前,打开第 1 个终端,执行下面的命令:

```
[kingbase@dbsvr ~]$ ls -l /u00/Kingbase/ES/V9/data/sys_tblspc
total 0
[kingbase@dbsvr ~]$
```

从输出可以看到,创建第 1 个用户表空间前,目录/u00/Kingbase/ES/V9/data/sys_tblspc 是空目录。

(2) 在第 2 个终端,创建目录/u01/Kingbase/ES/V9/userts/data_ts,并创建表空间 data_ts:

```
[kingbase@dbsvr ~]$ ksql -d test -U system
system@test=# \! mkdir -p /u01/Kingbase/ES/V9/data/userts/data_ts
system@test=# CREATE TABLESPACE data_ts LOCATION '/u01/Kingbase/ES/V9/data/
```

```
userts/data_ts';
CREATE TABLESPACE
system@test=# SELECT oid, spcname FROM sys_tablespace WHERE spcname = 'data_ts';
  oid  | spcname
-------+---------
 16628 | data_ts
(1 row)
system@test=#
```

从输出可以看到,新创建的表空间 data_ts 的 OID 是 16628。

(3) 在第 1 个终端上,执行下面的命令:

```
[kingbase@dbsvr sys_tblspc]$ ls -l /u00/Kingbase/ES/V9/data/sys_tblspc
total 0
lrwxrwxrwx. 1 kingbase dba 39 Feb 18 12:25 16628 -> /u01/Kingbase/ES/V9/data/userts/data_ts
[kingbase@dbsvr sys_tblspc]$ cd /u01/Kingbase/ES/V9/data/userts/data_ts
[kingbase@dbsvr data_ts]$ ls -l
total 0
drwx------. 2 kingbase dba 6 Feb 18 12:25 SYS_12_202305151
[kingbase@dbsvr data_ts]$ ls -l /u01/Kingbase/ES/V9/data/userts/data_ts/SYS_12_202305151
total 0
[kingbase@dbsvr data_ts]$
```

从输出可以看到,新建一个表空间 data_ts,将在数据目录的 sys_tblspc 子目录下创建一个符号链接,它直接指向了表空间的实际存储目录/u01/Kingbase/ES/V9/data/userts/data_ts。实际上,表空间 data_ts 直接对应着操作系统目录/u01/Kingbase/ES/V9/data/userts/data_ts/SYS_12_202305151。

KingbaseES 数据库通过数据目录下的符号链接/u00/Kingbase/ES/V9/data/sys_tblspc/16628 导引到表空间的实际存储位置,其中 16628 是表空间 data_ts 的 OID。由于目前表空间 data_ts 还没有存储任何数据,因此其对应的目录现在还是一个空目录。

(4) 返回第 2 个终端,在表空间 data_ts 中,为数据库 test 创建表 t2:

```
system@test=# CREATE TABLE t2(col1 int, col2 int) TABLESPACE data_ts;
CREATE TABLE
system@test=# -- 查询表 t2 的 oid 和 relfilenode
system@test=# SELECT oid, relname, relfilenode FROM sys_class WHERE relname='t2';
  oid  | relname | relfilenode
-------+---------+-------------
 16629 | t2      |       16629
(1 row)
system@test=# -- 查询表 t2 所对应的操作系统文件
system@test=# SELECT sys_relation_filepath('t2');
           sys_relation_filepath
------------------------------------------------
 sys_tblspc/16628/SYS_12_202305151/14386/16629
(1 row)
system@test=#
```

从输出可以看到，表 t2 对应的操作系统文件为：

/u00/Kingbase/ES/V9/data/sys_tblspc/16628/SYS_12_202305151/14386/16629

其中的/u00/Kingbase/ES/V9/data/sys_tblspc/16628 是一个符号链接：

```
system@test=# \! ls -ld  /u00/Kingbase/ES/V9/data/sys_tblspc/16628
lrwxrwxrwx. 1 kingbase dba 39 Feb 18 12:25 /u00/Kingbase/ES/V9/data/sys_tblspc/16628 -> /u01/Kingbase/ES/V9/data/userts/data_ts
system@test=#
```

因此，表 t2 真正对应的操作系统文件应该是：

```
/u01/Kingbase/ES/V9/data/userts/data_ts/SYS_12_202305151/14386/16629
system@test=# \! ls -l /u01/Kingbase/ES/V9/data/userts/data_ts/SYS_12_202305151/14386/16629
-rw-------. 1 kingbase dba 0 Feb 18 12:28 /u01/Kingbase/ES/V9/data/userts/data_ts/SYS_12_202305151/14386/16629
system@test=#
```

新创建的用户表空间，不能位于 KingbaseES 数据库的数据目录下，只能将新创建的表空间，创建在 KingbaseES 数据库的数据目录之外。例如：

```
system@test=# \! mkdir -p /u00/Kingbase/ES/V9/data/userts/newdata_ts
system@test=# CREATE TABLESPACE newdata_ts LOCATION '/u00/Kingbase/ES/V9/data/userts/newdata_ts';
ERROR:  tablespace location should not be inside the data directory
system@test=# \! rm -rf  /u00/Kingbase/ES/V9/data/userts
```

例 6.2　查看数据库和表空间之间的关系。

下面的示例可以验证一个数据库可以存放在多个表空间中，一个表空间可以存放多个数据库的数据。

（1）在 KingbaseES 数据库中创建表空间 userdata_ts、new1_ts 和 new2_ts：

```
system@test=# \! mkdir -p /u02/Kingbase/ES/V9/data/userts/userdata_ts
system@test=# CREATE TABLESPACE userdata_ts LOCATION '/u02/Kingbase/ES/V9/data/userts/userdata_ts';
system@test=# \! mkdir -p /u01/Kingbase/ES/V9/data/userts/new1_ts
system@test=# CREATE TABLESPACE new1_ts LOCATION '/u01/Kingbase/ES/V9/data/userts/new1_ts';
system@test=# \! mkdir -p /u02/Kingbase/ES/V9/data/userts/new2_ts
system@test=# CREATE TABLESPACE new2_ts LOCATION '/u02/Kingbase/ES/V9/data/userts/new2_ts';
```

（2）执行下面的 SQL 语句，创建数据库 newdb1 和 newdb2，默认的表空间为 userdata_ts：

```
system@test=# CREATE DATABASE newdb1  TABLESPACE userdata_ts;
system@test=# CREATE DATABASE newdb2  TABLESPACE userdata_ts;
```

（3）创建数据库用户 newuser，并允许它连接到数据库 newdb1、在数据库 newdb1 上创建数据库对象、将数据库 newdb1 的属主修改为 newuser、在表空间 data_ts 和 userdata_ts 上创建数据库对象：

```
system@test=# CREATE USER newuser PASSWORD 'newuser123';
system@test=# GRANT CONNECT ON DATABASE newdb1 TO newuser;
system@test=# GRANT CREATE ON DATABASE newdb1 TO newuser;
system@test=# ALTER DATABASE newdb1 OWNER TO newuser;
system@test=# GRANT CREATE ON TABLESPACE data_ts TO newuser;
system@test=# GRANT CREATE ON TABLESPACE userdata_ts TO newuser;
```

（4）允许数据库用户 newuser 连接到数据库 newdb2、在数据库 newdb2 上创建数据库对象、将数据库 newdb2 的属主修改为 newuser：

```
system@test=# GRANT CONNECT ON DATABASE newdb2 TO newuser;
system@test=# GRANT CREATE  ON DATABASE newdb2 TO newuser;
system@test=# ALTER DATABASE newdb2 OWNER TO newuser;
```

（5）将表空间 new2_ts 的属主修改为 newuser：

```
system@test=# ALTER TABLESPACE new2_ts OWNER TO newuser;
system@test=# \q
[kingbase@dbsvr ~]$
```

（6）使用数据库用户 newuser 为数据库 newdb1 在表空间 data_ts 中创建表 ttt1，为数据库 newdb1 在表空间 userdata_ts 中创建表 ttt2：

```
[kingbase@dbsvr ~]$ KINGBASE_PASSWORD=newuser123 ksql -d newdb1 -U newuser
newuser@newdb1=# CREATE TABLE ttt1 ( col1 int, col2 int, primary key(col1) ) tablespace data_ts;
newuser@newdb1=# CREATE TABLE ttt2 ( col1 int, col2 int, primary key(col1) ) tablespace userdata_ts;
newuser@newdb1=# \q
[kingbase@dbsvr ~]$
```

（7）查看所有数据库默认表空间的名字及其目录位置：

```
[kingbase@dbsvr ~]$ ksql -d test -U system
system@test=# SELECT d.datname as "Database Name",
test-#          t.spcname as "Default Tablespace",
test-#          CASE
test-#            WHEN t.spcname = 'sys_default' THEN '/u00/Kingbase/ES/V9/data/base'
test-#            WHEN t.spcname = 'sys_global'  THEN '/u00/Kingbase/ES/V9/data/global'
test-#            WHEN t.spcname = 'sysaudit'    THEN '/u00/Kingbase/ES/V9/data/sysaud'
test-#            ELSE sys_catalog.sys_tablespace_location(t.oid)
test-#          END AS "Location"
test-#   FROM sys_catalog.sys_database d
test-#        JOIN sys_catalog.sys_tablespace t ON d.dattablespace = t.oid;
 Database Name | Default Tablespace |                Location
---------------+--------------------+-----------------------------------------
 test          | sys_default        | /u00/Kingbase/ES/V9/data/base
 kingbase      | sys_default        | /u00/Kingbase/ES/V9/data/base
 template1     | sys_default        | /u00/Kingbase/ES/V9/data/base
 template0     | sys_default        | /u00/Kingbase/ES/V9/data/base
 security      | sys_default        | /u00/Kingbase/ES/V9/data/base
 newdb1        | userdata_ts        | /u02/Kingbase/ES/V9/data/userts/userdata_ts
```

```
 newdb2           | userdata_ts          | /u02/Kingbase/ES/V9/data/userts/userdata_ts
(7 rows)
system@test=#
```

（8）查看 KingbaseES 数据库集簇有哪些表空间，以及它们的目录位置：

```
system@test=#   SELECT t.spcname AS "Tablespace Name",
test-#          u.usename AS "Tablespace Owner",
test-#          CASE
test-#            WHEN t.spcname = 'sys_default' THEN '/u00/Kingbase/ES/V9/data/base'
test-#            WHEN t.spcname = 'sys_global'  THEN '/u00/Kingbase/ES/V9/data/global'
test-#            WHEN t.spcname = 'sysaudit'    THEN '/u00/Kingbase/ES/V9/data/sysaud'
test-#            ELSE sys_catalog.sys_tablespace_location(t.oid)
test-#          END AS "Location"
test-#   FROM sys_catalog.sys_tablespace t
test-#              JOIN sys_catalog.sys_user u ON t.spcowner = u.usesysid;
 Tablespace Name | Tablespace Owner |               Location
-----------------+------------------+----------------------------------------
 sys_default     | system           | /u00/Kingbase/ES/V9/data/base
 sys_global      | system           | /u00/Kingbase/ES/V9/data/global
 sysaudit        | system           | /u00/Kingbase/ES/V9/data/sysaud
 new1_ts         | system           | /u01/Kingbase/ES/V9/data/userts/new1_ts
 data_ts         | system           | /u01/Kingbase/ES/V9/data/userts/data_ts
 userdata_ts     | system           | /u02/Kingbase/ES/V9/data/userts/userdata_ts
 new2_ts         | newuser          | /u02/Kingbase/ES/V9/data/userts/new2_ts
(7 rows)
```

（9）查看指定名字的数据库（需要连接到这个数据库上）存储在哪些表空间上：

```
system@test=# SELECT p.spcname
test-#        FROM sys_database d  JOIN sys_tablespace p ON p.oid = d.dattablespace
test-#        WHERE d.datname = 'test'
test-#        UNION
test-#        SELECT t.spcname
test-#        FROM sys_class c JOIN sys_tablespace t ON c.reltablespace = t.oid
test-#        WHERE t.spcname IN ( SELECT spcname AS "Name"
test(#                             FROM sys_catalog.sys_tablespace
test(#                            )
test-#          AND c.relname NOT LIKE 'pg_%'
test-#          AND c.relname NOT LIKE 'kdb_%'
test-#          AND c.relname NOT LIKE 'sys_%'
test-#          AND c.relname NOT LIKE 'system_privilege_index'
test-#          AND t.spcname != 'sys_global';   -- 排除 sys_global 表空间;
   spcname
-------------
 data_ts
 sys_default
(2 rows)
```

从输出可以看到，数据库 test 当前存储在 data_ts 和 sys_default 两个表空间上。

（10）查看当前表空间 userdata_ts 存储了哪些数据库的数据：

```
system@test=# SELECT d.datname
test-#          FROM pg_database d, ( SELECT pg_tablespace_databases(oid) AS datoid
test(#                                  FROM pg_tablespace t
test(#                                  WHERE t.spcname='userdata_ts'
test(#                                ) t
test-#          WHERE t.datoid = d.oid;
 datname
---------
 newdb1
 newdb2
(2 rows)
```

从输出可以看到,表空间 userdata_ts 当前存储了数据库 newdb1 和 newdb2 的数据。

6.1.2 删除表空间

可以使用 SQL 语句 DROP TABLESPACE 来删除一个表空间。需要注意的是除了超级用户,只有表空间的属主才可以删除自己的表空间。

例 6.3 只有表空间的属主才可以删除自己的表空间。

```
[kingbase@dbsvr ~]$ KINGBASE_PASSWORD=newuser123 ksql -d newdb1 -U newuser
newuser@newdb1=> \db
                       List of tablespaces
    Name      | Owner   |              Location
--------------+---------+-------------------------------------------
 data_ts      | system  | /u01/Kingbase/ES/V9/data/userts/data_ts
 new1_ts      | system  | /u01/Kingbase/ES/V9/data/userts/new1_ts
 new2_ts      | newuser | /u02/Kingbase/ES/V9/data/userts/new2_ts
 sys_default  | system  |
 sys_global   | system  |
 sysaudit     | system  |
 userdata_ts  | system  | /u02/Kingbase/ES/V9/data/userts/userdata_ts
(7 rows)
newuser@newdb1=> DROP TABLESPACE new1_ts;
ERROR:  must be owner of tablespace new1_ts    --无法删除不属于自己的表空间
newuser@newdb1=> DROP TABLESPACE new2_ts;
DROP TABLESPACE                                --可以删除属于自己的表空间
newuser@newdb1=> -- 要彻底删除表空间 new1_ts,还要删除创建表空间前为该表空间创建的
                -- 目录
newuser@newdb1=> \! rmdir /u02/Kingbase/ES/V9/data/userts/new2_ts
newuser@newdb1=> \q
[kingbase@dbsvr ~]$
```

从输出可以看到,用户必须是表空间的属主才能删除表空间。

具有数据库 Superuser 权限的 DBA 可以删除不属于自己的表空间:

```
[kingbase@dbsvr ~]$ ksql -d test -U system
system@test=# -- 将 new1_ts 的属主修改为 newuser
system@test=# ALTER TABLESPACE new1_ts OWNER TO newuser;
system@test=# \db new1_ts
                       List of tablespaces
```

```
      Name    |  Owner   |             Location
 -------------+----------+------------------------------------
  new1_ts     | newuser  | /u01/Kingbase/ES/V9/data/userts/new1_ts
(1 row)
system@test=# DROP TABLESPACE new1_ts;
DROP TABLESPACE                 --system 是数据库超级用户,它可以删除不属于自己的表空间
system@test=# -- 要彻底删除表空间 new1_ts,还要删除创建表空间前为该表空间创建的目录
system@test=# \! rmdir /u01/Kingbase/ES/V9/data/userts/new1_ts
system@test=#
```

从输出可以看到,表空间 new1_ts 的属主不是用户 system,但由于 system 是超级用户,因此 system 可以删除表空间 new2_ts(属主为 newuser)。

例 6.4 删除非空的表空间。

如果表空间不为空,表空间不能被删除:

```
system@test=# DROP TABLESPACE userdata_ts;
ERROR:  tablespace "userdata_ts" is not empty
system@test=#
```

为了删除不为空的表空间 userdata_ts,首先执行下面的 ksql 命令和 SQL 语句,查看表空间 userdata_ts 目前保存了哪些数据库的数据:

```
system@test=# SELECT d.datname
test-#          FROM sys_database d, ( SELECT sys_tablespace_databases(oid) AS datoid
test(#                                   FROM sys_tablespace t
test(#                                  WHERE t.spcname='userdata_ts'
test(#                                ) t
test-#         WHERE t.datoid = d.oid;
  datname
-----------
 newdb1
 newdb2
(2 rows)
system@test=# \l+ newdb1
                                        List of databases
  Name  | Owner  | Encoding | Collate     |   Ctype     |  Access privileges  | Size  | Tablespace  | Description
--------+--------+----------+-------------+-------------+---------------------+-------+-------------+------------
 newdb1 | newuser| UTF8     | zh_CN.UTF-8 | zh_CN.UTF-8 | =Tc/newuser        +| 14 MB | userdata_ts |
        |        |          |             |             | newuser=CTc/newuser |       |             |
(1 row)
system@test=# \l+ newdb2
                                        List of databases
  Name  | Owner  | Encoding | Collate     |   Ctype     |  Access privileges  | Size  | Tablespace  | Description
--------+--------+----------+-------------+-------------+---------------------+-------+-------------+------------
 newdb2 | newuser| UTF8     | zh_CN.UTF-8 | zh_CN.UTF-8 | =Tc/newuser        +| 14 MB | userdata_ts |
        |        |          |             |             | newuser=CTc/newuser |       |             |
(1 row)
```

从输出可以看到,表空间 userdata_ts 保存了数据库 newdb1 和 newdb2 的数据,并且这两个数据库的默认表空间都是 userdata_ts。

假设不再需要数据库 newdb2 的数据,那么可以执行下面的 SQL 语句,直接删除数据库 newdb2:

```
system@test=# DROP DATABASE newdb2;
```

假设数据库 newdb1 的数据还有用，那么需要将数据库 newdb1 保存在 userdata_ts 表空间的数据库对象移动到其他的表空间（如表空间 data_ts）。

打开第 1 个终端，执行下面的 ksql 命令和 SQL 语句，查看当前连接的数据库 newdb1 有哪些表，它们存储在哪些表空间中：

```
[kingbase@dbsvr ~]$ KINGBASE_PASSWORD=newuser123 ksql -d newdb1 -U newuser
newuser@newdb1=> WITH DefaultTablespace AS (
newdb1(>            SELECT spcname
newdb1(>            FROM sys_tablespace
newdb1(>            WHERE oid = (SELECT dattablespace FROM sys_database WHERE datname = 'newdb1')
newdb1(>           )
newdb1->  SELECT schemaname,tablename,
newdb1->         COALESCE(tablespace,(SELECT spcname FROM DefaultTablespace)) AS tablespace
newdb1->  FROM sys_catalog.sys_tables
newdb1->  WHERE schemaname NOT IN
newdb1->        ('pg_catalog','sys_catalog','sys_hm','sys','sysmac','information_schema');
 schemaname | tablename | tablespace
------------+-----------+-------------
 public     | ttt1      | data_ts
 public     | ttt2      | userdata_ts
(2 rows)
```

从输出可以看到，数据库 newdb1 的表 ttt1 保存在表空间 data_ts 中。

需要将所有不在数据库 newdb1 的默认表空间 userdata_ts 的表，移动回默认表空间 userdata_ts。执行下面的 SQL 语句，将表 ttt1 移动到表空间 userdata_ts：

```
newuser@newdb1=> ALTER TABLE ttt1 SET TABLESPACE userdata_ts;
```

再次查看当前连接的数据库 newdb1 有哪些表，它们存储在哪些表空间中：

```
newuser@newdb1=> WITH DefaultTablespace AS (
newdb1(>            SELECT spcname
newdb1(>            FROM sys_tablespace
newdb1(>            WHERE oid = (SELECT dattablespace FROM sys_database WHERE datname = 'newdb1')
newdb1(>           )
newdb1->  SELECT schemaname,tablename,
newdb1->         COALESCE(tablespace,(SELECT spcname FROM DefaultTablespace)) AS tablespace
newdb1->  FROM sys_catalog.sys_tables
newdb1->  WHERE schemaname NOT IN
newdb1->        ('pg_catalog','sys_catalog','sys_hm','sys','sysmac','information_schema');
 schemaname | tablename | tablespace
------------+-----------+-------------
 public     | ttt2      | userdata_ts
 public     | ttt1      | userdata_ts
```

(2 rows)

从输出可以看到，数据库 newdb1 所有的表都保存在表空间 userdata_ts 中了。

打开第 2 个终端，执行下面的 ksql 命令和 SQL 语句，将数据库移动到表空间 data_ts：

```
[kingbase@dbsvr ~]$ ksql -d test -U system
system@test=# ALTER DATABASE newdb1 SET TABLESPACE data_ts;
ERROR:  database "newdb1" is being accessed by other users
DETAIL:  There is 1 other session using the database.
system@test=#
```

从输出可以看到，由于还有 1 个会话连接到了数据库 newdb1，此时无法将数据库 newdb1 的数据库对象移动到新的表空间。

回到第 1 个终端，执行元命令 \q，退出与数据库 newdb2 的会话连接：

```
newuser@newdb2=> \q
[kingbase@dbsvr ~]$
```

回到第 2 个终端，执行下面的 SQL 语句，将数据库 newdb1 移动到表空间 data_ts：

```
system@test=# ALTER DATABASE newdb1 SET TABLESPACE data_ts;
ALTER DATABASE
system@test=#
```

回到第 1 个终端，再次执行下面的 ksql 命令和 SQL 语句，查看当前连接的数据库 newdb1 有哪些表，它们存储在哪些表空间中：

```
[kingbase@dbsvr ~]$ KINGBASE_PASSWORD=newuser123 ksql -d newdb1 -U newuser
newuser@newdb1=> WITH DefaultTablespace AS (
newdb1(>          SELECT spcname
newdb1(>          FROM sys_tablespace
newdb1(>          WHERE oid = (SELECT dattablespace FROM sys_database WHERE datname = 'newdb1')
newdb1(>          )
newdb1-> SELECT schemaname,tablename,
newdb1->          COALESCE(tablespace,(SELECT spcname FROM DefaultTablespace)) AS tablespace
newdb1-> FROM sys_catalog.sys_tables
newdb1-> WHERE schemaname NOT IN
newdb1->          ('pg_catalog','sys_catalog','sys_hm','sys','sysmac','information_schema');
 schemaname | tablename | tablespace
------------+-----------+------------
 public     | ttt2      | data_ts
 public     | ttt1      | data_ts
(2 rows)
newuser@newdb1=> \l+ newdb1
                                        List of databases
  Name  | Owner   | Encoding | Collate     | Ctype       | Access privileges      | Size  | Tablespace | Description
--------+---------+----------+-------------+-------------+------------------------+-------+------------+------------
 newdb1 | newuser | UTF8     | zh_CN.UTF-8 | zh_CN.UTF-8 | =Tc/newuser           +| 14 MB | data_ts    |
        |         |          |             |             | newuser=CTc/newuser    |       |            |
(1 row)
newuser@newdb2=> \q
[kingbase@dbsvr ~]$
```

从输出可以看到，数据库 newdb1 的所有表，现在被移动到新的默认表空间 data_ts 中了。

回到第 2 个终端,删除表空间 userdata_ts:

```
newuser@newdb1=> \db userdata_ts
                List of tablespaces
   Name       | Owner  |              Location
--------------+--------+------------------------------------------
 userdata_ts  | system | /u02/Kingbase/ES/V9/data/userts/userdata_ts
(1 row)
system@test=# DROP TABLESPACE userdata_ts;
DROP TABLESPACE
system@test=# \! rmdir /u02/Kingbase/ES/V9/data/userts/userdata_ts
system@test=#
```

6.2 用户数据的存储规划

6.2.1 表空间的分类

表空间可以分为系统表空间和用户表空间。创建 KingbaseES 数据库集簇的时候,系统会自动创建系统表空间 sys_global 和 sysaudit,其中 sys_global 专门用于存储 KingbaseES 数据库的数据字典表(共享的系统目录),sysaudit 用来存储 KingbaseES 数据库的审计信息。还会创建存放用户数据的表空间 sys_default,在数据库中创建表、索引和临时文件时,如果没有给出 tablespace 子句,则默认保存在表空间 sys_default 中。数据库用户使用 SQL 语句 CREATE TABLESPACE 创建新的表空间存储用户数据。

表空间根据存储数据的持久性特点还可以分为永久表空间和临时表空间。**永久表空间**用于存储系统或者用户数据,系统自动创建的系统表空间和 DBA 创建的用于存储用户数据的用户表空间,都是永久表空间。**临时表空间**用于存储临时数据,临时表或临时表的索引,以及执行 SQL 语句时可能产生的临时文件(如排序、聚合、哈希等)。

为了获得更好的性能,建议将临时表空间配置在固态硬盘(solid state disk,SSD)或者 IOPS(input/output operations per second)比较高的磁盘阵列上。临时表空间中的数据通常在会话结束时或事务结束时被清除,不会永久存储。

可以为 KingbaseES 数据库配置多个临时表空间,方法是使用逗号分隔不同的临时表空间名。如果没有配置 temp_tablespaces 参数,临时表空间将默认使用系统表空间 sys_default。

6.2.2 表空间的规划

建议 KingbaseES 数据库用户将用户数据分门别类存储在不同的用户表空间中。

1. 规划用户表空间的存储目录位置

用户表空间不可以创建在 KingbaseES 数据库的数据目录下,在 KingbaseES 数据库安装示例中,即位于/u00/Kingbase/ES/V9/data/base 目录下。为了运维的简单,将用户表空间规划在文件系统/u01、/u02 等目录上。如果 KingbaseES 数据库服务器有多个磁盘,或者 KingbaseES 数据库服务器有一个专有的存储系统并划分了多个 LUN,那么可以将这些磁盘和 LUN 挂载到/u01、/u02 等目录上。然后可以为用户创建以下的目录：

```
/u01/Kingbase/ES/V9/data/userts
/u02/Kingbase/ES/V9/data/userts
…
```

并将用户表空间创建在这些目录下。这样规划的好处是可以用下面的两条命令来完成 KingbaseES 数据库集簇的冷备份或者热备份：

```
cd /                                          #转到根目录
tar cf dbcluster.tar u0?/Kingbase/ES/V9/data  #备份 KingbaseES 数据库集簇
```

之前为本章实战环境创建的表空间 data_ts、userdata_ts、new1_ts 和 new2_ts 就是按照这个规划来创建的。

为了防止将用户数据存储在系统默认的表空间 sys_default 中，通过在系统配置文件 kingbase.conf 中设置参数 default_tablespace 来为 KingbaseES 数据库指定一个新的系统默认表空间，用它来存储用户数据。

2. 规划用户表空间的存储数据

建议 KingbaseES 数据库 DBA 按如下的维度（或者维度组合）来规划表空间的数据存储。

（1）动态数据与静态数据分离存储：经常变更修改的数据表（动态数据）存储在单独的表空间中；一般不变更修改的数据表（静态数据），存储在单独的表空间中。

（2）大表数据与小表数据分离存储：大表数据存储在单独的表空间中；小表数据存储在单独的表空间中。

3. 创建专门的临时表空间

KingbaseES 数据库中的临时表空间主要用于存储查询执行过程中生成的临时数据。这些临时数据包括排序操作、哈希表、临时表和其他需要大量内存但可能溢出到磁盘的数据结构。

通过将临时表空间放置在不同的物理磁盘或 SSD 上，可以并行化数据访问，从而提高 I/O 性能。这对于执行大型排序操作或包含复杂连接和聚合的查询特别有用。

由于临时数据不需要持久保存，将这些数据与持久存储的数据分开，可以减少对持久数据的干扰，特别是在会生成大量的临时数据的计算时。

在系统配置文件 kingbase.conf 中设置参数 temp_tablespaces，为 KingbaseES 数据库指定一个或者多个临时表空间。

数据库用户可以在事务开始时，执行下面的语句，设置事务使用的临时表空间：

```
SET temp_tablespaces = 'temp_space1,temp_space2';
```

例 6.5 规划数据库服务器中的数据存储。

在 KingbaseES 数据库服务器，规划如下。

（1）KingbaseES 数据库的二进制程序安装在目录/u00/Kingbase/ES/V9/kingbase 下。

（2）KingbaseES 数据库的数据保存在目录/u00/Kingbase/ES/V9/data 下。

现在规划其中用户数据的存储目录。

（1）KingbaseES 数据库的用户表空间位于以下目录。

① /u01/Kingbase/ES/V9/data/userts。

② /u02/Kingbase/ES/V9/data/userts。

③ /u03/Kingbase/ES/V9/data/userts。

（2）KingbaseES 数据库的临时表空间为 temp_ts，位于目录 /u03/Kingbase/ES/V9/data/userts。

（3）KingbaseES 数据库的默认表空间为 user_ts，位于目录 /u02/Kingbase/ES/V9/data/userts。

首先执行下面的 ksql 命令和 SQL 语句，创建表空间 user_ts 和临时表空间 temp_ts：

```
[kingbase@dbsvr ~]$ ksql -d test -U system
system@test=# \! mkdir -p /u02/Kingbase/ES/V9/data/userts/user_ts
system@test=# CREATE TABLESPACE user_ts LOCATION '/u02/Kingbase/ES/V9/data/userts/user_ts';
system@test=# \! mkdir -p /u03/Kingbase/ES/V9/data/userts/temp_ts
system@test=# CREATE TABLESPACE temp_ts LOCATION '/u03/Kingbase/ES/V9/data/userts/temp_ts';
system@test=# \q
[kingbase@dbsvr ~]$
```

然后使用 vi 编辑器，修改系统配置文件 kingbase.conf 中的参数 default_tablespace 和 temp_tablespaces，修改如下：

```
default_tablespace = 'user_ts'
temp_tablespaces = 'temp_ts'
```

最后执行 sys_ctl restart 命令重新启动数据库：

```
[kingbase@dbsvr ~]$ sys_ctl restart
```

6.3 管理表中数据

6.3.1 表的填充因子

在 KingbaseES 数据库中，填充因子（fillfactor）是一个与表和索引存储相关的参数。它指定了表或索引页中预留给更新操作的空间百分比。填充因子的值可以在 10~100（包括两端的值），默认值通常是 100。

执行下面的 SQL 语句，创建表 example，为表的的索引指定填充因子的值为 65，为表本身指定填充因子的值为 70：

```
[kingbase@dbsvr ~]$ ksql -d test -U system
system@test=# CREATE TABLE example(
test(#          did     integer,
test(#          name    varchar(40),
test(#          UNIQUE(name) WITH (fillfactor=65)
test(#      )
test-#      WITH (fillfactor=70);
```

```
CREATE TABLE
system@test=#
```

当为表设置 fillfactor=70,这意味着数据页在初始填充时只填充到其容量的 70%。剩余的 30%空间保留用于表中的行进行更新操作。这样,当行被更新且新数据比旧数据占用更多空间时,行可能在原始的数据页中被重新放置,而不是移动到其他数据页。这有助于减少表碎片化,提高更新操作的效率。

当为索引设置 fillfactor=65,该值决定了索引页初始填充时的数据量。一个较低的填充因子在索引上留有更多的空间用于未来的插入或更新操作,这可以减少索引的重排次数。

填充因子值对 KingbaseES 数据库的性能有如下的影响。

(1) **更新性能**:较低的填充因子允许表或索引页有更多的空间处理更新和插入操作,从而减少了页分裂的可能性。在一个已满的页需要为新的或更新的行腾出空间时,它会降低性能并增加 I/O 负载。

(2) **读取性能**:虽然较低的填充因子可以提高更新性能,但它也意味着数据在物理存储上更加分散。这可能导致读取操作需要访问更多的数据页,从而降低查询性能。

(3) **空间使用**:较低的填充因子会占用更多的存储空间,因为相同数量的数据被存储在更多的页中。

对于如何设置填充因子的值,有如下的建议。

(1) 对于**频繁更新的表或索引**,设置一个低于默认值的填充因子可以改善性能。

(2) 对于**经常执行读操作很少更新的表**,可以使用默认的填充因子,或者将其设置得更高,以优化读取性能并减少所需的存储空间。

在实际使用中,合理地设置填充因子可以平衡读写性能。通常需要根据实际的工作负载和表的使用模式来调整这个参数。

6.3.2 生成批量测试数据

执行下面的 SQL 语句,可以生成一个 1000 万行的大表:

```
system@test=# DROP TABLE IF EXISTS test_orders;
system@test=# CREATE TABLE test_orders (
test(#                  o_orderkey SERIAL PRIMARY KEY,
test(#                  o_custkey INTEGER NOT NULL,
test(#                  o_orderstatus CHAR(1) NOT NULL,
test(#                  o_totalprice NUMERIC(12, 2) NOT NULL,
test(#                  o_description VARCHAR(64),
test(#                  o_value BIGINT
test(#                  );
system@test=# INSERT INTO test_orders (o_custkey, o_orderstatus, o_totalprice,
test-#                      o_description,o_value)
test-#          SELECT (RANDOM() * 9900 + 100)::INTEGER,  -- 为列 o_custkey 生成随机整数值
test-#              CASE
test-#                  WHEN RANDOM() < 1/3 THEN 'A'
test-#                  WHEN RANDOM() < 2/3 THEN 'F'
test-#                  ELSE 'O'
test-#              END,                      -- 为列 o_orderstatus 生成随机字符值
```

```
test-#                  ROUND((RANDOM() * 9900+100)::NUMERIC, 2),
                                        -- 为列 o_totalprice 生成随机浮点数
test-#         MD5(RANDOM()::text || clock_timestamp()::text)::varchar(64),
test-#                                  -- 为 o_description 生成随机文本
test-#          (FLOOR(RANDOM() * 1000000000))::BIGINT
                                        -- 为列 o_value 生成一个随机的 BIGINT 值
test-#          FROM generate_series(1, 10000000); -- 生成 10000000 条记录
INSERT 0 10000000
system@test=# SELECT count(*) FROM test_orders;
  count
----------
 10000000
(1 row)
```

其中，

（1）RANDOM() 是 KingbaseES 数据库中的一个函数，用于生成一个介于 0 和 1 之间的随机浮点数（包含 0，但不包含 1）。这个函数不接受任何参数，并且每次调用时都会返回一个新的随机值。例如：表达式（RANDOM() * 9900 + 100)::INTEGER 将生成 [100, 9999] 上的随机整数。

（2）表达式(FLOOR(RANDOM() * 1000000000))::BIGINT 的作用是生成一个随机的、[0, 1 000 000 000) 内的、类型为 BIGINT 的整数。

（3）表达式 MD5(RANDOM()::text || clock_timestamp()::text)::varchar(64) 的作用是生成一个基于随机数和当前时间戳的 MD5 散列值，并将其转换为最多 64 个字符的可变长度字符类型(VARCHAR(64))。

（4）表达式 ROUND((RANDOM() * 9900+100)::NUMERIC, 2) 的作用是生成一个随机的、[100, 10 000) 内的随机浮点数，并将其四舍五入到小数点后两位。

（5）generate_series 是 KingbaseES 数据库中的一个非常有用的集合返回函数，它用于生成一系列连续的数值。generate_series(1,10000000) 生成了一个从 1 到 10 000 000 的数值序列。每一个生成的数值都会作为一行出现在结果集中，这样就可以使用 INSERT 语句，为表生成 1000 万行的测试数据。

6.3.3 使用 COPY 命令导入/导出数据库表

1. 使用 COPY 命令将数据库表导出为 CSV 文件

执行下面的 SQL 语句，将数据库 test 中的表 orders 导出为一个 CSV 格式的文件：

```
system@test=# COPY tpch.orders TO '/home/kingbase/orders.csv' WITH CSV HEADER;
COPY 1500000
system@test=# \q
[kingbase@dbsvr ~]$ ls -l /home/kingbase/orders.csv
-rw-r--r--. 1 kingbase dba 194354495 Jan 27 16:05 /home/kingbase/orders.csv
[kingbase@dbsvr ~]$ head -3 /home/kingbase/orders.csv
o_orderkey,o_custkey,o_orderstatus,o_totalprice,o_orderdate,o_orderpriority,o_clerk,o_shippriority,o_comment
1,36901,O,173665.47,1996-01-02 00:00:00,5-LOW          ,Clerk#000000951,0,nstructions sleep furiously among
```

```
2,78002,O,46929.18,1996-12-01 00:00:00,1-URGENT         ,Clerk#000000880,0,"
foxes. pending accounts at the pending, silent asymptot"
[kingbase@dbsvr ~]$
```

导出的 CSV 文件,第 1 列是表的列名,从第 2 列开始是表的数据。

2. 使用 COPY 命令将 CSV 文件导入数据库表

要将 CSV 文件的数据导入数据库,首先要数据库 test 的 tpch 模式中,创建一个与 CSV 文件的列名相对应的表 myorders:

```
[kingbase@dbsvr ~]$ ksql -d test -U system
system@test=# CREATE TABLE tpch.myorders (
test(#            o_orderkey          integer NOT NULL,
test(#            o_custkey           integer NOT NULL,
test(#            o_orderstatus       char(1) NOT NULL,
test(#            o_totalprice        numeric NOT NULL,
test(#            o_orderdate         timestamp without time zone NOT NULL,
test(#            o_orderpriority     char(15) NOT NULL,
test(#            o_clerk             char(15) NOT NULL,
test(#            o_shippriority      integer NOT NULL,
test(#            o_comment           varchar(79) NOT NULL,
test(#            PRIMARY KEY (o_orderkey),
test(#            FOREIGN KEY (o_custkey) REFERENCES tpch.customer(c_custkey)
test(#      );
system@test=#
```

然后执行下面的命令,将 CSV 文件的数据导入数据库表:

```
system@test=# COPY tpch.myorders   FROM '/home/kingbase/orders.csv' WITH CSV HEADER;
COPY 1500000
system@test=#
```

执行下面的 SQL 语句,验证已经成功导入数据:

```
system@test=# SELECT o_orderkey,o_custkey,o_orderstatus,o_totalprice FROM tpch.myorders LIMIT 3;
 o_orderkey | o_custkey | o_orderstatus | o_totalprice
------------+-----------+---------------+--------------
          1 |     36901 | O             |    173665.47
          2 |     78002 | O             |     46929.18
          3 |    123314 | F             |    193846.25
(3 rows)
system@test=#
```

6.3.4 使用 sys_bulkload 装载大表

sys_bulkload 是一个 KingbaseES 数据库用于高效加载数据的工具。它主要用于快速导入大量数据,特别是在数据仓库和其他需要快速数据迁移或导入的场景中。

sys_bulkload 支持以下几种文件格式进行数据导入:CSV、文本、二进制和自定义格式(通过编写自定义的输入函数来实现)。

在使用 sys_bulkload 导入数据时，通常需要提供一个控制文件，这个文件描述了数据文件的格式，包括字段分隔符、文本限定符、转义字符、空值表示法等。确保数据文件符合指定的格式是成功导入数据的关键。

1. 安装 sys_bulkload 插件

执行下面的 SQL 语句，安装扩展插件 sys_bulkload：

```
system@test=# CREATE EXTENSION sys_bulkload;
CREATE EXTENSION
system@test=#
```

2. 使用 sys_bulkload 将 CSV 文件导入到表 myorders

（1）要将 CSV 文件的数据导入数据库，首先要创建一个与 CSV 文件的列名相对应的表 tpch.myorders。由于表 tpch.myorders 目前已经存在，因此只需要执行下面的 SQL 语句，清空表 tpch.myorders 的数据：

```
system@test=# truncate table tpch.myorders;
```

（2）使用 vi 编辑器，为 sys_bulkload 创建控制文件 /home/kingbase/myorders.ctl，将以下内容编辑到文件中：

```
TYPE = CSV
INPUT = /home/kingbase/orders.csv
OUTPUT = tpch.myorders
DELIMITER = ,
ENCODING = UTF8
SKIP = 1
```

其中跳过（skip）标识跳过 CSV 文件的第 1 行（标题行）。

（3）执行下面的 sys_bulkload 命令，将 CSV 文件导入到表 myorders 中：

```
[kingbase@dbsvr ~]$ sys_bulkload  -U system -d test myorders.ctl
NOTICE: BULK LOAD START
NOTICE: BULK LOAD END
    1 Rows skipped.
    1500000 Rows successfully loaded.
    0 Rows not loaded due to parse errors.
    0 Rows not loaded due to duplicate errors.
    0 Rows replaced with new rows.
    log path: myorders.log
    parse error path: /home/kingbase/orders.bad
    duplicate error path: /home/kingbase/orders.dupbad
    ctrl file path: /home/kingbase/myorders.ctl
    data file path: /home/kingbase/orders.csv
    Run began on 2024-01-27 16:59:18.472701+08
    Run ended on 2024-01-27 16:59:22.515040+08
[kingbase@dbsvr ~]$
```

（4）执行下面的 SQL 语句，验证已经成功导入了数据：

```
system@test=# SELECT o_orderkey,o_custkey,o_orderstatus,o_totalprice FROM tpch.
myorders LIMIT 3;
 o_orderkey | o_custkey | o_orderstatus | o_totalprice
------------+-----------+---------------+--------------
          1 |     36901 | O             |     173665.47
          2 |     78002 | O             |      46929.18
          3 |    123314 | F             |     193846.25
(3 rows)
```

6.3.5 大表模糊查询

1. 大表数据抽样查询

在 SQL 语句中，TABLESAMPLE 子句用于从表中随机抽取一定比例的行。这个子句在处理非常大的表时特别有用，因为它允许用户在不扫描整个表的情况下，快速获得样本数据，可以大大加快查询速度，付出的代价是查询结果只是近似的，而不是精确的。

抽样方法有 2 种。

（1）伯努力（Bernoulli）抽样：抽取更均匀更具有代表性的样本。

（2）系统（system）抽样：考虑更多的是查询速度而不是每个样本的随机独立性。

下面的 SQL 语句，采用伯努力抽样，抽样的比例为 1%，查询显示表 orders 的 o_totalprice 列的平均值：

```
system@test=# SELECT AVG(o_totalprice) FROM tpch.orders TABLESAMPLE BERNOULLI(1);
        avg
---------------------
 150475.821718873910
(1 row)
```

下面的 SQL 语句，采用系统抽样，抽样的比例为 1%，查询显示表 orders 的 o_totalprice 列的平均值：

```
system@test=# SELECT AVG(o_totalprice) FROM tpch.orders TABLESAMPLE SYSTEM(1);
        avg
---------------------
 151606.680418814843
(1 row)
```

2. 大表行估算

执行下面的 SQL 语句，可以直接计算出一个大表精确的总行数：

```
system@test=# SELECT count(*) FROM tpch.orders;
  count
---------
 1500000
(1 row)
```

执行下面的 SQL 语句，可以估算（不是精确计算）出一个大表大概的总行数：

```
system@test=# SELECT (CASE
```

```
test(#                    WHEN reltuples > 0 THEN sys_relation_size(oid) * reltuples/
(8192 * relpages)
test(#                    ELSE 0
test(#               END)::bigint AS estimated_row_count
test-#          FROM sys_class
test-#          WHERE oid = 'tpch.orders'::regclass;
 estimated_row_count
---------------------
             1500000
(1 row)
```

这条语句使用系统表 sys_class 中的统计信息来进行估算。表 sys_class 中的 reltuples 字段提供了表中行的近似数目,而 relpages 字段提供了表占用的页数。这个查询通过这些统计数据和页的平均大小(通常是 8KB,即 8192B)来估算行数。对于大型表来说,由于不需要扫描整个表,这个方法通常更快,代价是精度较低,仅提供一个估算值,而不是精确的行数。并且这个估算值的准确性依赖于 KingbaseES 数据库的统计信息的更新程度。如果表最近经历了大量的插入、删除或更新操作,而统计信息尚未刷新,这个估算值可能不太准确。

6.3.6 数据存储空间查询

1. 查看指定名字的数据库占用的磁盘存储空间的大小

执行下面的 SQL 语句,查看数据库 test 占用的存储空间的大小:

```
system@test=# SELECT sys_size_pretty( sys_database_size('test') );
 sys_size_pretty
-----------------
 3297 MB
(1 row)
```

2. 查看数据库集簇中每一个数据库占用的磁盘存储空间的大小

执行下面的 SQL 语句,查看数据库集簇中每一个数据库占用的存储空间的大小:

```
system@test=# SELECT datname, sys_size_pretty(sys_database_size(oid)) FROM sys_
database;
  datname   | sys_size_pretty
------------+-----------------
 test       | 3297 MB
 kingbase   | 14 MB
 template1  | 14 MB
 security   | 14 MB
 template0  | 14 MB
(5 rows)
```

3. 查看数据库集簇占用的磁盘存储空间的大小

执行下面的 SQL 语句,查看 KingbaseES 数据库集簇占用的磁盘存储空间的大小:

```
system@test=# SELECT  'KingbaseES Database Cluster(Total)' AS "",
test-#                sys_size_pretty(sum(sys_database_size(oid)))  AS "SIZE"
test-#          FROM sys_database;
```

```
                                    | size
--------------------------------+---------
 KingbaseES Database Cluster(Total)  | 3365 MB
(1 row)
```

4. 查看指定名字的表空间占用磁盘存储的大小

执行下面的 SQL 语句，查看表空间 user_ts 占用磁盘存储的大小：

```
system@test=# SELECT SYS_SIZE_PRETTY( SYS_TABLESPACE_SIZE('user_ts') );
 sys_size_pretty
-----------------
 1352 MB
(1 row)
```

5. 查看一个表所占的存储空间大小

执行下面的 SQL 语句，查看表 orders 所占的存储空间大小：

```
system@test=# SELECT sys_size_pretty( sys_total_relation_size('tpch.orders') );
 sys_size_pretty
-----------------
 294 MB
(1 row)
```

6. 查看数据库中最大的前 n 个表

执行下面的 SQL 语句，查看当前连接的数据库中最大的 3 个表：

```
system@test=# SELECT relname AS "Table",sys_size_pretty(sys_total_relation_size
(relid)) AS "Size"
test-#          FROM sys_catalog.sys_statio_user_tables
test-#          ORDER BY sys_total_relation_size(relid) DESC
test-#          LIMIT 3;
    Table    |  Size
-------------+---------
 lineitem    | 1364 MB
 test_orders | 1103 MB
 orders      | 294 MB
(3 rows)
```

第 7 章

事务与并发控制

数据库是一个共享资源,如果允许多个用户并发访问同一个数据库,那么就有可能带来数据不一致的问题。大型的数据库管理系统必须提供事务处理和并发控制来协调多个用户的并发操作,在系统出现故障时进行故障恢复,并确保事务可以正确运行。

本章主要介绍 KingbaseES 数据库的事务处理和并发控制技术。

7.1 事务的基本概念

事务(transaction)是用户定义的一个有限的数据库操作序列,这些操作要么全做,要么全不做,是一个不可分割的工作单位。事务是一系列数据操作,这些操作将数据库从一个一致性状态转换到另一个一致性状态。

数据库的事务具有以下四个属性。

(1)**原子性**(atomicity):事务作为一个整体被执行,事务所包含的对数据库的操作,要么全部被执行(commit),要么都不执行(roll back)。

(2)**一致性**(consistency):事务应确保数据库的状态从一个一致状态转变为另一个一致状态。一致状态的含义是对数据库中数据的变更,都满足完整性约束以及组织机构的业务规则。

(3)**隔离性**(isolation):多个事务并发地执行,一个事务的执行不会影响其他事务的执行,好像当前数据库管理系统就只在执行一个事务一样。

(4)**持久性**(durability):已提交事务对数据库的变更将永久地保存到数据库中,即使发生数据库故障也能确保这一点。

事务的这四个属性简称为**事务的 ACID**(取自这 4 个属性的英文首字母)。保证事务 ACID 特性是事务管理的重要任务。事务 ACID 特性可能遭到破坏的因素有:

(1)多个事务并行运行时,不同事务的操作交叉执行;

(2)事务在运行过程中被强行停止。

在第一种情况下,数据库管理系统必须保证多个事务的交叉运行不影响这些事务的原子性;在第二种情况下,数据库管理系统必须保证被强行终止的事务对数据库和其他事务没有任何影响。

并发控制技术和故障恢复技术保证事务的 ACID 特性。事务是系统恢复和并发控制的基本单位。数据库管理系统有责任和义务保证事务的 ACID 特性,这也是数据库管理系统的关键核心技术。

KingbaseES 数据库提供日志系统保证了事务的原子性(A)和持久性(D);采用锁机制和 MVCC 相结合的并发控制机制来保证事务的一致性(C)和隔离性(I)。

 ## 7.2 事务处理模型

7.2.1 显式事务与隐式事务

在 KingaseES 数据库中,所有 SQL 语句都在事务中执行。KingaseES 数据库支持隐式事务和显式事务。

显式事务需要明确指定事务开始语句和结束语句,在它们之间的 SQL 语句形成一个事务块。一个事务块以 BEGIN 开始,以 COMMIT(事务提交)或 ROLLBACK(事务回滚)结束。COMMIT 是指成功完成事务的所有操作,将事务中所有对数据库的更新写回到磁盘上的物理数据库中去,事务正常结束。提交事务的更新使数据库进入一个新的一致状态。ROLLBACK 是指在事务运行的过程中发生了某种故障,事务不能继续执行,系统将事务中对数据库的所有已完成的操作全部撤销,回滚到事务开始时的状态。

对于不在事务块中的 SQL 语句,KingaseES 数据库会自动在内部开启一个事务,执行该 SQL 语句,执行正确则自动提交,执行失败则自动回滚,称为**隐式事务**,也称单语句事务。

下面是隐式事务的例子:

```
[kingbase@dbsvr ~]$ ksql -d test -U system
system@test=# CREATE TABLE t1(col1 int  primary key, col2 int);
CREATE TABLE
system@test=# INSERT INTO t1 VALUES(1,1);
INSERT 0 1
system@test=# INSERT INTO t1 VALUES(2,1);
INSERT 0 1
system@test=# SELECT * FROM t1;
 col1 | col2
------+------
    1 |    1
    2 |    1
(2 rows)
```

下面是显式事务的例子:

```
system@test=# BEGIN;
BEGIN
system@test=# INSERT INTO t1 VALUES(3,1);
INSERT 0 1
system@test=# INSERT INTO t1 VALUES(4,1);
INSERT 0 1
system@test=# SELECT * FROM t1;
 col1 | col2
------+------
```

```
    1 |    1
    2 |    1
    3 |    1
    4 |    1
(4 rows)
```

如果此时回滚事务,则该事务中所有的操作都会撤销:

```
system@test=# ROLLBACK;
ROLLBACK
system@test=# SELECT * FROM t1;
 col1 | col2
------+------
    1 |    1
    2 |    1
(2 rows)
```

KingbaseES 数据库的事务块中不支持语句级的回滚,因此,在事务块中任何一个 SQL 语句出错,后面的 SQL 语句不能继续执行,整个事务块必须回滚,通过回滚结束失败的事务:

```
system@test=# BEGIN;
BEGIN
system@test=# INSERT INTO t1 VALUES(5,1);
INSERT 0 1
system@test=# SELECT * FROM t1;
 col1 | col2
------+------
    1 |    1
    2 |    1
    5 |    1
(3 rows)
system@test=# INSERT INTO t1 VALUES(1,1);
ERROR:  duplicate key value violates unique constraint "t1_pkey"
DETAIL:  Key (col1)=(1) already exists.
system@test=# INSERT INTO t1 VALUES(6,1);
ERROR:  current transaction is aborted, commands ignored until end of transaction block
system@test=# ROLLBACK;
ROLLBACK
system@test=# SELECT * FROM t1;
 col1 | col2
------+------
    1 |    1
    2 |    1
(2 rows)
```

除了通过 ROLLBACK 命令显式结束一个事务,还可以通过执行 COMMIT 或者 END 命令来显式结束事务:

```
system@test=# BEGIN;
```

```
BEGIN
system@test=# INSERT INTO t1 VALUES(7,1);
INSERT 0 1
system@test=# COMMIT;
COMMIT
system@test=# SELECT * FROM t1;
 col1 | col2
------+------
    1 |    1
    2 |    1
    7 |    1
(3 rows)
system@test=# BEGIN;
BEGIN
system@test=# INSERT INTO t1 VALUES(8,1);
INSERT 0 1
system@test=# END;
COMMIT
system@test=# SELECT * FROM t1;
 col1 | col2
------+------
    1 |    1
    2 |    1
    7 |    1
    8 |    1
(4 rows)
```

7.2.2　DDL 语句与事务

一些数据库管理系统(如 Oracle 数据库)，一旦 DDL 语句执行完成后会自动执行一条 COMMIT 语句，KingbaseES 数据库系统与此不同，在事务块内的 DDL 语句不会自动提交。事务回滚时，其中的 DDL 语句是一起回滚的：

```
system@test=# BEGIN;
system@test=# CREATE TABLE t2(col1 int);
CREATE TABLE
system@test=# INSERT INTO t2 VALUES(10);
INSERT 0 1
system@test=# SELECT * FROM t2;
 col1
------
   10
(1 row)
system@test=# ROLLBACK;
ROLLBACK
system@test=# SELECT * FROM t2;
ERROR:  relation "t2" does not exist
```

```
LINE 1: SELECT * FROM t2;
                ^
system@test=# \q
[kingbase@dbsvr ~]$
```

7.2.3 事务隔离级别

计算机系统对并发事务的操作调度是随机的,不同的调度会产生不同的结果。几个事务被并发调度执行,当且仅当这几个事务的并行执行的结果,与这几个事务的某个串行执行结果相同时,称这些事务是**可串行化调度**的。可串行化是并发事务正确性的准则。多个事务并发操作时,可串行化调度保证了数据的一致性,一个用户事务的执行不受其他事务的干扰。

实现可串行化调度可能会影响系统的并发度,对于有些应用来说,可以容忍一些数据库的不一致来换取系统的高性能和高并发度。因此,SQL 语句标准中定义了事务不同的隔离级别,以满足不同的应用场景。

事务并发有可能导致的数据不一致包括如下几种情况。

(1)**丢失修改(lost update)**:指两个事务 T1 和 T2 读入同一数据,各自进行修改,T2 提交的结果破坏了 T1 提交的结果,导致 T1 的修改被丢失,又称脏写。

(2)**脏读(dirty read)**:指事务 T1 修改某一数据还未提交,事务 T2 读取同一数据。

(3)**不可重复读(nonrepeatable read)**:指事务 T1 读取数据后,事务 T2 执行更新或删除操作并提交,当事务 T1 再次读该数据时,得到与前一次不同的值或该数据被删除。

(4)**幻读(phantom read)**:指事务 T1 按一定条件从数据库中读取某些数据记录后,事务 T2 插入了一些记录,当 T1 再次按相同条件读取数据时,发现多了一些记录。

在 SQL 语句标准中定义了事务的四类隔离级别,由低到高分别是:READ UNCOMMITTED(读未提交)、READ COMMITTED(读已提交)、REPEATABLE READ(可重复读)、SERIALIZABLE(可串行化)。所有的隔离级别都不允许出现丢失修改的数据异常,但对其他数据一致性的保障程度各异。

注意:可串行化隔离级别的定义是事务执行顺序是可串行化的,因此,在该级别下可以规避丢失修改、脏读、不可重复读和幻读这些数据异常,但是,只规避这些数据异常,并不能保证做到了可串行化。

表 7-1 给出了事务四个隔离级别与数据异常的关系。

表 7-1 事务隔离级别与数据异常的关系

隔离级别	丢失修改	脏读	不可重复读	幻读
READ UNCOMMITTED	不允许	允许	允许	允许
READ COMMITTED	不允许	不允许	允许	允许
REPEATABLE READ	不允许	不允许	不允许	允许
SERIALIZABLE	不允许	不允许	不允许	不允许

在 KingbaseES 数据库可以设置 4 种隔离级别,如表 7-2 所示。

表 7-2 Kingbase 数据库的事务隔离级别

隔离级别	丢失修改	脏读	不可重复读	幻读	串行化异常
READ UNCOMMITTED	不允许	不可能	可能	可能	可能
READ COMMITTED	不允许	不可能	可能	可能	可能
REPEATABLE READ	不允许	不可能	不可能	不可能	可能
SERIALIZABLE	不允许	不可能	不可能	不可能	不可能

KingbaseES 数据库默认的事务隔离级别是 READ COMMITTED。KingbaseES 数据库使用系统配置参数 default_transaction_isolation 来定义事务的隔离级别,数据库用户可以在系统级、会话级和事务级上设置该参数的值。

（1）系统级：在系统配置文件 kingbase.conf 中的设置该参数的值。

（2）会话级：在会话中执行 SET SESSION 语句,可以设置会话的事务隔离级别。

（3）事务级：在会话中使用 BEGIN TRANSACTION ISOLATION LEVEL 语句,可以为会话中的当前事务设置事务隔离级别。

请看下面的示例：

```
[kingbase@dbsvr ~]$ ksql -d test -U system
system@test=# -- 查看系统配置文件中参数 default_transaction_isolation 的值
system@test=# SHOW default_transaction_isolation;
 default_transaction_isolation
-------------------------------
 read committed
(1 row)
system@test=# -- 显式当前的事务隔离级别(系统级)
system@test=# SHOW transaction_isolation;
 transaction_isolation
-----------------------
 read committed
(1 row)
system@test=# -- 在会话级设置事务隔离级别
system@test=# SET SESSION CHARACTERISTICS AS TRANSACTION ISOLATION LEVEL SERIALIZABLE;
SET
system@test=# -- 显式当前的事务隔离级别(会话级)
system@test=# SHOW transaction_isolation;
 transaction_isolation
-----------------------
 serializable
(1 row)
system@test=# -- 开启一个新事务,并设置该事务的隔离级别
system@test=# BEGIN TRANSACTION ISOLATION LEVEL REPEATABLE READ;
BEGIN
system@test=# -- 显式当前的事务隔离级别(事务级)
system@test=# SHOW transaction_isolation;
 transaction_isolation
-----------------------
```

```
    repeatable read
(1 row)
system@test=# -- 结束会话中的事务,将恢复会话级的事务隔离级别
system@test=# COMMIT;
COMMIT
system@test=# SHOW transaction_isolation;
 transaction_isolation
-----------------------
 serializable
(1 row)
system@test=# -- 连接到另外一个数据库,将开启一个新的会话,这将恢复为系统默认的事务隔
离级别
system@test=# \c kingbase
You are now connected to database "kingbase" as userName "system".
system@kingbase=# SHOW transaction_isolation;
 transaction_isolation
-----------------------
 read committed
(1 row)
system@kingbase=# \q
[kingbase@dbsvr ~]$
```

7.2.4 事务并发控制机制

多用户共享是数据库的一个显著特点,而多用户同时访问同一数据对象可能会破坏事务的隔离性,例如,前面介绍的并发操作带来的丢失修改、脏读、不可重复读、幻读等数据异常。并发控制机制就是要用正确的方式调度并发操作,使一个用户事务的执行不受其他事务的干扰,从而避免造成数据的不一致性。

封锁是实现并发控制的一个非常重要的技术,所谓**封锁**就是事务 T 在对某个数据对象记录等操作之前,先向系统发出请求,对其加锁。加锁后事务 T 就对该数据对象有了一定的控制,在事务 T 释放它的锁之前,其他事务不能更新或读取此数据对象。

理论已经证明,如果所有的事务均遵守**两段锁协议**(**two-phase locking**,**2PL**),则这些事务的所有交叉调度都是可串行化的。2PL 协议规定所有的事务应遵守下面的规则。

(1) 在对任何数据进行读、写操作之前,事务首先要获得对该数据的封锁。

(2) 在释放了一个封锁之后,事务不再获得其他任何封锁。

2PL 在对任何数据进行读、写操作之前,需要对该数据加锁。在封锁相容矩阵中,共享锁(shared lock,S 锁)和排他锁(exclusive lock,X 锁)是不相容的,因此,当事务 T1 正对数据 A 进行读操作(加 S 锁)时,事务 T2 想要对数据进行写操作(加 X 锁),那么事务 T2 必须等待事务 T1 释放数据 A 上的 S 锁,才能继续执行。

为了提高系统的并发度,提出了 MVCC 策略,通过在数据库中维护数据对象的多个版本信息来实现事务的高效并发。

在多版本机制中,每个写操作都创建该数据项的一个新版本,在数据库中,一个逻辑数据项在物理上会存储多个版本,读操作可以读取事务开始时该数据项的最新版本。因此,一个事务在修改某个数据项时,其他并发的事务可以读取该数据项的旧版本,从而避免数据库

中数据对象读和写操作的冲突,有效地提高了系统的性能。

MVCC 虽然提高了并发度,但也带来了维护多个版本的存储开销。

也就是说,KingbaseES 数据库使用 MVCC 和多种锁相结合来维护数据的一致性。

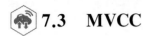 7.3　MVCC

多版本并发控制不仅是并发控制协议,它还影响着一个数据库管理系统的多方面。

(1) 多个版本怎么存放。

(2) 无效的版本怎么回收。

(3) 怎么做到事务的可串行化。

7.3.1　事务号与事务状态

KingbaseES 数据库给每个可能修改数据库中数据的事务分配一个事务号(事务 ID)。事务号是一个 32 位的无符号数。事务号可以比较大小,事务号比较大的是开始比较晚的事务。

事务号是一个 32 位的无符号数,只能标识约 40 亿(2^{32})个事务。为了减少事务号的消耗,KingbaseES 数据库对于只读事务先不分配事务号,而是使用虚拟事务号来唯一标识该事务。虚拟事务号由该会话的标识号和内部维护的一个局部的事务号组成,可以唯一标识一个事务。

使用下面的内置函数可以获得当前事务的事务号。

(1) txid_current():该函数返回当前事务的事务号,如果当前事务还没有分配事务号,则会给它分配事务号。

(2) txid_current_if_assigned():该函数返回当前事务的事务号,如果当前事务还没有分配事务号,则返回 null。

为了避免浪费事务号,建议使用 txid_current_if_assigned()函数来获取当前事务的事务号:

```
[kingbase@dbsvr ~]$ ksql -d test -U system
system@test=# BEGIN;
system@test=# SELECT * FROM t1;
--省略了输出
system@test=# SELECT txid_current_if_assigned();
 txid_current_if_assigned
--------------------------

(1 row)                                    --输出为空白,标识此时还未分配事务号
system@test=# INSERT INTO t1 VALUES(9,1);
INSERT 0 1
system@test=# SELECT txid_current_if_assigned();
 txid_current_if_assigned
--------------------------
                     1190
(1 row)                                    --执行了 INSERT 语句,此时分配了事务号 1190
```

```
system@test=# COMMIT;
COMMIT
```

事务有三种状态：提交、回滚和正在运行。KingbaseES 数据库的数据目录下有一个 sys_xact 子目录，其中的文件中记录系统中事务的状态信息，每个事务使用 2 位保存。可以使用 txid_status() 函数来获取指定事务号的状态：

```
system@test=# SELECT txid_status('1190');
 txid_status
-------------
 committed
(1 row)
```

7.3.2 元组的结构

KingbaseES 数据库中元组的多个版本都是存储在数据库中。每条元组都会在其头部记录版本信息，通过这些信息可以标识出元组的每个版本。

KingbaseES 数据库给每个可能修改数据库中数据的事务分配一个事务号，一个事务可以包含多个 SQL 命令，每个 SQL 命令都有一个序号，称为命令号。在当前事务中，命令的序号，从 0 开始。KingbaseES 数据库通过在元组头部标注事务号和命令号来标识元组的每个版本。

KingbaseES 数据库的每个表都有多个由系统隐式定义的系统列，通过这些列可以查询到每个元组头部的这些信息。系统有以下的隐藏列，可以使用 SELECT 语句查询。

(1) **tableoid**：包含该元组的表 oid。
(2) **xmin**：插入该元组的事务 ID。
(3) **xmax**：删除该元组的事务 ID。
(4) **cmin**：插入该元组的命令号。
(5) **cmax**：删除该元组的命令号。
(6) **ctid**：该元组所处的表内物理位置，其形式为 (x,y)，其中 x 是该元组所在的页面编号，y 是该元组所在页面内的编号。

执行下面的 SQL 语句，创建测试表 t_mvcc，插入一条数据，然后查看这些隐藏列：

```
system@test=# CREATE TABLE t_mvcc(id serial primary key,usercol varchar(20));
system@test=# INSERT INTO t_mvcc(usercol) VALUES('1');
system@test=# SELECT tableoid,ctid,xmin,xmax,cmin,cmax, * FROM t_mvcc;
 tableoid | ctid  | xmin | xmax | cmin | cmax | id | usercol
----------+-------+------+------+------+------+----+---------
    16638 | (0,1) | 1192 |    0 |    0 |    0 |  1 | 1
(1 row)
```

从输出可以看出，表 t_mvcc 中有一个元组，ctid 的值是 (0,1)，存放在第一个页面(页面号为 0)的第 1 个槽中，是由事务号为 1192 的事务插入的。

KingbaseES 数据库自带一个扩展插件 pageinspect，可以用来直接检查数据库物理文件的数据页面的具体内容。执行下面的 SQL 语句，安装这个扩展模块：

```
system@test=# CREATE EXTENSION pageinspect;
```

安装完扩展插件 pageinspect 后,可以执行下面的 SQL 语句,创建视图 v_pageinspect,用于查看表 t_mvcc 的存储情况:

```
system@test=# CREATE OR REPLACE VIEW v_pageinspect AS
test-#           SELECT '(0,' || lp || ')' AS ctid,
test-#                  CASE lp_flags
test-#                      WHEN 0 THEN ' Unused '
test-#                      WHEN 1 THEN ' Normal '
test-#                      WHEN 2 THEN ' Redirect to' || lp_off
test-#                      WHEN 3 THEN ' Dead '
test-#                  END,
test-#                  t_xmin::text::int8 AS xmin,
test-#                  t_xmax::text::int8 AS xmax,
test-#                  t_ctid
test-#           FROM heap_page_items (get_raw_page ('t_mvcc', 0))
test-#           ORDER BY lp;
```

使用视图 v_pageinspect,查看表 t_mvcc 的存储情况:

```
system@test=#   SELECT * FROM v_pageinspect;
  ctid  |  case  | xmin | xmax | t_ctid
--------+--------+------+------+--------
 (0,1)  | Normal | 1192 |    0 | (0,1)
(1 row)
```

从输出可以看到,表 t_mvcc 当前只有一行数据。

(1) ctid 的值为(0,1),表示这行数据位于表的第 0 个物理块上第 1 行位置。

(2) case 是该行的当前状态(Normal)。

(3) xmin 表示行的最小事务号,即创建该行的事务号,它标识了行的起始事务,也被称为行的插入事务号。

(4) xmax 表示行的最大事务号,即删除或更新该行的事务号。如果 xmax=0,表示没有任何事务对该行进行过更新或删除操作,它仍然处于原始的插入状态。

(5) t_ctid 保存的是指向元组自身或者指向更新版本的元组标识符 ctid,本例指向了元组自身。

7.3.3 元组的增、删、改

1. 插入元组

插入新元组时,将新元组的 xmin 设置为执行插入该元组的事务标识 txid,xmax 设置为 0。

在第 1 个终端执行如下的 SQL 语句:

```
system@test=# --开始一个新的事务
system@test=# BEGIN;
system@test=# SELECT txid_current();
 txid_current
--------------
         1195
```

```
(1 row)
system@test=# INSERT INTO t_mvcc(usercol) VALUES('2');
system@test=# SELECT * FROM v_pageinspect;
 ctid  | case   | xmin | xmax | t_ctid
-------+--------+------+------+--------
 (0,1) | Normal | 1192 |    0 | (0,1)
 (0,2) | Normal | 1195 |    0 | (0,2)
(2 rows)
```

从输出可以看到,插入新行后,新行的 xmin 列填充插入新行的事务号(本例 xmin＝1195),表示该行是由这个事务插入表中的;xmax 列被设置为 0;ctid=(0,2)表示该行目前在表的第 0 个物理块的第 2 行位置。因为这个新插入的行到目前为止只有一个版本,所以 t_ctid 的值指向元组自身,与 ctid 的值一样,都是(0,2)。

继续在第 1 个终端执行下面的 SQL 语句,然后结束事务,紧接着开启一个新的事务,并查看这个新事务的事务号:

```
system@test=# SELECT xmin,xmax,ctid,* FROM t_mvcc;
 xmin | xmax | ctid  | id | usercol
------+------+-------+----+---------
 1192 |    0 | (0,1) |  1 | 1
 1195 |    0 | (0,2) |  2 | 2
(2 rows)
system@test=# END;
COMMIT
system@test=# BEGIN;
BEGIN
system@test=# SELECT txid_current();
 txid_current
--------------
         1196
(1 row)
```

请保持第 1 个终端不要退出。

2. 删除元组

删除表中的行只需要把被删除行的 xmax 值修改为执行删除操作的事务的事务号。
打开第 2 个终端窗口,并执行下面的 ksql 命令和 SQL 语句:

```
[kingbase@dbsvr ~]$ ksql -U system -d test
system@test=# BEGIN;
system@test=# SELECT txid_current();
 txid_current
--------------
         1197
(1 row)
system@test=# DELETE FROM t_mvcc WHERE usercol='2';
system@test=# SELECT xmin,xmax,ctid,* FROM t_mvcc;
 xmin | xmax | ctid  | id | usercol
------+------+-------+----+---------
 1192 |    0 | (0,1) |  1 | 1
```

```
(1 row)
system@test=# SELECT * FROM v_pageinspect;
  ctid  |  case  | xmin | xmax | t_ctid
--------+--------+------+------+--------
  (0,1) | Normal | 1192 |    0 | (0,1)
  (0,2) | Normal | 1195 | 1197 | (0,2)
(2 rows)
```

从输出可以看到,删除一行,将用执行删除的事务的事务号填充该行的 xmax 值(本例删除的是位于(0,2)的行,xmax 值被设置为1197)。删除后,在删除行所在的事务立刻看不到该行。此刻还未提交事务1197。

继续在第1个终端窗口下执行下面的命令和 SQL 语句:

```
system@test=# SELECT xmin,xmax,ctid,* FROM t_mvcc;
 xmin | xmax | ctid  | id | usercol
------+------+-------+----+---------
 1192 |    0 | (0,1) |  1 | 1
 1195 | 1197 | (0,2) |  2 | 2
(2 rows)
```

由于第2个终端上的删除事务还未被提交,因此在第1个终端上还能看到旧的数据,但是旧的数据已经被打上了删除标记(xmax 值=1197)。

返回到第2个终端,执行下面的 SQL 语句:

```
system@test=# COMMIT;
COMMIT
```

再转回第1个终端,执行下面的 SQL 语句:

```
system@test=# SELECT xmin,xmax,ctid,* FROM t_mvcc;
 xmin | xmax | ctid  | id | usercol
------+------+-------+----+---------
 1192 |    0 | (0,1) |  1 | 1
(1 row)
```

从输出可以看到,在第1个终端已经看不到这行被删除的数据了。

继续在第1个终端,执行下面的 SQL 语句:

```
system@test=# SELECT * FROM v_pageinspect;
  ctid  |  case  | xmin | xmax | t_ctid
--------+--------+------+------+--------
  (0,1) | Normal | 1192 |    0 | (0,1)
  (0,2) | Normal | 1195 | 1197 | (0,2)     --被删除的行,由事务1195创建,由事务1197删除
(2 rows)
```

从输出可以看到,在表 t_mvcc 的存储中,还保留了被删除行的数据。

3. 修改元组

在 KingbaseES 数据库中修改表中的元组,相当于删除旧元组,插入新元组。因此,会为表添加新元组,该新元组的的 xmin 设置为执行更新操作的事务的事务号,xmax 值设置为0。对于旧元组,将其 xmax 值设置为执行更新操作的事务的 txid 值,t_ctid 字段的值指

向新行的位置 ctid。

继续在第 2 个终端执行下面的命令和 SQL 语句：

```
system@test=# BEGIN;
system@test=# SELECT txid_current();
 txid_current
--------------
         1198
(1 row)
system@test=# UPDATE t_mvcc SET usercol='3' WHERE usercol='1';
UPDATE 1
system@test=# SELECT xmin,xmax,ctid, * FROM t_mvcc;
 xmin | xmax | ctid  | id | usercol
----+----+-----+--+--------
 1198 |    0 | (0,3) |  1 | 3
(1 row)
system@test=# SELECT * FROM v_pageinspect;
  ctid  |  case  | xmin | xmax | t_ctid
------+-------+----+----+--------
 (0,1) | Normal | 1192 | 1198 | (0,3)     更新前的行
 (0,2) | Normal | 1195 | 1197 | (0,2)     被删除的行
 (0,3) | Normal | 1198 |    0 | (0,3)     更新后的行
(3 rows)
system@test=#
```

从输出可以看到，事务 1198 修改了数据行，在这个事务中，使用 SELECT 语句查询该表时，返回更新后的数据（usercol='3'）。修改数据行只是将旧元组的 xmax 值设置为修改该元组的事务的事务号 1198，旧元组的 t_ctid 被修改为新元组的位置标识(0,3)，并创建一个新元组(0,3)，将新元组的 xmin 设置为修改这个数据的事务号（本例是 1198），xmax 值设置为 0，t_ctid 指向自己(0,3)。

返回到第 1 个终端，执行下面的 SQL 语句：

```
system@test=# SELECT xmin,xmax,ctid, * FROM t_mvcc;
 xmin | xmax | ctid  | id | usercol
----+----+-----+--+--------
 1192 | 1198 | (0,1) |  1 | 1
(1 row)
system@test=#
```

从输出可以看到，被修改的原始行，xmin 的值（1192）不变，xmax 的值被修改为 1198，是发起修改操作的那个事务的事务号 txid。由于当前的事务隔离级别是读提交，所以此时在第 1 个终端的事务中，还看不到刚才在第 2 个终端上修改的数据（事务还未提交），数据依旧是 usercol=1。

在第 2 个终端上，结束并提交事务：

```
system@test=# COMMIT;
```

返回到第 1 个终端，执行下面的 SQL 语句：

```
system@test=# SELECT xmin,xmax,ctid, * FROM t_mvcc;
 xmin | xmax | ctid  | id | usercol
------+------+-------+----+---------
 1198 |    0 | (0,3) |  1 | 3
(1 row)
```

从输出可以看到,由于在第 2 个终端上的修改已经提交,因此在第 1 个终端上的事务,可以看到修改后的新版本数据(usercol=3)。

7.3.4 元组的访问

KingbaseES 数据库中存储了元组的多个版本,如何来判断查询应该返回元组的哪个版本?

KingbaseES 数据库中的查询都是基于某个**快照**(**snapshot**),使用快照来判断数据库中的可见元组。快照表示系统某一时刻的数据库状态,依据某个快照查询数据库中的数据,结果集只能返回在该快照时刻数据库中已经提交的数据。因此,快照给出了在某个特定时刻数据库中数据的一个一致性视图。

根据不同的事务隔离级别,查询使用快照的方式也不同。

(1) 对于隔离级别是 READ COMMITTED 的事务,**在每个查询开始时生成一个快照**,其查询结果与查询开始时的数据库状态保持一致,可以看到该查询开始时所有提交事务的更新。

(2) 对于隔离级别是 REPEATABLE READ 和 SERIALIZABLE 的事务,**在事务开始时生成一个快照**,该事务的所有查询都使用该快照。事务中查询结果与事务开始时的数据库状态保持一致,可以看到该事务开始时所有提交事务的更新。

例 7.1 在第 1 个终端上,以 READ COMMITTED 的事务隔离级别,执行转账事务;在第 2 个终端,以 READ COMMITTED 的事务隔离级别,来观察第 1 个终端事务开始前、执行中和提交后,可以看到的用户账号金额。

(1) 首先在第 1 个终端上执行如下的 ksql 命令和 SQL 语句,创建测试表 account 并初始化测试数据:

```
[kingbase@dbsvr ~]$ ksql -d test -U system
system@test=# DROP TABLE IF EXISTS account;
system@test=# CREATE TABLE account(name char(20),balance number(12,2));
system@test=# INSERT INTO account VALUES ('Tom',20000);
system@test=# INSERT INTO account VALUES ('Jack',30000);
system@test=# \q
[kingbase@dbsvr ~]$
```

(2) 然后在两个终端上,按照表 7-3 的操作顺序,分别执行 ksql 命令和 SQL 语句:

表 7-3 操作顺序 1

时刻	第 1 个终端	第 2 个终端
T0	kingbase@dbsvr ~]$ ksql -d test -U system system@test=#	kingbase@dbsvr ~]$ ksql -d test -U system system@test=#

续表

时刻	第 1 个终端	第 2 个终端
T1		system@test=# SELECT * FROM ACCOUNT; 　　name　　｜ balance ----------------+---------- 　Tom　　　　｜ 20000.00 　Jack　　　　｜ 30000.00 (2 rows) --此时看到的是两个账号的初始值 system@test=#

T2	system @ test = # BEGIN TRANSACTION ISOLATION LEVEL Read Committed;
	BEGIN
	system @ test = # UPDATE account SET balance=balance-10000 WHERE name='Tom';
	UPDATE 1
	system@test=#	

T2		system @ test = # BEGIN TRANSACTION ISOLATION LEVEL Read Committed;
		BEGIN
		system@test=# SELECT * FROM ACCOUNT;
		　　name　　｜ balance
		----------------+----------
		　Tom　　　　｜ 20000.00
		　Jack　　　　｜ 30000.00
		(2 rows)
		--此时两个账号的值不变
		system@test=#

T3	system @ test = # UPDATE account SET balance=balance+10000 WHERE name='Jack'; UPDATE 1 1 system@test=#	
T4		system@test=# SELECT * FROM ACCOUNT; 　　name　　｜ balance ----------------+---------- 　Tom　　　　｜ 20000.00 　Jack　　　　｜ 30000.00 (2 rows) --此时两个账号的值不变 system@test=#
T5	system@test=# COMMIT; COMMIT system@test=# \q [kingbase@dbsvr ~]$	

续表

时刻	第 1 个终端	第 2 个终端
| T6 | | ```
system@test=# SELECT * FROM ACCOUNT;
    name         | balance
-----------------+----------
 Tom             | 10000.00
 Jack            | 40000.00
(2 rows)
--此时虽然终端2上的事务尚未结束,但是由于
--终端2上的事务隔离级别是读提交,因此仍然
--可以看到终端1上已经提交的转账事务的结果
system@test=# COMMIT;
COMMIT
system@test=# \q
[kingbase@dbsvr ~]$
``` |

从输出可以看出,在 READ COMMITTED 事务隔离级别下,会为每个查询开始时生成一个快照。

例 7.2 在第 1 个终端上,以 READ COMMITTED 的事务隔离级别,执行转账事务;在第 2 个终端,以 Repeatable Read 的事务隔离级别,来观察第 1 个终端事务开始前、执行中和提交后,可以看到的用户账号金额:

(1) 在第 1 个终端上执行如下的 ksql 命令和 SQL 语句,重新创建测试表 account 并初始化测试数据:

```
[kingbase@dbsvr ~]$ ksql -d test -U system
system@test=# DROP TABLE IF EXISTS account;
system@test=# CREATE TABLE account(name char(20),balance number(12,2));
system@test=# INSERT INTO account VALUES ('Tom',20000);
system@test=# INSERT INTO account VALUES ('Jack',30000);
system@test=# \q
[kingbase@dbsvr ~]$
```

(2) 在两个终端上,按照表 7-4 的操作顺序,执行 ksql 命令和 SQL 语句:

表 7-4 操作顺序 2

时刻	第 1 个终端	第 2 个终端
| T0 | ```
kingbase@dbsvr ~]$ ksql -d test -U system
system@test=#
``` | ```
kingbase@dbsvr ~]$ ksql -d test -U system
system@test=#
``` |
| T1 | | ```
system@test=# SELECT * FROM ACCOUNT;
    name         | balance
-----------------+----------
 Tom             | 20000.00
 Jack            | 30000.00
(2 rows)
--终端1上转账事务开始前,在终端2上可以看
--到的两个账号的初始值
system@test=#
``` |

续表

时刻	第 1 个终端	第 2 个终端

T2	system @ test = # BEGIN TRANSACTION ISOLATION LEVEL Read Committed;
	BEGIN
	system @ test = # UPDATE account SET balance=balance-10000 WHERE name='Tom';
	UPDATE 1
	system@test=#	

T2		system @ test = # BEGIN TRANSACTION ISOLATION LEVEL Repeatable Read;
		BEGIN
		system@test=# SELECT * FROM ACCOUNT;
		    name | balance
		----------------+----------
		 Tom | 20000.00
		 Jack | 30000.00
		(2 rows)
		--终端 1 上转账事务在进行中,终端 2 上的事务
		--在 Repeatable Read 事务隔离级别下,可以
		--看到的两个账号的值,还是转账前的初始值
		system@test=#

T3	system @ test = # UPDATE account SET balance=balance+10000 WHERE name='Jack'; UPDATE 1 1 system@test=#	
T4		system@test=# SELECT * FROM ACCOUNT; name | balance ----------------+---------- Tom | 20000.00 Jack | 30000.00 (2 rows) --终端 1 上转账事务在进行中,终端 2 上的事务 --在 Repeatable Read 事务隔离级别下,可以 --看到的两个账号的值,还是转账前的初始值 system@test=#
T5	system@test=# COMMIT; COMMIT system@test=# \q [kingbase@dbsvr ~]$	

续表

时刻	第 1 个终端	第 2 个终端
| T6 | | ```
system@test=# SELECT * FROM ACCOUNT;
    name         | balance
-----------------+----------
 Tom             | 20000.00
 Jack            | 30000.00
(2 rows)
--由于终端2上事务的隔离级别是Repeatable
--Read,终端2上的事务还未结束,因此在终端2
--上还看不到终端1上已经提交的转账事务的
--结果值!看到的仍然是事务开始时的值
system@test=# COMMIT;
COMMIT
system@test=# \q
[kingbase@dbsvr ~]$
``` |

从输出可以看出,在 REPEATABLE READ 事务隔离级别下,会在事务开始前,为事务生成一个快照。

7.3.5 元组的并发更新

KingbaseES 数据库使用了 MVCC 并发控制策略,实现了对同一个数据库对象读写操作的不冲突,从而提高了数据库并发处理的性能,但是当多个用户对同一个数据库对象进行写操作时,会是什么效果?

对于不同隔离级别的事务,当出现写操作冲突时,系统采用的处理方式也不同。

(1) 对于 **READ COMMITTED 隔离级别**的事务,当并发更新同一条元组时,第一个获取锁的事务(假设是 T1)对元组进行更新,其他事务则等待事务 T1 结束。如果事务 T1 提交,则下一个获取锁的事务继续更新该元组;如果事务 T1 回滚,则下一个获取锁的事务继续更新该元组。

(2) 对于**隔离级别是 REPEATABLE READ 和 SERIALIZABLE** 的事务,当并发更新同一条元组时,第一个获取锁的事务(假设是 T1)对元组进行更新,其他事务等待事务 T1 结束。如果事务 T1 提交,则其他事务回滚;如果事务 T1 回滚,则下一个获取锁的事务继续更新该元组。

例 7.3 在第 1 个终端上,以 READ COMMITTED 的事务隔离级别,更新用户 tom 的余额为 50 000;在第 2 个终端,以 READ COMMITTED 的事务隔离级别,更新用户 tom 的余额为 60 000;观察这两个事务的并发更新情况。

(1) 首先在第 1 个终端上执行如下的 ksql 命令和 SQL 语句,创建测试表 account 并初始化测试数据:

```
[kingbase@dbsvr ~]$ ksql -d test -U system
system@test=# DROP TABLE IF EXISTS account;
system@test=# CREATE TABLE account(name char(20),balance number(12,2));
```
```

```
system@test=# INSERT INTO account VALUES ('Tom',20000);
system@test=# INSERT INTO account VALUES ('Jack',30000);
system@test=# \q
[kingbase@dbsvr ~]$
```

（2）在两个终端上，按照表 7-5 的操作顺序，执行 ksql 命令和 SQL 语句：

表 7-5　操作顺序 3

时刻	第 1 个终端	第 2 个终端
T0	`kingbase@dbsvr ~]$ ksql -d test -U system` `system@test=#`	`kingbase@dbsvr ~]$ ksql -d test -U system` `system@test=#`

T1		`system@test=# SELECT * FROM ACCOUNT;`
		` name       | balance`
		`----------------+----------`
		` Tom        | 20000.00`
		` Jack       | 30000.00`
		`(2 rows)`
		`--查看并发更新前的初始值`
		`system@test=#`

T2	`system@test=# BEGIN TRANSACTION ISOLATION LEVEL Read Committed;`
	`BEGIN`
	`system@test=# UPDATE account SET balance=balance-10000 WHERE name='Tom';`
	`UPDATE 1`
	`system@test=#`	

T2		`system@test=# BEGIN TRANSACTION ISOLATION LEVEL Read Committed;`
		`BEGIN`
		`system@test=# UPDATE account SET balance=balance+20000 WHERE name='Tom';`
		`--执行等待阻塞中……`
		`--终端 1 和终端 2 同时并发更新，终端 2 上的更新等`
		`--待终端 1 上的事务结束`

T3	`system@test=# COMMIT;`
	`COMMIT`
	`system@test=# \q`
	`[kingbase@dbsvr ~]$`	

续表

时刻	第 1 个终端	第 2 个终端

T6		--阻塞解除,继续执行继续
		UPDATE 1
		system@test=#
		--由于终端 1 上事务以提交的方式结束,因此终端 2
		--上的并发更新事务获得继续执行需要的锁,执行
		--更新
		system@test=# COMMIT;
		COMMIT
		system@test=# SELECT * FROM ACCOUNT;
		    name | balance
		----------------+----------
		 Jack | 30000.00
		 Tom | 30000.00
		(2 rows)
		system@test=#
		--可以看到,Tom 的账号余额先被减去 10 000 然后
		--再被加上 20 000(实际增加了 10 000)
		system@test=# \q
		[kingbase@dbsvr ~]$

从输出可以看出,在 READ COMMITTED 事务隔离级别下的并发更新,如果第 1 个事务的更新提交,那么第 2 个更新将会继续进行。

例 7.4 在第 1 个终端上,以 REPEATABLE READ 的事务隔离级别,更新用户 tom 的余额为 50 000;在第 2 个终端,以 REPEATABLE READ 的事务隔离级别,更新用户 tom 的余额为 60 000;观察这两个事务的并发更新情况。

(1) 首先在第 1 个终端上执行如下的 ksql 命令和 SQL 语句,创建测试表 account 并初始化测试数据:

```
[kingbase@dbsvr ~]$ ksql -d test -U system
system@test=# DROP TABLE IF EXISTS account;
system@test=# CREATE TABLE account(name char(20),balance number(12,2));
system@test=# INSERT INTO account VALUES ('Tom',20000);
system@test=# INSERT INTO account VALUES ('Jack',30000);
system@test=# \q
[kingbase@dbsvr ~]$
```

(2) 在两个终端上,按照表 7-6 的操作顺序,执行 ksql 命令和 SQL 语句:

表 7-6 操作顺序 4

时刻	第 1 个终端	第 2 个终端
T0	kingbase@dbsvr ~]$ ksql -d test -U system system@test=#	kingbase@dbsvr ~]$ ksql -d test -U system system@test=#

续表

时刻	第 1 个终端	第 2 个终端
T1		system@test=# SELECT * FROM ACCOUNT; 　　　　name　　　　| balance ----------------+---------- 　Tom　　　　　　| 20000.00 　Jack　　　　　 | 30000.00 (2 rows) --查看并发更新前的初始值。 system@test=#

T2	system@test=# BEGIN TRANSACTION ISOLATION LEVEL Repeatable Read;
	BEGIN
	system@test=# UPDATE account SET balance=balance-10000 WHERE name='Tom';
	UPDATE 1
	system@test=#	

T2		system@test=# BEGIN TRANSACTION ISOLATION LEVEL Repeatable Read;
		BEGIN
		system@test=# UPDATE account SET balance=balance+20000 WHERE name='Tom';
		--执行等待阻塞中……
		--终端 1 和终端 2 同时并发更新,终端 2 上的更新等
		--待终端 1 上的事务结束

T3	system@test=# COMMIT; COMMIT system@test=# \q [kingbase@dbsvr ~]$	

T6		--解除阻塞继续执行
		ERROR:　could not serialize access due to concurrent update
		system@test=#
		--由于终端 1 上事务以提交的方式结束,因此终端 2
		--上的并发更新事务将无法继续进行
		system@test=# SELECT * FROM ACCOUNT;
		ERROR:　current transaction is aborted, commands ignored until end of transaction block
		system@test=#
		--终端 2 上的并发更新事务已经失败,只能回滚
		system@test=# ROLLBACK;
		ROLLBACK
		system@test=# SELECT * FROM ACCOUNT;

续表

时刻	第1个终端	第2个终端
| T6 | | ```
         name         | balance
----------------------+----------
 Jack                 | 30000.00
 Tom                  | 10000.00
(2 rows)
--可以看到,Tom 的账号余额只是在终端 1 的事务中
--被减去 10 000,当前值为 10 000
system@test=# \q
[kingbase@dbsvr ~]$
``` |

从输出可以看出,在 REPEATABLE READ 事务隔离级别下的并发更新,如果第 1 个事务的更新提交,那么第 2 个更新将失败并且只能回滚。

7.4 管理元组的多版本

KingbaseES 数据库采用 MVCC 的并发控制机制,随着数据库中事务的运行,当对某些表执行了大量的增、删、改等 DML 操作后,在数据库中存在元组的多个版本,从而造成表膨胀的问题。

KingbaseES 数据库事务中的查询都是基于快照的,对于数据库中保存每个元组的多个版本,有些版本对于所有活动的后台服务进程事务快照都是不可见的,称为**无效元组**。需要定期清理无效元组,回收其所占用的数据库空间。

7.4.1 手工清理无效元组

KingbaseES 数据库提供 VACUUM 命令来清理数据库中的无效元组。VACUUM 命令根据其是否有 FULL 选项,有两种运行方式。

(1) 不带 FULL 的 VACUUM 命令会删除表文件中每个页面中的无效元组,其他事务可以在其运行期间继续访问该表,称为并发清理,用于日常无效元组的清理。

(2) 带 FULL 的 VACUUM 命令不仅会移除整个文件中所有的无效元组,还会对文件进行碎片整理,其他事务可能在其运行期间无法访问该表。

例 7.5 观察表(索引)膨胀现象,并执行 VACUUM 和 VACUMM FULL 命令来清理无效元组。

(1) 创建一个表 vacuumtest,在表 vacuumtest 上创建索引 idx_vacuumtest_col1 和索引 idx_vacuumtest_col2:

```
[kingbase@dbsvr ~]$  ksql -d test -U system
system@test=# create table vacuumtest(col1 varchar(100),col2 varchar(100),col3 varchar(100));
system@test=# create index idx_vacuumtest_col1 on vacuumtest(col1);
system@test=# create index idx_vacuumtest_col2 on vacuumtest(col2);
```
```

（2）向表 vacuumtest 插入 500 万条测试数据，并查看此时表和索引所占存储空间的大小：

```
system@test=# insert into vacuumtest
test-#              select id::varchar,md5(id::varchar),md5(md5(id::varchar))
test-#              from generate_series(1,5000000) as id;
system@test=# SELECT sys_size_pretty(sys_total_relation_size('vacuumtest'));
 sys_size_pretty
----------------
 1071 MB
system@test=# SELECT sys_size_pretty( sys_total_relation_size('idx_vacuumtest_col1') );
 sys_size_pretty
----------------
 186 MB
system@test=# SELECT sys_size_pretty( sys_total_relation_size('idx_vacuumtest_col2') );
 sys_size_pretty
----------------
 365 MB
```

（3）删除表 vacuumtest 中的 400 万条数据后再次查看表和索引的大小：

```
system@test=# DELETE FROM vacuumtest WHERE col1::int4 > '1000000';
system@test=# SELECT sys_size_pretty(sys_total_relation_size('vacuumtest'));
 sys_size_pretty
----------------
 1071 MB
system@test=# SELECT sys_size_pretty( sys_total_relation_size('idx_vacuumtest_col1') );
 sys_size_pretty
----------------
 186 MB
system@test=# SELECT sys_size_pretty( sys_total_relation_size('idx_vacuumtest_col2') );
 sys_size_pretty
----------------
 365 MB
```

在 KingbaseES 数据库中，表中元组的删除只是在元组头上设置删除该元组的事务号，并不把元组从数据页面上清除，因此，删除表 vacuumtest 的大部分记录（在总的 500 万行中删除了 400 万行）之后，表 vacuumtest 本身及其索引的大小没有发生任何改变。

（4）执行 VACUUM 命令，并再次查看表和索引的大小。

```
system@test=# VACUUM;
system@test=# SELECT sys_size_pretty(sys_total_relation_size('vacuumtest'));
 sys_size_pretty
----------------
 654 MB
system@test=# SELECT sys_size_pretty( sys_total_relation_size('idx_vacuumtest_col1') );
```

```
 sys_size_pretty
-----------------
 186 MB
system@test=# SELECT sys_size_pretty( sys_total_relation_size('idx_vacuumtest_
col2') );
 sys_size_pretty
-----------------
 365 MB
```

VACUUM 命令用于回收数据库表中已经删除或更新的行所占用的空间,使得这部分空间可以被新的数据重用,它**不会减少物理文件的大小**,但会使得文件内部的空间得到重用。**VACUUM 命令是非锁定的**,意味着它在运行时不会阻塞对表的读写操作。执行完 VACUUM 命令之后,可以看到,表 vacuumtest 收缩变小了,但是表 vacuumtest 上面的索引并没有收缩变小。

(5)执行 VACUUM FULL 命令,并再次查看表和索引的大小。

```
system@test=# VACUUM FULL;
system@test=# SELECT sys_size_pretty(sys_total_relation_size('vacuumtest'));
 sys_size_pretty
-----------------
 181 MB
system@test=# SELECT sys_size_pretty( sys_total_relation_size('idx_vacuumtest_
col1') );
 sys_size_pretty
-----------------
 22 MB
system@test=# SELECT sys_size_pretty( sys_total_relation_size('idx_vacuumtest_
col2') );
 sys_size_pretty
-----------------
 56 MB
```

VACUUM FULL 命令会重写表到一个新的磁盘文件中,从而回收更多的空间,尽可能**减少表占用的磁盘空间**。**VACUUM FULL 命令会锁定表**,直到整个过程完成,这可能会影响数据库的可用性,特别是对于大表来说。因为它需要额外的磁盘空间来创建新的表副本,并在重写完成后删除旧的文件,所以在磁盘空间紧张的情况下使用 VACUUM FULL 命令需要谨慎。执行完 VACUUM FULL 命令之后,可以看到,表 vacuumtest 进一步收缩变小,表 vacuumtest 上面的索引也收缩变小了。

注意:执行 vacuum full 语句会锁定表,因此需要在系统不忙的时候执行,否则会造成业务卡顿,甚至酿成运维事故。

对于一些长时间执行的 VACUUM 命令,可执行下面的 SQL 语句,监控 VACUUM 命令的进度:

```
SELECT *, relid :: regclass,
       heap_blks_scanned / heap_blks_total :: float * 100 "% scanned",
       heap_blks_vacuumed / heap_blks_total :: float * 100 "% vacuumed"
  FROM sys_stat_progress_vacuum;
```

7.4.2 自动清理无效元组

KingbaseES 数据库提供了自动清理无效元组的机制。如果系统配置参数 AutoVacuum 设置为 true,则系统启动时会启动自动清理启动进程,该进程一直运行,由它定期启动自动清理工作进程来完成该项工作。

自动清理工作进程主要负责对有过大量更新操作的表进行空间清理和更新统计信息,即包括 Vacuum(清理)和 Analyze(分析)两部分工作。Vacuum 负责对表进行空间清理,将过期的元组清除,使其空间可以被其他数据使用;Analyze 负责将表的统计信息进行更新,使得查询在生成查询计划时,可以根据比较准确的统计信息选择较优的查询计划。

自动清理工作进程根据系统的统计信息和系统配置参数来决定对哪些表进行 Vacuum,对哪些表进行 Analyze,在系统表 sys_stat_all_tables 中记录了对表更新的统计信息。

如果表中无效元组的个数超出一定的阈值,将自动对表执行 vacuum 操作,具体条件如下:

```
sys_stat_all_tables.n_dead_tup >
    autovacuum_vacuum_threshold + autovacuum_vacuum_scale_factor × pg_class.reltuples
```

其中:

(1) autovacuum_vacuum_threshold 的默认值是 50;

(2) autovacuum_vacuum_scale_factor 的默认值是 0.2,即无效元组超过元组总数的 20%,需要进行 Vacuum。

如果表中增、删、改的元组个数超出一定的阈值应该对表进行 analyze:

```
sys_stat_all_tables.n_mod_since_analyze >
    autovacuum_analyze_threshold + autovacuum_analyze_scale_factor × pg_class.reltuples
```

其中:

(1) autovacuum_ analyze_threshold 的默认值是 50;

(2) autovacuum_ analyze_scale_factor 的默认值是 0.1,即更新的元组数超过元组总数的 10%,需要进行 Analyze。

自动清理启动进程根据系统配置参数 autovacuum_naptime 的取值定期唤醒,由它为每个活动的数据库创建一个自动清理工作进程来进行 Vacuum 的相关工作,因此最多会创建 autovacuum_max_workers 个自动清理工作进程。自动清理工作进程在每个数据库上对满足条件的表进行 Vacuum 和 Analyze 操作。

总结一下,系统配置参数定制了自动清理工作进程的行为,相关的系统配置参数包括以下几种。

(1) 参数 autovacuum 设置是否启动自动清理操作。

(2) 参数 autovacuum_vacuum_threshold 和 autovacuum_vacuum_scale_factor 设置选择 Vacuum 表的阈值。

(3) 参数 autovacuum_analyze_threshold 和 autovacuum_analyze_scale_factor 设置选择 Analyze 表的阈值。

(4) 参数 autovacuum_naptime 设置自动清理工作进程的启动周期。

(5) 参数 autovacuum_max_workers 设置自动清理工作进程的最大个数。

DBA 可以根据生产数据库的工作负载，在系统配置文件 Kingbase.conf 中配置这些参数。

DBA 还可以为特定的表，设置这些自动清理参数，语法如下：

```
ALTER TABLE table_name SET (autovacuum_enabled = true/false);
ALTER TABLE table_name SET (autovacuum_vacuum_threshold = value);
ALTER TABLE table_name SET (autovacuum_vacuum_scale_factor = value);
```

7.5 管理事务号

7.5.1 自动冻结事务号

KingbaseES 数据库使用元组头上的事务号来标识元组的版本，并根据查询使用的快照来决定访问哪个版本，元组头上的事务号跟元组一起存储在数据磁盘上。

KingbaseES 数据库中的事务号是 32 位的无符号数，最多可以存放 $2^{32}-1$ 个事务，随着系统的长时间的运行，就会出现事务号不够用的情况。对于该问题，目前 KingbaseES 数据库采用的策略是事务号循环使用，称为**事务号的回卷**（**wraparound**），就像日常生活中的钟表，指针跨过 12 点后就变成了 1 点。但是，系统中是根据事务号先后来判断元组的可见性，如果事务号发生了回卷，就不能根据其大小来判断事务发生的先后，这就是事务 ID 回卷的问题。

KingbaseES 数据库引入**事务年龄**（**age**）的概念来判断事务发生的先后。一个事务的年龄定义为其事务号与系统最新下一个事务号之间的距离。使用 age()函数可以获得一个事务的年龄。通过事务的年龄可以判断事务发生的先后，先发生的事务其年龄更老。

因为在 KingbaseES 数据库中事务号是 32 位循环使用的，事务的最大年龄是 $2^{31}-1$，系统对特别老的事务采用**冻结**（**freezing**）的方式，表明该事务号比当前所有的事务号都老，不参与正常的比较运算，称为**事务号的冻结**。KingbaseES 数据库通过自动清理工作进程定期对数据库中元组头上的老事务号冻结，预防事务号回卷带来的问题。

系统配置参数 vacuum_freeze_min_age 定义了事务号冻结的最小年龄，默认值为 50 000 000（5000 万），即如果元组头上的事务号的年龄大于该值，则冻结该事务号。

系统表 sys_class 中 relfrozenxid 字段的值记录了每个表中元组事务号的冻结点，即比该点老的事务都已经进行冻结，可以使用下面的 SQL 语句查询每个表的冻结点：

```
system@test=# SELECT relfrozenxid, age(relfrozenxid) FROM sys_class WHERE relname = 't_mvcc';
 relfrozenxid | age
--------------+-----
         1251 | 113
(1 row)
```

下面的 SQL 语句可以查询出当前连接的数据库中所有表的事务号的冻结点，以及它们的事务年龄：

```
system@test=# SELECT n.nspname AS schema_name,  c.relname AS table_name,
test-#             c.relfrozenxid,   age(c.relfrozenxid) AS age
test-#       FROM  sys_class c  JOIN sys_namespace n ON c.relnamespace = n.oid
test-#       WHERE c.relkind = 'r'  AND n.nspname NOT IN
test-#               ('pg_catalog', 'information_schema', 'anon',
test(#                'dbms_sql', 'perf','sys','sys_catalog', 'src_restrict',
test(#                'sys_hm','sysaudit', 'sysmac', 'xlog_record_read')
test-#       ORDER BY  schema_name,age DESC;
 schema_name | table_name | relfrozenxid | age
-------------+------------+--------------+-----
 public      | t1         |         1250 | 114
--省略了许多输出
(12 rows)
```

系统表 sys_database 中 datfrozenxid 字段记录了本数据库中的元组事务号的冻结点，可以使用下面的 SQL 语句进行查询：

```
system @ test = # SELECT datname, datfrozenxid, age (datfrozenxid) FROM sys_database;
  datname   | datfrozenxid | age
------------+--------------+-----
 test       |         1012 | 352
 kingbase   |         1012 | 352
 template1  |         1012 | 352
 template0  |         1012 | 352
 security   |         1012 | 352
(5 rows)
```

如果由于各种原因没有定期对表进行冻结操作，或者有些数据库长久不使用，就有可能发生事务号回卷的问题，从而造成数据的丢失。为了规避这种现象，数据库中元组头上的事务号的年龄老到一定程度，KingbaseES 数据库会强制启动自动清理工作进程进行数据库中元组头上事务号的冻结。系统参数 autovacuum_freeze_max_age 定义了强制启动自动清理工作进程的年龄，默认值是 200 000 000(2亿)，即数据库中的元组头上的事务号的年龄如果超过 2 亿，则强制启动自动清理工作进程。

如果强制启动自动清理工作进程后也没有解决问题，数据库中元组头上的事务号年龄越来越大，当距离回卷点还剩 40 000 000(4000万)个事务号时，系统会抛出警告，提示执行全数据库的 VACUUM。

如果忽略了该警告或系统原因没能冻结数据库中的老事务号，则当距离回卷点还剩 3 000 000(300万)个事务号时，系统会自动关闭，防止事务号回卷导致数据的丢失。

KingbaseES 数据库中事务号是循环使用的，图 7-1 给出了事务号冻结的几个关键节点。

假设某个数据库的冻结点的事务号是 A，回卷事务号为 F，F-A 表示它们之间的距离，即相隔的事务号个数，则 F-A=2^{31}(约 20 亿)，F-D=4000 万，F-E=300 万。当前事务号是

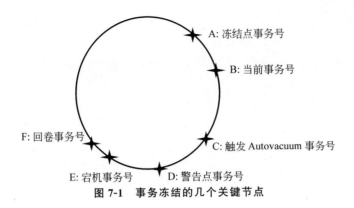

图 7-1　事务冻结的几个关键节点

B,则冻结点年龄就是 B－A。

默认情况下,当前事务号 B 到达 C 点时,系统启动自动清理工作进程,也就是说,C 点是自动清理进程执行事务号冻结操作的触发点,它由系统参数 autovacuum_freeze_max_age 定义(C－A＝2 亿)。

如果自动清理工作进程冻结了年龄大于 5000 万(由系统参数 vacuum_freeze_min_age 定义)的事务号,则 A 点会向前移动,从而降低其年龄,后面所有的关键点也会相应向前移动相同的距离,从而规避事务回卷。

7.5.2　手工冻结事务号

KingbaseES 数据库自动清理工作进程会在执行清理操作的时候自动执行事务号冻结,在必要时,也可以通过在 ksql 中执行下面的命令来手工冻结表的事务号,语法如下:

```
vacuum   (freeze,verbose) [TableName];        -- 手工冻结单个表 TableName
vacuum   (freeze,verbose) ;                   -- 手工冻结当前连接的数据库的所有表
```

在开始示例之前,执行下面的步骤,创建一个以很高速率消耗 n 个数据库事务号的存储过程 insert_data_loop(n)。

(1) 执行下面的 SQL 语句,创建一个测试表 test_table:

```
system@test=# CREATE TABLE test_table ( id SERIAL PRIMARY KEY,data TEXT );
```

(2) 执行下面的 SQL 语句,基于测试表 test_table,创建一个名叫 insert_data_loop 存储过程:

```
system@test=# CREATE OR REPLACE PROCEDURE insert_data_loop(max_iterations_param INT) AS $$
test$#          DECLARE
test$#              i INT := 0;
test$#              data TEXT;
test$#          BEGIN
test$#              -- 使用循环插入数据
test$#              LOOP
test$#                  EXIT WHEN i >= max_iterations_param;
test$#
test$#                  -- 构建数据字符串
```

```
test$#                    data := 'Transaction ' || i;
test$#
test$#                    -- 执行插入语句
test$#                    INSERT INTO test_table (data) VALUES (data);
test$#
test$#                    -- 提交事务消耗事务号
test$#                    commit;
test$#
test$#                    -- 递增计数器
test$#                    i := i + 1;
test$#            END LOOP;
test$#
test$#        END;
test$#   $$ LANGUAGE plsql;
```

（3）执行下面的 SQL，测试存储过程 insert_data_loop()：

```
system@test=# SELECT txid_current();
 txid_current
--------------
         1366
system@test=# CALL insert_data_loop(10000);
system@test=# SELECT txid_current();
 txid_current
--------------
        11367
```

从输出可以看到，存储过程 CALL insert_data_loop(10 000) 以很高的速率消耗了 10 000 个事务号，可以仿真生产数据库持续运行了一段时间。

示例的目的是观察持续时间很长不提交的最古老事务是如何影响冻结操作的。先说结论：冻结表或者数据库的操作，不可能回收早于最古老活动事务之前的事务号。也就是说，如果一个事务 T 一直不提交，随着生产数据库的运行（以高速率消耗事务号的存储过程 insert_data_loop(10 000)来仿真），将影响冻结表或者数据库的操作，在最严重的情况下，会导致 KingbaseES 数据库事务号回卷问题的发生，示例的过程如下。

（1）首先要确保测试系统没有其他的活动事务。然后在第 1 个终端，在 ksql 中，创建表 t1 后，开始事务 T1、插入 1 行后暂停，不要结束事务 T1，期望事务 T1 称为最古老的活动事务：

```
[kingbase@dbsvr ~]$ ksql -U system -d test
system@test=# DROP TABLE IF EXISTS t1;
system@test=# CREATE TABLE t1(id int);
system@test=# BEGIN;
system@test=# INSERT INTO t1 VALUES (1);
system@test=# SELECT txid_current();
 txid_current
--------------
        11370
(1 row)
```

事务 T1 的事务号是 11370。

（2）在第 2 个终端，在 ksql 中，使用隐式事务，创建表 t2 后，为表 t2 新增 1 条记录，这条新增的记录是在最老的事务之后产生的。

```
[kingbase@dbsvr ~]$ ksql -U system -d test
system@test=#    DROP TABLE IF EXISTS t2;
system@test=#    CREATE TABLE t2(id int);
system@test=#    INSERT INTO t2 VALUES(100);
```

（3）在第 3 个终端，执行下面的命令和存储过程，消耗 10 000 个事务号，仿真生产数据库持续运行了一段时间：

```
[kingbase@dbsvr ~]$ ksql -U system -d test
system@test=# CALL insert_data_loop(10000);
```

此时第 1 个终端上的事务 T1 是一个运行时间很长的（基于事务号来衡量）最古老的活动事务。

（4）在第 2 个终端，开始事务 T2、插入 1 行后暂停，不要结束事务 T2：

```
system@test=# BEGIN;
system@test=# INSERT INTO t1 VALUES (2);
system@test=# SELECT txid_current();
 txid_current
--------------
        21374
(1 row)
```

（5）在第 3 个终端，执行存储过程 insert_data_loop(10 000)，消耗 10 000 个事务号：

```
system@test=# CALL insert_data_loop(10000);
CALL
```

（6）在第 3 个终端，执行下面的命令和 SQL 语句，查看表 t2 的事务年龄：

```
system@test=# SELECT age(relfrozenxid), relname, pg_size_pretty(pg_total_
relation_size(oid))
test-#          FROM sys_class
test-#          WHERE relname='t2';
  age  | relname | pg_size_pretty
-------+---------+----------------
 20006 | t2      | 8192 bytes
```

（7）在第 3 个终端，执行下面的 SQL 语句，查看所连接的数据库中，属于某个用户（如用户 system）的每个表的事务年龄：

```
system@test=# SELECT n.nspname AS schema_name,   c.relname AS table_name,
test-#          c.relfrozenxid,   age(c.relfrozenxid) AS age
test-#       FROM  sys_class c   JOIN sys_namespace n ON c.relnamespace = n.oid
test-#       WHERE  c.relkind = 'r'  AND n.nspname NOT IN
test-#              ( 'pg_catalog', 'information_schema', 'anon',
test(#                'dbms_sql', 'perf','sys','sys_catalog', 'src_restrict',
test(#                'sys_hm','sysaudit' , 'sysmac', 'xlog_record_read')
```

```
test-#          ORDER BY  schema_name,age DESC;
 schema_name | table_name | relfrozenxid |  age
-------------+------------+--------------+-------
 public      | t_mvcc     |         1251 | 30125
 public      | account    |         1325 | 30051
 public      | vacuumtest |         1342 | 30034
 public      | test_table |         1364 | 30012
 public      | t1         |        11369 | 20007
 public      | t2         |        11370 | 20006
 tpch        | nation     |         1238 | 30138
--省略了一些输出
(14 rows)
```

(8) 在第 3 个终端，执行下面的 SQL 语句，查看 Kingbase 数据库集簇中，每个数据库的事务年龄：

```
system@test=# SELECT datname, age(datfrozenxid) FROM sys_database;
  datname  |  age
-----------+-------
 test      | 30365
 kingbase  | 30365
 template1 | 30365
 template0 | 30365
 security  | 30365
(5 rows)
```

(9) 在第 3 个终端，执行下面的 SQL 语句，手工冻结表 t2 的事务号，并再次查看表 t2 的事务年龄：

```
system@test=# vacuum (freeze,verbose) t2;
INFO:   aggressively vacuuming "public.t2"
INFO:   "t2": found 0 removable, 1 nonremovable row versions in 1 out of 1 pages
DETAIL:  0 dead row versions cannot be removed yet, oldest xmin: 11370
There were 0 unused item identifiers.           --执行冻结操作时的最古老活动事务号
Skipped 0 pages due to buffer pins, 0 frozenTup pages.
0 pages are entirely empty.
CPU: user: 0.00 s, system: 0.00 s, elapsed: 0.00 s.
VACUUM
system@test=# SELECT age(relfrozenxid),relname,pg_size_pretty(pg_total_relation_size(oid))
test-#          FROM sys_class
test-#          WHERE relname='t2';
  age  | relname | pg_size_pretty
-------+---------+----------------
 20008 | t2      | 40 kB
```

由于表 t2 的行是在最古老的活动事务(第 1 个终端上的事务，事务号是 11370)之后产生的，因此必须保留其版本，不能被标记为 freezing，因此无法将表 t2 的事务年龄降下来。

(10) 在第 1 个终端，提交事务 T1(事务号是 11370)：

```
system@test=# COMMIT;
```

COMMIT

在第 2 个终端的事务 T2 成了最古老的活动事务(事务号是 21374)。

(11) 在第 3 个终端，执行 SQL 语句，再次手工冻结表 t2，并查看表 t2 的事务年龄：

```
system@test=# vacuum (freeze,verbose) t2;
INFO:  aggressively vacuuming "public.t2"
INFO:  "t2": found 0 removable, 1 nonremovable row versions in 1 out of 1 pages
DETAIL:  0 dead row versions cannot be removed yet, oldest xmin: 21374
There were 0 unused item identifiers.            --执行冻结操作时的最古老活动事务号
Skipped 0 pages due to buffer pins, 0 frozenTup pages.
0 pages are entirely empty.
CPU: user: 0.00 s, system: 0.00 s, elapsed: 0.00 s.
VACUUM
system@test=# SELECT age(relfrozenxid),relname,sys_size_pretty(sys_total_
relation_size(oid))
test-#          FROM sys_class
test-#          WHERE relname='t2';
  age  | relname | pg_size_pretty
-------+---------+----------------
 10005 | t2      | 40 kB
(1 row)
system@test=# SELECT txid_current();
 txid_current
--------------
        31379
(1 row)
```

从输出可以看出，此时表 t2 的事务年龄并没有降到 1，而是 10005(第 3 个终端的事务号 31379 减去第 2 个终端的事务号 21374，值为 10005)，原因是需要保留最古老的活动事务(此时第 2 个终端上的事务，事务号是 21374)以后产生的行。

(12) 在第 2 个终端，提交事务 T2(事务号是 21374)：

```
system@test=# COMMIT;
COMMIT
```

(13) 在第 3 个终端执行 SQL 语句，再次手工冻结表 t2，并查看表 t2 的事务年龄：

```
system@test=# vacuum (freeze,verbose) t2;
INFO:  aggressively vacuuming "public.t2"
INFO:  "t2": found 0 removable, 1 nonremovable row versions in 1 out of 1 pages
DETAIL:  0 dead row versions cannot be removed yet, oldest xmin: 31380
--省略了一些输出             --执行冻结操作时的最古老活动事务号,此时是最新的事务号
system@test=# SELECT age(relfrozenxid),relname,pg_size_pretty(pg_total_
relation_size(oid))
test-#          FROM sys_class
test-#          WHERE relname='t2';
 age | relname | pg_size_pretty
-----+---------+----------------
   1 | t2      | 40 kB
```

从输出可以看到,这一次表 t2 的事务年龄已经降到 1。

7.6 数据库锁

KingbaseES 数据库采用 MVCC 的并发控制策略,读数据项可以不加锁,实现了数据库的读写不冲突。对于数据库的 DDL 操作,写写冲突等情况还提供了多种锁模式用于控制对表中数据的并发访问。KingbaseES 数据库自动封锁,无须用户干涉,缺省的封锁机制是在保证数据完整性的前提下尽量提高数据的并发性。KingbaseES 数据库遵循严格的两段锁协议,在命令的执行过程中根据需求自动对访问的数据库对象加锁,在事务结束时释放锁。

对于数据库的锁,通常从以下几个角度描述:封锁对象类型(粒度)、封锁对象标识、封锁模式以及持锁时间。具体来说,就是对谁加锁、加什么样的锁、什么时候释放锁。

7.6.1 表级锁

KingbaseES 数据库对表的任何操作(表上的 DDL/DML/SELECT 等)都需要首先加表级锁,根据不同的操作加不同模式的锁。封锁模式决定了对数据项的封锁程度,即锁的冲突程度。KingbaseES 数据库提供多种封锁类型来减少操作间冲突。

表 7-7 给出了 KingbaseES 数据库的锁模式,以及使用场合。

表 7-7 KingbaseES 数据库的封锁模式以及使用场合

序号	锁 模 式	对 应 操 作	与之冲突的模式
1	AccessShareLock	SELECT	8
2	RowShareLock	SELECT FOR UPDATE/SHARE	7,8
3	RowExclusiveLock	INSERT/UPDATE/DELETE	5,6,7,8
4	ShareUpdateExclusiveLock	VACUUM(no full),ANALYZE,CREATE INDEX CONCURRENTLY,REINDEX CONCURRENTLY,CREATE STATISTICS,	4,5,6,7,8
5	ShareLock	CREATE INDEX(WITHOUT CONCURRENTLY)	3,4,6,7,8
6	ShareRowExclusiveLock	CREATE TRIGGER,ALTER TABLE(部分)	3,4,5,6,7,8
7	ExclusiveLock	REFRESH MATERIALIZED VIEW CONCURRENTLY	2,3,4,5,6,7,8
8	AccessExclusiveLock	DROP TABLE/TRUNCATE/ALTER TABLE(部分)/REINDEX/CLUSTER/VACUUM FULL	全部

从表 7-7 可以看出,1 号锁 AccessShareLock 最宽容,只跟 8 号锁冲突,8 号锁则与所有的锁冲突。锁的序号越大,封锁程度越强。

在数据库系统运行期间,用户可以通过系统视图 sys_locks 来观察系统中常规锁的状态,其中的每一行表示系统中目前存在的一个锁状态。

```
system@test=# SELECT pid , locktype , relation:: regclass, virtualxid, transactionid,
mode, granted
```

```
test-#         FROM sys_locks
test-#         ORDER BY pid;
  pid  | locktype   | relation  | virtualxid | transactionid |       mode       | granted
-------+------------+-----------+------------+---------------+------------------+--------
 69475 | relation   | pg_locks  |            |               | AccessShareLock  | t
 69475 | relation   | sys_locks |            |               | AccessShareLock  | t
 69475 | virtualxid |           | 4/10454    |               | ExclusiveLock    | t
(3 rows)
```

SQL 语句的输出中：pid 列是服务进程的进程号；locktype 列是封锁对象的类型，它的值可以是关系表、事务号等；locktype 列后面的几列（relation、virtualxid）是具体的封锁对象，根据 locktype 的值来决定使用哪个列；mode 是封锁模式；granted 是一个布尔值，表示该封锁是否授予，如果是 t 表示获得锁；如果是 f 表示未获得锁需要等待。

下面通过 SQL 语句来观察 KingbaseES 数据库中各种操作对锁的使用方式。测试过程将使用 3 个终端，在其中的 2 个终端中并发运行 SQL 语句事务，另外 1 个终端用来观察系统的封锁情况。

测试之前需要首先创建测试表 locktest，并插入两条记录：

```
system@test=# CREATE TABLE locktest(col1 int,col2 int);
system@test=# INSERT INTO locktest VALUES (1,1);
system@test=# INSERT INTO locktest VALUES (2,2);
```

1. SELECT 语句与 DML 语句的并发执行

在第 1 个终端，开始事务，执行 SELECT 语句：

```
[kingbase@dbsvr ~]$ ksql -U system -d test
system@test=# SELECT sys_backend_pid();
 sys_backend_pid
-----------------
            4969
(1 row)
system@test=# BEGIN;
system@test=# SELECT * FROM locktest ;
 col1 | col2
------+------
    1 |    1
    2 |    2
(2 rows)
```

从输出可以看到，第 1 个会话的服务进程的 PID 是 4969，开启一个事务，查询表 locktest 中的所有元组。

在第 2 个终端，开始事务，执行 UPDATE 语句：

```
[kingbase@dbsvr ~]$ ksql -U system -d test
system@test=# SELECT sys_backend_pid();
 sys_backend_pid
-----------------
            5223
(1 row)
```

```
system@test=# BEGIN;
system@test=# UPDATE locktest SET col2=11 WHERE col1=1;
```

从输出可以看到，第 2 个会话的服务进程的 PID 是 5223，开启一个事务，修改表 locktest 中的第 1 个元组。虽然修改的是第 1 个会话查询的元组，但是读写不冲突，可以同时进行。

在第 1 个终端，再次执行 SELECT 语句：

```
system@test=# SELECT * FROM locktest ;
 col1 | col2
------+------
    1 |    1
    2 |    2
(2 rows)
```

SELECT 语句可以并发执行，并且查询的结果还是原来的结果集，即读取的该元组的老版本。

在第 3 个终端，执行下面的 SQL 语句，查看当前活动的锁：

```
[kingbase@dbsvr ~]$ ksql -d test -U system
system@test=# SELECT sys_backend_pid();
 sys_backend_pid
-----------------
            5282
(1 row)
system@test=# SELECT pid ,locktype ,relation::regclass,virtualxid, transactionid, mode, granted
test-#          FROM sys_locks
test-#          ORDER BY pid;
 pid  | locktype      | relation  | virtualxid | transactionid |      mode        | granted
------+---------------+-----------+------------+---------------+------------------+---------
 4969 | relation      | locktest  |            |               | AccessShareLock  | t
 4969 | virtualxid    |           | 4/10716    |               | ExclusiveLock    | t
 5223 | relation      | locktest  |            |               | RowExclusiveLock | t
 5223 | virtualxid    |           | 5/124      |               | ExclusiveLock    | t
 5223 | transactionid |           |            |         31359 | ExclusiveLock    | t
--省略了一些输出
(8 rows)
```

从输出可以看到，第 1 个会话执行的 SELECT 查询语句，向系统申请了表 locktest 上的 AccessShareLock，并且已经成功地获取了该锁（granted＝t）。第 2 个会话执行的 UPDATE 语句向系统申请了表 locktest 上的 RowExclusiveLock，并且已经成功地获取了该锁（granted＝t）。

由此可以看出 KingbaseES 数据库中 SELECT 语句和 DML 语句不冲突，可以并发执行。

在第 2 个终端，提交当前的事务：

```
system@test=# COMMIT;
```

2. SELECT 语句与 DDL 语句的并发执行

在第 2 个终端开启一个新事务后，在表 locaktest 上执行 ALTER TABLE 语句（DDL）：

```
system@test=# BEGIN;
system@test=# ALTER table locktest ADD col3 int;
--语句执行阻塞中……
```

从输出可以看到,在表 locktest 上执行 ALTER TABLE 语句将被阻塞,不能继续执行。在第 3 个终端,查看当前活动的锁:

```
system@test=# SELECT pid,locktype,relation::regclass,virtualxid, transactionid, mode, granted
test-#          FROM sys_locks
test-#          ORDER BY pid;
 pid  | locktype      | relation  | virtualxid | transactionid |      mode           | granted
------+---------------+-----------+------------+---------------+---------------------+--------
 4969 | relation      | locktest  |            |               | AccessShareLock     | t
 4969 | virtualxid    |           | 4/10716    |               | ExclusiveLock       | t
 5223 | virtualxid    |           | 5/125      |               | ExclusiveLock       | t
 5223 | relation      | locktest  |            |               | AccessExclusiveLock | f
 5223 | transactionid |           |            |         31360 | ExclusiveLock       | t
--省略了一些输出
(8 rows)
```

第 2 个会话执行的 DDL 语句 ALTER TABLE,向 KingbaseES 数据库申请了对表 locktest 的 AccessExclusiveLock 锁,但是暂时还没有成功获得该锁(granted=f),因为它需要的锁 AccessExclusiveLock 与第 1 个会话对表 locktest 持有的锁 AccessShareLock 冲突。

转到第 1 个终端,执行下面的 COMMIT 语句提交事务,这将会释放表 locktest 上的 AccessShareLock 锁:

```
system@test=# COMMIT;
```

转到第 2 个终端,可以看到,被阻塞的 ALTER TABLE 语句解除阻塞并继续执行成功了:

```
--解除阻塞继续执行
ALTER TABLE
system@test=#
```

转到第 3 个终端,查看当前活动的锁:

```
system@test=# SELECT pid,locktype,relation::regclass,virtualxid, transactionid, mode, granted
test-#          FROM sys_locks
test-#          ORDER BY pid;
 pid  | locktype      | relation  | virtualxid | transactionid |      mode           | granted
------+---------------+-----------+------------+---------------+---------------------+--------
 5223 | virtualxid    |           | 5/125      |               | ExclusiveLock       | t
 5223 | relation      | locktest  |            |               | AccessExclusiveLock | t
 5223 | transactionid |           |            |         31360 | ExclusiveLock       | t
 5282 | relation      | pg_locks  |            |               | AccessShareLock     | t
 5282 | relation      | sys_locks |            |               | AccessShareLock     | t
 5282 | virtualxid    |           | 6/20265    |               | ExclusiveLock       | t
(6 rows)
```

从输出可以看到,第 1 个会话中已经没有任何锁了,第 2 个会话在表 locktest 上申请的 AccessExclusiveLock 已授予,并且由于事务还没有完成,所以一直持有这个锁。

由此可以看出 KingbaseES 数据库中 SELECT 语句和有些 DDL 语句是冲突的,不可以并发执行。

3. SELECT 语句与 VACUUM 语句的并发执行

在第 1 个终端,再次开始一个新事务后,执行 SELECT 语句:

```
system@test=# BEGIN;
system@test=# SELECT * FROM locktest;
--语句执行阻塞中……
```

从输出可以看到,SELECT 语句目前被阻塞,原因是第 2 个会话的事务中在表 locktest 上执行了 DDL 语句,还持有 AccessExclusiveLock 锁,阻止了在第 1 个会话上的 SELECT 语句申请 AccessShareLock 锁。

转到第 2 个终端,提交该事务:

```
system@test=# COMMIT;
```

转到第 1 个终端,可以看到 SELECT 语句已经解除阻塞并继续成功执行完成:

```
--语句解除阻塞继续执行
col1 | col2 | col3
----+----+------
   2|   2 |
   1|  11 |
(2 rows)
```

转到第 2 个终端,执行下面的 VACUUM 语句:

```
system@test=# VACUUM locktest;
VACUUM
system@test=# VACUUM FULL locktest;
--语句执行阻塞中……
```

从输出可以看到,VACUUM 语句可以正常执行,但是 VACUUM FULL locktest 语句被阻塞了,处于等待状态。

在第 3 个终端,查看当前活动的锁:

```
system@test=# SELECT pid,locktype,relation::regclass,virtualxid,transactionid,mode,granted
test-#       FROM sys_locks
test-#       ORDER BY pid;
 pid  | locktype      | relation  | virtualxid | transactionid | mode                | granted
------+---------------+-----------+------------+---------------+---------------------+--------
 4969 | relation      | locktest  |            |               | AccessShareLock     | t
 4969 | virtualxid    |           | 4/10717    |               | ExclusiveLock       | t
 5223 | virtualxid    |           | 5/130      |               | ExclusiveLock       | t
 5223 | relation      | locktest  |            |               | AccessExclusiveLock | f
 5223 | transactionid |           |            | 31362         | ExclusiveLock       | t
--省略了一些输出
(8 rows)
```

从输出可以看到,第 2 个会话申请表 locktest 上的 AccessExclusiveLock 锁没有授予,与第 1 个会话持有的表 locktest 上的 AccessShareLock 冲突。

转到第 1 个终端，提交第 1 个会话的事务：

```
system@test=# COMMIT;
```

马上转到第 2 个终端，可以看到 VACUUM FULL 语句解除阻塞并继续成功执行：

```
--语句解除阻塞继续执行
VACUUM
system@test=#
```

由此可以看出 KingbaseES 数据库中 SELECT 语句和 VACUUM 语句是不冲突的，但是跟 VACUUM FULL 语句冲突。

4. DML 语句与 DML 语句的并发执行

在第 1 个终端，重新开启 1 个新事务，执行 UPDATE 语句：

```
system@test=# BEGIN;
system@test=# UPDATE locktest SET col2=111 WHERE col1=1;
UPDATE 1
```

从输出可以看到，第 1 个会话上的事务成功修改了表 locktest 中 col1＝1 的元组。

在第 2 个终端，重新开启 1 个新事务，执行 UPDATE 语句：

```
system@test=# BEGIN;
system@test=# UPDATE locktest SET col2=222 WHERE col1=2;
UPDATE 1
```

从输出可以看到，第 2 个会话上的事务成功修改了表 locktest 中 col1＝2 的元组。

在第 3 个终端，查看当前活动的锁：

```
system@test=# SELECT pid,locktype,relation::regclass,virtualxid,transactionid,mode,granted
test-#        FROM sys_locks
test-#        ORDER BY pid;
 pid  | locktype      | relation | virtualxid | transactionid | mode             | granted
------+---------------+----------+------------+---------------+------------------+--------
 4969 | virtualxid    |          | 4/10718    |               | ExclusiveLock    | t
 4969 | transactionid |          |            | 31363         | ExclusiveLock    | t
 4969 | relation      | locktest |            |               | RowExclusiveLock | t
 5223 | relation      | locktest |            |               | RowExclusiveLock | t
 5223 | virtualxid    |          | 5/132      |               | ExclusiveLock    | t
 5223 | transactionid |          |            | 31364         | ExclusiveLock    | t
--省略了一些输出
(9 rows)
```

从输出可以看到，第 1 个会话和第 2 个会话在表 locktest 上都持有 RowExclusiveLock，该锁之间是不冲突的。

因此可以看出 KingbaseES 数据库中 2 个事务并发修改同一张表，如果修改的是不同的元组，是不会发生冲突的。

在继续之前，在第 1 个终端和第 2 个终端上执行 COMMIT 语句，提交事务：

```
system@test=# COMMIT;
```

7.6.2 事务锁

KingbaseES 数据库采用 MVCC 的并发控制策略,实现了数据库的读写不阻塞,如何解决写写冲突(更新同一个数据项的 DML 语句的并发)? KingbaseES 数据库采用元组上的锁标识来实现对同一个元组的并发更新操作,即写写冲突。每个元组的元组头的 XMAX 表示删除/修改该元组的事务号,如果有并发事务同时更新同一个元组产生的写写冲突,则必然是当前事务和该元组头上 XMAX 表示的事务,KingbaseES 数据库采用对事务号加锁的办法来解决对同一元组的写写冲突问题。

后台服务进程每次给事务分配事务号时,就在该事务号上加 ExclusiveLock,直到事务结束。在上面例子中查询封锁信息时,可以看到每个会话都会持有对事务号的封锁,没有分配事务号的,在虚拟事务号上持锁;已经分配事务号的,在本事务号上持锁。

当事务 T1 更新某个元组 tuple_x 时,首先会检查该元组头上 xmax 表示的事务 T2 是否结束,如果 T2 不是一个合法的事务号,则说明 tuple_x 没有并发的更新,可以直接修改;如果 T2 是一个合法的事务号并且还在运行,则说明 T2 正在更新该元组,事务 T1 则会对 T2 事务号加 ShareLock 进行等待,T2 结束时,唤醒 T1 继续操作。T1 获得该共享锁后,随即释放该锁,继续后续的操作。

使用前面例子中的 3 个终端来观察更新同一个元组,出现写写冲突时 KingbaseES 数据库的封锁策略。

在第 1 个终端,再开启事务,执行 UPDATE 语句:

```
system@test=# BEGIN;
system@test=# UPDATE locktest SET col2=1111 WHERE col1=1;
UPDATE 1
system@test=# SELECT txid_current();
 txid_current
--------------
        31304
(1 row)
```

从输出可以看到,第 1 个会话中事务号为 31304 的事务成功地修改了表 locktest 中 col1=1 的元组。

在第 3 个终端,查看当前活动的锁:

```
system@test=# SELECT pid,locktype,relation::regclass,virtualxid, transactionid, mode, granted
test-#       FROM sys_locks
test-#       ORDER BY pid;
 pid  | locktype      | relation  | virtualxid | transactionid | mode             | granted
------+---------------+-----------+------------+---------------+------------------+--------
 4969 | relation      | locktest  |            |               | RowExclusiveLock | t
 4969 | virtualxid    |           | 4/10719    |               | ExclusiveLock    | t
 4969 | transactionid |           |            | 31304         | ExclusiveLock    | t
 5282 | relation      | pg_locks  |            |               | AccessShareLock  | t
 5282 | relation      | sys_locks |            |               | AccessShareLock  | t
 5282 | virtualxid    |           | 6/20268    |               | ExclusiveLock    | t
(6 rows)
```

可以看到第 1 个会话在事务号 31304 上持有 ExclusiveLock 锁、在表 locktest 上持有 RowExclusiveLock 锁。

在第 2 个终端,再开启事务,执行 UPDATE 语句:

```
system@test=# BEGIN;
system@test=# SELECT txid_current();
 txid_current
--------------
        31366
(1 row)
system@test=# UPDATE locktest SET col2=1111 WHERE col1=1;
--语句执行阻塞中……
```

第 2 个会话中的事务尝试修改表 locktest 中 col1=1 的元组,目前处于阻塞等待中。

在第 3 个终端,查看当前活动的锁:

```
system@test=# SELECT pid,locktype,relation::regclass,virtualxid, transactionid, mode, granted
test-#        FROM sys_locks
test-#        ORDER BY pid;
 pid  | locktype      | relation  | virtualxid | transactionid | mode              | granted
------+---------------+-----------+------------+---------------+-------------------+---------
 4969 | virtualxid    |           | 4/10719    |               | ExclusiveLock     | t
 4969 | transactionid |           |            |         31304 | ExclusiveLock     | t
 4969 | relation      | locktest  |            |               | RowExclusiveLock  | t
 5223 | relation      | locktest  |            |               | RowExclusiveLock  | t
 5223 | virtualxid    |           | 5/133      |               | ExclusiveLock     | t
 5223 | tuple         | locktest  |            |               | ExclusiveLock     | t
 5223 | transactionid |           |            |         31304 | ShareLock         | f
 5223 | transactionid |           |            |         31366 | ExclusiveLock     | t
 5282 | relation      | sys_locks |            |               | AccessShareLock   | t
 5282 | virtualxid    |           | 6/20269    |               | ExclusiveLock     | t
 5282 | relation      | pg_locks  |            |               | AccessShareLock   | t
(11 rows)
```

从输出可以看到,第 2 个会话中的事务在事务号 31304 上持有 ExclusiveLock 锁、在表 locktest 上持有 RowExclusive 锁,并申请对事务号 31304 的 ShareLock,但是目前还没法获得该锁,处于等待状态。因为目前第 1 个会话中的事务号正是 31304,还处于正在运行状态,持有自己事务号的 ExclusiveLock 锁,出现冲突。

KingbaseES 数据库采用元组上的锁标识来实现概念上的行级锁,避免了在全局锁表中为每个元组存放锁对象,即提高了系统的并发度,又避免了巨大的锁管理开销。

在第 1 个终端,提交事务:

```
system@test=# COMMIT;
```

到第 2 个终端,可以看到第 2 个会话中事务的 UPDATE 语句解除阻塞继续执行完成:

```
--语句解除阻塞继续执行
UPDATE 1
system@test=#
```

在第 3 个终端,查看当前活动的锁:

```
system@test=# SELECT pid,locktype,relation::regclass,virtualxid, transactionid, mode, granted
test-#          FROM sys_locks
test-#          ORDER BY pid;
 pid  |   locktype    | relation  | virtualxid | transactionid |      mode       | granted
------+---------------+-----------+------------+---------------+-----------------+---------
 5223 | relation      | locktest  |            |               | RowExclusiveLock | t
 5223 | virtualxid    |           | 5/10478    |               | ExclusiveLock    | t
 5223 | transactionid |           |            | 31305         | ExclusiveLock    | t
 5282 | relation      | pg_locks  |            |               | AccessShareLock  | t
 5282 | relation      | sys_locks |            |               | AccessShareLock  | t
 5282 | virtualxid    |           | 6/20193    |               | ExclusiveLock    | t
(6 rows)
```

因为第 2 个会话还没有提交，依然持有事务号 31305 的 ExclusiveLock 锁和表 locktest 上的 RowExclusive 锁。

在第 2 个终端，提交该事务：

```
system@test=# COMMIT;
```

在第 3 个终端，查看当前活动的锁：

```
system@test=# SELECT pid,locktype,relation::regclass,virtualxid, transactionid, mode, granted
test-#          FROM sys_locks
test-#          ORDER BY pid;
 pid  | locktype   | relation  | virtualxid | transactionid |      mode       | granted
------+------------+-----------+------------+---------------+-----------------+---------
 5282 | relation   | pg_locks  |            |               | AccessShareLock  | t
 5282 | relation   | sys_locks |            |               | AccessShareLock  | t
 5282 | virtualxid |           | 6/20271    |               | ExclusiveLock    | t
(3 rows)
```

从输出可以看到，第 2 个会话上的事务提交后，所有的锁都释放了。

由此可以看出，KingbaseES 数据库采用元组上的锁标识来实现概念上的行级锁，避免了在全局锁表中为每个元组存放锁对象，即提高了系统的并发度，又避免了巨大的锁管理开销。

表级锁和事务锁是最常使用的锁，KingbaseES 数据库内部偶尔还会在其他对象上封锁，如页面等，这里不再赘述。

7.6.3 死锁

两个或多个事务都已封锁了一些数据对象，然后请求已被其他事务封锁的数据对象加锁，事务相互等待对方先释放资源，从而出现死等待，这就是死锁。KingbaseES 数据库自动检测死锁，检测到死锁后，回滚其中的一个语句。

继续使用前面例子中的 3 个终端，观察死锁现以及 KingaseES 数据库的封锁策略。
在第 1 个终端开启一个事务，执行 UPDATE 命令，修改表 locktest 中 col1=1 的元组：

```
system@test=# BEGIN;
system@test=# UPDATE locktest SET col2=1 WHERE col1=1;
UPDATE 1
```

在第 2 个终端开启一个事务，执行 UPDATE 命令，修改表 locktest 中 col1=2 的元组：

```
system@test=# BEGIN;
system@test=# UPDATE locktest SET col2=22 WHERE col1=2;
UPDATE 1
```

在第 1 个终端继续执行 UPDATE 命令，修改表 locktest 中 col1＝2 的元组：

```
system@test=# UPDATE locktest SET col2=21 WHERE col1=2;
--语句阻塞等待中…
```

在第 3 个终端，查看当前活动的锁：

```
system@test=# SELECT pid,locktype,relation::regclass,virtualxid, transactionid, mode, granted
test-#        FROM sys_locks
test-#        ORDER BY pid;
 pid  |  locktype     | relation  | virtualxid | transactionid |      mode        | granted
------+---------------+-----------+------------+---------------+------------------+---------
 4969 | transactionid |           |            |     31307     | ShareLock        | f
 4969 | tuple         | locktest  |            |               | ExclusiveLock    | t
 4969 | relation      | locktest  |            |               | RowExclusiveLock | t
 4969 | virtualxid    |           | 4/163      |               | ExclusiveLock    | t
 4969 | transactionid |           |            |     31306     | ExclusiveLock    | t
 5223 | relation      | locktest  |            |               | RowExclusiveLock | t
 5223 | virtualxid    |           | 5/10479    |               | ExclusiveLock    | t
 5223 | transactionid |           |            |     31307     | ExclusiveLock    | t
 5282 | relation      | pg_locks  |            |               | AccessShareLock  | t
 5282 | relation      | sys_locks |            |               | AccessShareLock  | t
 5282 | virtualxid    |           | 6/20195    |               | ExclusiveLock    | t
(11 rows)
```

从输出可以看出，第 1 个会话申请事务号 31307 上的 ShareLock 未被授予，因为第 2 个会话中的事务的事务号就是 31307，事务 31307 还正在运行，持有其自己事务号 31307 上的 ExclusiveLock，这两者存在锁冲突。

在第 2 个终端继续执行 UPDATE 命令，修改 locktest 表中 col1＝1 的元组：

```
system@test=# UPDATE locktest SET col2=12 WHERE col1=1;
ERROR:   deadlock detected
DETAIL:  Process 5223 waits for ShareLock on transaction 31306; blocked by process 4969.
Process 4969 waits for ShareLock on transaction 31307; blocked by process 5223.
HINT:   See server log for query details.
CONTEXT:  while updating tuple (0,6) in relation "locktest"
system@test=#
```

从输出可以看到，KingbaseES 数据库检测到系统发生了死锁，于是结束第 2 个会话中的事务，释放了这个会话上持有的锁。

此时转到第 1 个终端上，可以看到其中的事务解除了阻塞等待，正常执行完成：

```
--语句解除阻塞继续执行
UPDATE 1
system@test=#
```

系统配置参数 deadlock_timeout 定义了在发生死锁时系统等待的时间，如果在指定的超时时间内无法获得封锁，系统就会检查是否出现了死锁。如果出现死锁，回滚其中一个事务以解除死锁。参数 deadlock_timeout 的默认值为 1s。参数 deadlock_timeout 的值应该根据系统的性能和并发负载进行适当的调整。设置过长的超时时间可能导致事务长时间等待，影响系统性能，而设置过短的超时时间可能增加死锁的发生频率。

死锁产生的原因通常是数据库中的事务对于数据库对象的更新交叉进行。预防死锁的一个办法就是事务尽可能按照同一顺序对数据库对象加锁，例如，所有的应用开发人员在开发应用程序时，都遵守下面的规则，当需要修改主表和细表时，按一定的顺序进行。

7.7 故障恢复机制

7.7.1 故障恢复概述

系统可能发生各种各样的故障，每种故障需要不同的方法来处理。数据库系统主要会遇到 4 种故障，分别是事务故障、系统故障、介质故障和用户错误。

事务故障是指事务的运行没有到达预期的终点就被终止。例如，SQL 语句在执行语句的过程中出现违反完整性约束的情况，这时事务就需要回滚。事务故障由系统自动处理，DBA 通常不需要进行干预。

系统故障是指造成系统停止运转的任何事件，使得系统要重新启动。例如，特定类型的硬件错误（CPU 故障）、操作系统故障、数据库管理系统代码错误、系统断电、导致系统崩溃的计算机病毒等。发生系统故障时，可能造成数据库处于不正确的状态。系统故障通常在下次启动时自动恢复。

介质故障指外存故障，如磁盘损坏、破坏磁盘数据的计算机病毒等。介质故障比前两类故障发生的可能性小得多，但破坏性最大。介质故障需要借助存储在其他地方的数据备份来恢复数据库，通常需要 DBA 进行干预。

对于上述 4 种故障的恢复，KingbaseES 数据库都是采用日志系统来保证事务的原子性和持久性。

用户在使用数据库的过程中可能会出现一些误操作，例如，误删了表中的数据行、误删除系统中的表、用户提交了错误数据等，对于这类问题，KingbaseES 数据库提供了逻辑备份与还原、闪回，以及时间点恢复（point-in-time-recovery，PITR）技术来帮助用户找回误操作带来的数据损失。

7.7.2 日志系统组件

KingbaseES 数据库日志系统包括以下相关组件：
（1）共享内存中的日志缓冲区；
（2）外存上的日志文件；
（3）外存上的控制文件；
（4）检查点进程、后台写进程、日志写进程、后台服务进程、日志归档进程等。

数据库服务进程处理客户的请求，更新数据库的内容，所有对数据库的更新都会形成日志写到共享内存中的日志缓冲区中，日志写进程定期把共享内存中的日志缓冲区的内容写到外存的日志文件中,特别是事务提交时一定保证该事务的所有日志记录都写到外存。检查点进程和后台写进程定期把数据缓存区的内容写到外存的数据文件,以减少系统出现故障时进行系统恢复的时间,检查点的信息保存在系统外存的控制文件中。日志系统的这些组件相互协作,保证事务的原子性和持久性。

为保证数据库是可恢复的,登记日志文件时必须遵循 WAL 协议。KingbaseES 数据库中把对数据的修改写到数据库的数据文件中和把表示这个修改的日志记录写到日志文件中是两个不同的操作。WAL 协议保证了事务提交的时候,对应的日志必须刷到磁盘上,即使系统出错,也可以通过日志的恢复操作完成事务的操作,这就保证了事务的持久性。KingbaseES 数据库采用多版本并发控制策略,对于写到磁盘上的未提交事务所做的修改,用户根据快照规则是看不到的,因此,不需要事务的回滚操作。

7.7.3 WAL 文件

KingbaseES 数据库的 WAL 文件位于目录/u00/Kingbase/ES/V9/data/sys_wal 下,默认每个文件大小是 16MB(可以在系统初始化时通过系统配置参数 wal_segment_size 指定):

```
[kingbase@dbsvr ~]$ cd /u00/Kingbase/ES/V9/data/sys_wal
[kingbase@dbsvr sys_wal]$ ls -l
total 1032192
-rw-------. 1 kingbase dba 16777216 Mar  2 15:18 000000010000000100000087
-rw-------. 1 kingbase dba 16777216 Mar  2 14:45 000000010000000100000088
#省略了一些输出
drwx------. 2 kingbase dba        6 Feb 29 21:27 archive_status
[kingbase@dbsvr sys_wal]$
```

执行下面的 SQL 语句,可以列出所有的 WAL 文件:

```
[kingbase@dbsvr ~]$ ksql -U system -d test
system@test=# SELECT * FROM sys_ls_waldir() ORDER BY  modification ASC;
           name           |   size   |      modification
--------------------------+----------+------------------------
 000000010000000100000089 | 16777216 | 2024-03-02 14:45:04+08
 00000001000000010000008A | 16777216 | 2024-03-02 14:45:04+08
--省略了一些输出
(63 rows)
```

数据库的日志文件详细记录了数据库服务进程对数据库的操作过程。数据库的日志逻辑上是一个连续的文件,但物理上分割成一个个独立的小文件。日志文件的文件名由 24 个字符组成,每个字符以十六进制表示,其中前 8 个字符是事务日志的时间线(timeline),中间 8 个字符是事务日志的逻辑 ID,每个逻辑 ID 默认是 $256 \times 16MB$,即每个逻辑 ID 分为 256 个日志段,每个日志段对应一个日志文件,每个日志文件默认 16M,最后的 8 个字符则表示日志段的编号。

日志段文件的文件名从 000000010000000000000001 开始,如果该日志段写满后,就会创建第二个日志段文件 000000010000000000000002,后续的文件名使用升序。在日志段文件 0000000100000000000000FF 填满后,就使用下一个文件 000000010000000100000000。

通过这种方式，每当最后两个数字要进位时，中间 8 位数字就会加 1，例如，0000000100000001000000FF 被填满后，就会开始使用 00000001000000020000000，以此类推。

每个日志页面中有多条日志记录顺序排放，每条日志记录的开始位置可以由日志序列号(log sequence nunber, LSN)唯一标识。日志序列号由日志的逻辑 ID、段号以及段内的偏移量组成，LSN 是一个单调递增的数值，标识了每条日志记录的位置。当数据库页面被修改，生成一条日志记录时，该页面头上会记录该 LSN 的值，KingbaseES 数据库系统使用 LSN 来跟踪每个数据库页面的状态。

例如，sys_current_wal_insert_lsn()函数可以返回当前日志的插入点，即下一条日志记录插入的位置：

```
test=# select sys_current_wal_insert_lsn();
 sys_current_wal_insert_lsn
----------------------------
 1/87275C70
(1 row)
```

使用 sys_walfile_name()函数可以查询日志序列号所对应的日志段文件名：

```
system@test=# select sys_walfile_name('1/87275C70');
     sys_walfile_name
--------------------------
 000000010000000100000087
(1 row)
```

使用命令行工具 sys_waldump 可以查看日志文件的内容：

```
system@test=# create table t_wal(col1 int,col2 char(10));
CREATE TABLE
system@test=# select sys_current_wal_insert_lsn();
 sys_current_wal_insert_lsn
----------------------------
 1/87285F28
(1 row)
system@test=# insert into t_wal values(1,'111');
INSERT 0 1
system@test=# select sys_current_wal_insert_lsn();
 sys_current_wal_insert_lsn
----------------------------
 1/87285FC8
(1 row)
system@test=# \q
[kingbase@dbsvr ~]$ sys_waldump -p /u00/Kingbase/ES/V9/data  -s 1/87285F28  -e 1/87285FC8
rmgr: Heap        len (rec/tot):     70/    70, tx:      31309, lsn: 1/87285F28,
prev 1/87285EF8, desc: INSERT+INIT off 1 flags 0x00, blkref #0: rel 1663/14386/17257 blk 0
rmgr: Transaction len (rec/tot):     34/    34, tx:      31309, lsn: 1/87285F70,
prev 1/87285F28, desc: COMMIT 2024-03-02 15:39:56.998871 CST
```

```
rmgr: Standby       len (rec/tot):     46/     46, tx:          0, lsn: 1/87285F98,
prev 1/87285F70, desc: RUNNING_XACTS nextXid 31310 latestCompletedXid 5480763
oldestRunningXid 31309
[kingbase@dbsvr sys_wal]$
```

可以看出在 INSERT 语句前后共有 3 条日志记录，分别是：插入一条元组、事务提交、生成一个事务号。重点插入元组的日志记录，起点在 1/87285F28，结束点在 1/87285F70，即下一条日志记录的起点。

通常情况下，数据库日志文件是顺序使用的，一个日志文件写满，会自动使用下一个日志文件，称为日志切换。也可以使用 sys_switch_wal()函数手工切换日志文件：

```
[kingbase@dbsvr ~]$ ksql -d test -U system
system@test=# SELECT sys_switch_wal();
 sys_switch_wal
----------------
 5/4BD99FC0
(1 row)
```

7.7.4　检查点机制

虽然把所有的操作都记录到了 WAL 文件，但是当实例故障发生之后，应该从哪里开始执行恢复操作？如果没有**检查点**（**Checkpoint**），那么只能从 KingbaseES 数据库启动的时刻开始恢复，这需要大量的空间来保存自数据库启动以来的 WAL 文件，此外通过这些 WAL 文件进行恢复的时间也不可接受，这就是引入检查点的动机。

检查点是一个事件，KingbaseES 数据库在执行检查点操作就是把系统缓冲区中的数据写回到磁盘，然后在 WAL 文件中写入一条检查点日志记录，并且在检查点操作成功后，在控制文件中记录检查点的信息。

系统出现故障数据库，需要恢复的最主要原因就是数据库缓冲区的数据与磁盘数据文件不一致，检查点将缓冲区的脏数据写入磁盘，保证了在检查点开始时刻内存和磁盘中的数据是一致的，因此系统故障恢复可以从最后一个成功的检查点日志开始。有了检查点之后，当发生**系统故障**后的恢复过程，如图 7-2 所示。

KingbaseES 数据库的恢复从最后一个成功的检查点开始，到故障点发生时的 WAL 文件，对于恢复操作来说是必不可少的。KingbaseES 数据库的恢复只需要从控制文件中读出最后一个成功的检查点位置，然后从这个位置开始重做 WAL 文件，一直到故障点就算完成了。

检查点完成之后，检查点之前的 WAL 文件都没用了，KingbaseES 数据库根据检查点的情况自动回收 WAL 文件的空间。

KingbaseES 数据库执行检查点的时机由以下几个因素决定。

（1）时间间隔：KingbaseES 数据库有一个**检查点进程**，定期自动执行检查点操作。根据参数 checkpoint_timeout 设置的时间间隔，周期性地执行一个检查点操作，即使在这段时间内数据库活动不高，KingbaseES 数据库也会定期进行检查点操作。

（2）WAL 文件数据量。

① 参数 checkpoint_completion_target 用来设置在检查点周期内完成检查点的目标时间，以减少检查点对数据库性能的影响。

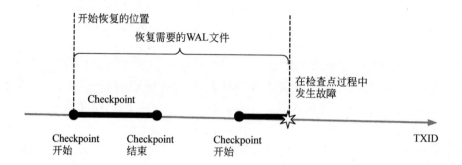

图 7-2　检查点与实例恢复

② 参数 max_wal_size 可以设置在触发检查点之前 WAL 文件能够增长到的最大尺寸。

③ 参数 min_wal_size 设置了在自动清理（回收或删除）WAL 文件之前保留的最小 WAL 文件的总大小，用于保证有足够的 WAL 文件空间可用于正常的数据库操作，特别是在数据库负载较高时。

（3）以 fast 或者 smart 方式关闭 KingbaseES 数据库时，将执行一个检查点操作。

（4）KingbaseES 数据库内部操作，如大型批量数据加载或索引创建等，可能会触发检查点，以确保数据的持久性，减少恢复时间。

（5）DBA 手工执行 CHECKPOINT 语句时，将执行一个检查点操作。

```
system@test=# CHECKPOINT;
CHECKPOINT
system@test=#
```

执行下面的命令，可以查看控制文件中关于检查点的信息：

```
[kingbase@dbsvr ~]$ sys_controldata /u00/Kingbase/ES/V9/data
sys_control version number:            1201
Catalog version number:                202305151
Database system identifier:            7341015102399125796
Database cluster state:                in production
sys_control last modified:             Sat 02 Mar 2024 03:46:13 PM CST
Latest checkpoint location:            1/88000088
Latest checkpoint's REDO location:     1/88000058
Latest checkpoint's REDO WAL file:     000000010000000100000088
Latest checkpoint's WalTimeLineID:     1
Latest checkpoint's PrevTimeLineID:    1
Latest checkpoint's full_page_writes:  on
```

```
Latest checkpoint's NextXID:                 0:31310
Latest checkpoint's NextOID:                 24663
Latest checkpoint's NextMultiXactId:         1
Latest checkpoint's NextMultiOffset:         0
Latest checkpoint's oldestXID:               1012
Latest checkpoint's oldestXID's DB:          1
Latest checkpoint's oldestActiveXID:         31310
Latest checkpoint's oldestMultiXid:          1
Latest checkpoint's oldestMulti's DB:        1
Latest checkpoint's oldestCommitTsXid:       0
Latest checkpoint's newestCommitTsXid:       0
Time of latest checkpoint:                   Sat 02 Mar 2024 03:46:13 PM CST
#省略了一些输出
[kingbase@dbsvr ~]$
```

其中，

（1）Latest checkpoint location：数据库异常停止后再重新启动时，需要做实例恢复，实例恢复的过程是从 WAL 文件中，找到最后一次的检查点，然后读取这个点之后的 WAL 文件，重新应用这些日志。最后一次的检查点的信息记录在 Latest checkpoint 项中。

（2）Latest checkpoint's REDO location：记录数据库日志文件上检查点。

（3）Latest checkpoint's REDO WAL file：记录 WAL 文件名，在目录 sys_wal 下可以查到这个日志文件。

7.7.5 配置 WAL 文件

1. 配置 WAL 文件的内容

在 KingbaseES 数据库的系统配置文件 kingbase.conf 中，参数 wal_level 用来确定将多少信息写入 WAL 文件。从低到高的三个级别分别是 minimal、replica、logical。

（1）**wal_level＝minimal**：写入最少的 WAL 文件，此时 WAL 文件只包含从崩溃或者立即关机后，进行恢复所需的 WAL 文件信息。

（2）**wal_level＝replica**：写入足够的数据以支持 WAL 文件归档和复制，包括在备用服务器上运行只读查询。

（3）**wal_level＝logical**：除了包含 replica 日志级别的所有信息，还会附加逻辑解码所需的信息（用于逻辑复制）。

生产数据库一般将日志级别参数 wal_level 设置为 replica。

与 WAL 文件的内容相关的另一个参数是 full_page_write。当前 KingbaseEs 数据库的默认页面大小是 8KB，当操作系统在向磁盘写页面时系统出现故障，就很有可能出现数据库页面前半部分和后半部分数据不一致的情况，造成数据页的损坏。为了解决这个问题，KingbaseES 数据库采用 Full Page Write 机制，即在检查点后对数据页的第一次修改需要在日志文件中记录整个数据页面。Full Page Write 机制增加了系统的可靠性，但是会使日志量增大，从而影响系统的性能，KingbaseES 数据库提供参数 full_page_write 控制是否开启该机制。生产数据库一般设置该参数为 on。

2. 配置 WAL 文件的写盘方式

当事务提交时，对应的日志缓冲区中的日志记录必须写到外存的日志文件中，包括操作系统的缓存也要一并写到外存（称为 sync），以保证事务的原子性和持久性。但是这是非常耗时的操作，当负载很大时，对系统的吞吐量影响很大。

WAL 文件写盘相关的参数会影响系统的性能。

（1）fsync：表示当事务提交时，是否需要把操作系统的缓存也一并写到外存。当设置为 off 时，不需要写出操作系统的缓存，可以提高性能，但是存在一定的风险。系统默认是 on。

（2）wal_sync_method：设置操作系统缓存写盘的方式，可以选的值包括：open_datasync、open_sync、fsync 和 fsync_writethrough。不同的操作系统可能提供不同的方式，有的操作系统会提供多种方式，可以根据应用的负载测试选择合适的写盘方式。默认是 fsync。

3. 配置 WAL 文件的成组提交

成组提交是指在多用户的高并发环境中，在事务提交前日志刷盘的时候，等待并发的其他事务日志一起刷盘，可以减少日志刷盘的次数，从而提高系统的吞吐量。

KingbaseES 数据库提供了两个参数 commit_delay 和 commit_siblings 来设置成组提交的使用。系统在日志刷盘的时候，如果同时有 commit_siblings 个并发的事务存在，则系统会等待 commit_delay 微秒再刷盘。

4. 配置检查点的执行频率

检查点是进行系统故障恢复的起始点，经常执行检查点可以缩短系统崩溃后恢复的时间，提高系统可用性。但是，系统在执行检查点时会将数据缓冲区中所有的脏页面写到外存的数据文件中，会加大系统的 I/O 负载，影响系统的性能。

检查点的执行频率与参数 checkpoint_timeout 和 max_wal_size 有关。

（1）参数 checkpoint_timeout 定义了检查点进程启动的频率，默认值为 5min。

（2）max_wal_size 定义了日志文件的最大长度，默认值为 1GB，当系统的日志文件量到达该值时，会启动检查点进程执行检查点，删除不需要的日志文件。

如果发现系统中检查点的执行已经影响了系统的性能，则可以增加这两个参数的值。

7.7.6 归档日志模式

成功执行一个检查点后，该检查点之间的 WAL 文件不再用于数据库实例恢复，但是这些 WAL 文件需要用于数据库备份恢复。一般来说，生产数据库应该运行在归档文件模式下，这样 WAL 文件在被删除重用之前会被复制到归档日志目录下，归档日志的文件名与 WAL 文件的名字相同。

执行下面的 SQL 语句，查看当前 KingbaseES 数据库的归档设置：

```
system@test=# SELECT name,setting FROM sys_settings WHERE name like 'archive%' or name='wal_level';
          name             |   setting
---------------------------+-------------
 archive_cleanup_command   |
 archive_command           | (disabled)
 archive_dest              | (disabled)
```

```
 archive_mode                    | off
 archive_timeout                 | 0
 wal_level                       | replica
(6 rows)
system@test=#
```

从输出可以看到，当前 KingbaseES 数据库工作在非归档日志模式(archive_mode 的值为 off)。

执行下面的步骤，配置 KingbaseES 数据库运行在归档日志模式下。

(1) 关闭 KingbaseES 数据库：

```
[kingbase@dbsvr ~]$ sys_ctl stop
```

(2) 创建用于保存归档日志的目录：

```
[kingbase@dbsvr ~]$ mkdir -p /u04/Kingbase/ES/V9/archivelog
```

(3) 编辑 KingbaseES 数据库的系统配置文件 kingbase.conf，修改配置文件中关于 WAL 文件和归档设置的参数：

```
#  wal_level 可以取以下的值:minimal, replica,or logical
# 修改 wal_level 的值需要重新启动数据库
# 工作在归档模式下不能设置为 minimal,可以设置为除 minimal 之外的其他参数
wal_level = replica
# 修改 archive_mode 的值需要重新启动数据库
archive_mode=on
# 修改 archive_command 的值不需要重新启动数据库,只需要 reload
archive_command = 'cp %p /u04/Kingbase/ES/V9/archivelog/%f'
# 修改 archive_time:归档周期,900 表示每 900s(15min)切换一次
archive_timeout = 900
```

(4) 执行下面的命令，启动 KingbaseES 数据库：

```
[kingbase@dbsvr ~]$ sys_ctl start
```

(5) 执行下面的命令和 SQL 语句，查看当前 KingbaseES 数据库的归档设置：

```
[kingbase@dbsvr ~]$ ksql -d test -U system
system@test=# SELECT name,setting FROM sys_settings WHERE name like 'archive%' or name='wal_level';
          name            |              setting
--------------------------+------------------------------------
 archive_cleanup_command  |
 archive_command          | cp %p /u04/Kingbase/ES/V9/archivelog/%f
 archive_dest             |
 archive_mode             | on
 archive_timeout          | 900
 wal_level                | replica
(6 rows)
system@test=#
```

从输出可以看到，KingbaseES 数据库已经工作在归档模式下了(archive_mode 的值为 on)。

第 8 章

数据库日常运行监控

KingbaseES 数据库的运维目标是要做到"数据不丢失,服务不间断"。为了达到这个目标,DBA 需要做好 KingbaseES 数据库的日常运行监控。

KingbaseES 数据库服务器的日常运行监控,包括以下 3 部分的内容。

(1) KingbaseES 数据库服务器的运行维护日志。

(2) KingbaseES 数据库服务的器操作系统监控。

(3) KingbaseES 数据库的运行监控。

本章的实验环境可以使用 1.4.5 节制作的虚拟机备份文件 dbserver-3DB-BestPractice.rar。

8.1 数据库服务器的运行维护日志

KingbaseES 数据库的 DBA,每天首先要做的事情,就是阅读 KingbaseES 数据库的运行维护日志,然后针对阅读过程中发现的问题,逐个进行处理。

KingbaseES 数据库的运行维护日志位于数据目录的 sys_log 子目录中(本书安装示例,位于目录/u00/Kingbase/ES/V9/data/sys_log 下)。

KingbaseES 数据库的运行维护日志有如下两种。

(1) KingbaseES 数据库启动日志文件 startup.log。

(2) KingbaseES 数据库运行日志文件 kingbase.conf(文件格式由系统配置参数 log_filename 定义)。

8.1.1 启动日志文件 startup.log

KingbaseES 数据库启动日志文件 startup.log 通常记录了 KingbaseES 数据库启动过程中的日志信息,它对于诊断启动期间的问题至关重要,尤其是在 KingbaseES 数据库无法正常启动的情况下。

文件 startup.log 主要包括以下内容。

(1) **启动序列信息**:记录 KingbaseES 数据库服务启动的时间戳及启动过程中的各个步骤。

(2) **配置文件加载信息**:记录 kingbase.conf 和其他相关配置文件(如 sys_hba.conf 等)的加载情况。

（3）**系统初始化检查信息**：记录对数据目录的检查、共享内存的初始化、控制文件的读取等信息。

（4）**恢复过程信息**：如果数据库在启动时有崩溃恢复或归档恢复，这些信息将被记录在日志中。

（5）**连接和身份验证信息**：记录监听端口、连接、身份验证过程的初始信息。

（6）**错误和警告消息**：如果在启动过程中遇到问题，如配置错误、资源不足、文件无法访问等，相关的错误或警告信息将被记录在日志中。

（7）**版本信息**：记录 KingbaseES 数据库版本信息，有时也包括编译信息。

（8）**检查点信息**：记录 KingbaseES 数据库启动时的第一个检查点信息。

可以使用 vi 编辑器或 cat、more 命令来查看启动日志文件 startup.log。

8.1.2 运行日志文件

KingbaseES 数据库的运行日志文件会记录数据库运行期间产生的各种日志信息，具体内容取决于其他日志相关的配置参数，通常包括以下信息。

（1）**查询日志**：在启用状态下，记录所有或特定类型（如慢查询等）的 SQL 语句。

（2）**错误和警告消息**：记录系统错误、警告及其他重要的系统消息。

（3）**连接和断开信息**：记录客户端连接和断开服务器的消息。

（4）**事务信息**：在启用状态下，记录事务的开始、结束等信息。

（5）**系统事件信息**：记录自动清理事件、检查点创建等信息。

（6）**配置更改信息**：记录 KingbaseES 数据库运行时的配置更改信息。

控制 KingbaseES 数据库运行日志文件的配置参数，可以分为以下 4 类。

1. 日志收集器配置

参数 logging_collector 决定是否启用 KingbaseES 数据库日志收集器进程，其默认值是 on，启动 KingbaseES 数据库后，日志收集器进程会在后台运行，并负责收集和管理 KingbaseES 数据库的日志信息。如果该参数设置为 off，则不会启动日志收集器进程。修改参数 logging_collector 的值后，需要重新启动 KingbaseES 数据库。

2. 日志文件管理

参数 log_directory 用来指定 KingbaseES 数据库日志文件的保存位置，默认保存在 KingbaseES 数据库的数据目录的 sys_log 子目录中。

参数 log_filename 用来指定日志文件的文件名格式。

参数 log_rotation_age 用来控制 KingbaseES 日志文件的切换时间，如 1d 表示每天切换一次。

参数 log_rotation_size 用来控制 KingbaseES 日志文件的大小，如果日志超过该参数设置的大小，将进行日志切换。

3. 日志输出目标

参数 log_destination 用于指定日志的输出对象，它决定了日志消息将被输出到哪个对象上。该参数的值有以下几种。

（1）**stderr**：表示日志消息将被发送给标准错误流。

(2) csvlog：表示日志消息将以 CSV 格式记录到日志文件中。

(3) syslog：表示日志消息将被发送给系统运行日志。

4．日志内容控制

参数 log_min_duration_statement 用于设置一个阈值，其单位是毫秒(ms)，如果执行时间超过这个阈值的 SQL 语句，都应当被记录到日志中。该配置参数的值如下。

(1) －1(默认值)：表示不记录 SQL 语句。

(2) 0：表示记录所有的 SQL 语句。

(3) 大于 0：如 60 000 表示记录执行时间超过 60s 的 SQL 语句。

例 8.1 启用系统的日志收集进程，并配置日志文件的名字、位置、文件的大小及日志切换时间周期。

一般情况下，不需要在运行日志文件中记录执行的 SQL 语句，KingbaseES 数据库的 DBA，只需要使用 vi 编辑器，编辑文件/u00/Kingbase/ES/V9/data/kingbase.conf，配置如下几个参数：

```
logging_collector = on                    #启用 KingbaseES 数据库日志收集进程
log_directory = 'sys_log'                 #日志保存在数据目录的 sys_log 子目录下
log_filename = 'kingbase-%Y-%m-%d_%H%M%S.log'
                                          #日志文件名形如 kingbase-年月日时分秒.log
log_rotation_age = 1d                     #每天切换一次运行日志文件
log_rotation_size = 100MB                 #运行日志文件大于 100MB 时切换一次日志
```

修改完 kingbase.conf 文件后，执行下面的命令，关闭 KingbaseES 数据库：

```
[kingbase@dbsvr ~]$ sys_ctl stop
```

为了更清楚地显示运行日志的情况，执行下面的命令，先删除所有的日志文件，然后重新启动 KingbaseES 数据库服务器：

```
[kingbase@dbsvr ~]$ rm -f /u00/Kingbase/ES/V9/data/sys_log/*
[kingbase@dbsvr ~]$ su -
Password: #输入用户 root 的密码 root123
Last login: Sun Dec 31 12:48:05 CST 2023 on pts/0
[root@dbsvr ~]# reboot
```

使用用户 kingbase，重新登录 KingbaseES 数据库服务器，执行下面的命令查看 KingbaseES 数据库服务器的运行日志：

```
[kingbase@dbsvr ~]$ cd /u00/Kingbase/ES/V9/data/sys_log/
[kingbase@dbsvr sys_log]$ ls
kingbase-2024-02-04_000910.log  startup.log
[kingbase@dbsvr sys_log]$ cat kingbase-2024-02-04_000910.log
2024-02-04 00:09:10.954 CST [1683] LOG:  database system was shut down at 2024-02-04 00:08:29 CST
2024 - 02 - 04 00: 09: 10. 970 CST [1603] LOG:   database system is ready to accept connections
[kingbase@dbsvr sys_log]$
```

例 8.2 在运行日志文件中记录数据库用户的登录、退出及执行的 SQL 语句。

基于例 8.1 的配置，在 kingbase.conf 中配置以下几个参数：

```
log_connections = on           #在日志文件中记录用户登录信息
log_disconnections = on        #在日志文件中记录用户退出信息
log_statement = 'all'          #在日志文件中记录所有的 SQL 语句
```

其中，参数 log_statement 的值还可以是 none（不记录任何 SQL 语句）、ddl（只记录 DDL 语句）、mod（记录 DDL 语句、INSERT 语句、UPDATE 语句、DELETE 语句和 TRUNCATE 语句）。

修改系统参数值后需要重新启动 KingbaseES 数据库。

注意：虽然将参数 log_statement 的值设置为 all，会记录所有的 SQL 语句，但是不会记录 SQL 语句的执行时间。这里记录所有执行过的 SQL 语句，目的是进行安全审计。

例 8.3 在运行日志文件中记录系统的慢 SQL 语句（执行时间超过 60s）。

基于例 8.1 的配置，在 kingbase.conf 中配置如下系统参数值：

```
log_min_duration_statement = 60000    #记录运行时间超过 60s 的 SQL 语句
```

修改系统参数值后需要重新启动 KingbaseES 数据库：

```
[kingbase@dbsvr ~]$ cd /u00/Kingbase/ES/V9/data/sys_log/
[kingbase@dbsvr sys_log]$ sys_ctl stop
[kingbase@dbsvr sys_log]$ sys_ctl start
[kingbase@dbsvr sys_log]$ ls -l
total 12
-rw-------. 1 kingbase dba 2162 Feb  4 00:11 kingbase-2024-02-04_000910.log
-rw-------. 1 kingbase dba  186 Feb  4 00:11 kingbase-2024-02-04_001124.log
-rw-------. 1 kingbase dba  693 Feb  4 00:09 startup.log
[kingbase@dbsvr sys_log]$
```

可以看到重新启动数据库后，新生成了运行日志文件 kingbase-2024-02-04_001124.log。

使用用户 kingbase，执行下面的命令，模拟执行一条慢 SQL 语句：

```
[kingbase@dbsvr ~]$ ksql -U system -d test -c "select sys_sleep(70);"
Password for user system: #输入用户 system 的密码 Passw0rd
#执行中……等待 70s
 sys_sleep
-----------

(1 row)
[kingbase@dbsvr ~]$
```

当上面的语句执行完成之后，执行下面的命令，在 KingbaseES 数据库当前的运行日志中，发现了这条慢 SQL 语句：

```
[kingbase@dbsvr ~]$ cd /u00/Kingbase/ES/V9/data/sys_log
[kingbase@dbsvr sys_log]$ ls
kingbase-2024-02-04_000910.log  kingbase-2024-02-04_001124.log  startup.log
[kingbase@dbsvr sys_log]$ cat kingbase-2024-02-04_001124.log
2024-02-04 00:11:24.695 CST [3439] LOG:  database system was shut down at 2024-02-04 00:11:22 CST
```

```
2024-02-04 00:11:24.698 CST [3437] LOG:  database system is ready to
accept connections
2024-02-04 00:14:57.572 CST [3487] LOG:  duration: 70010.615 ms  statement:
select sys_sleep(70);
[kingbase@dbsvr sys_log]$
```

如果 DBA 想使用另外一个日志文件专门记录慢 SQL 语句，而不是在 KingbaseES 数据库当前的运行日志文件中记录慢 SQL 语句，可以在文件/u00/Kingbase/ES/V9/data/kingbase.conf 中，设置如下系统参数值：

```
log_destination = 'csvlog'              # 将 SQL 语句记录到单独的 CSV 文件
```

然后重新启动 KingbaseES 数据库：

```
[kingbase@dbsvr ~]$ cd /u00/Kingbase/ES/V9/data/sys_log/
[kingbase@dbsvr sys_log]$ sys_ctl stop
[kingbase@dbsvr sys_log]$ sys_ctl start
[kingbase@dbsvr sys_log]$ ls -l
total 20
-rw-------. 1 kingbase dba 2162 Feb  4 00:11 kingbase-2024-02-04_000910.log
-rw-------. 1 kingbase dba 2260 Feb  4 00:25 kingbase-2024-02-04_001124.log
-rw-------. 1 kingbase dba  496 Feb  4 00:25 kingbase-2024-02-04_002555.csv
-rw-------. 1 kingbase dba  166 Feb  4 00:25 kingbase-2024-02-04_002555.log
-rw-------. 1 kingbase dba  693 Feb  4 00:09 startup.log
[kingbase@dbsvr sys_log]$
```

可以看到重新启动数据库后，新生成了运行日志文件 kingbase-2024-02-04_002555.log 和用于记录 SQL 语句的日志文件 kingbase-2024-02-04_002555.csv。

使用用户 kingbase，执行下面的命令，模拟执行一条慢 SQL 语句：

```
[kingbase@dbsvr ~]$ ksql -U system -d test -c "select sys_sleep(80);"
Password for user system: #输入用户 system 的密码 Passw0rd
#执行中……等待 80s
 sys_sleep
-----------
 
(1 row)
[kingbase@dbsvr ~]$
```

当上面的语句执行完成之后，执行下面的命令，在一个单独的 CSV 日志文件中，发现了这条慢 SQL 语句：

```
[kingbase@dbsvr ~]$ cd /u00/Kingbase/ES/V9/data/sys_log
 [kingbase@dbsvr sys_log]$ ls -l
total 20
-rw-------. 1 kingbase dba 2162 Feb  4 00:11 kingbase-2024-02-04_000910.log
-rw-------. 1 kingbase dba 2260 Feb  4 00:25 kingbase-2024-02-04_001124.log
-rw-------. 1 kingbase dba  693 Feb  4 00:29 kingbase-2024-02-04_002555.csv
-rw-------. 1 kingbase dba  166 Feb  4 00:25 kingbase-2024-02-04_002555.log
-rw-------. 1 kingbase dba  693 Feb  4 00:09 startup.log
[kingbase@dbsvr sys_log]$ cat kingbase-2024-02-04_002555.csv
```

```
#省略了一些输出
2024-02-04 00:29:36.219 CST,"system","test",3930,"[local]",65be69aa.f5a,1,"
SELECT",2024-02-04 00:28:26 CST,4/0,0,LOG,00000,"duration: 70011.582 ms
statement: select sys_sleep(80);",,,,,,,,,,"ksql"
[kingbase@dbsvr sys_log]$
```

如果 DBA 想记录所有的 SQL 语句,可以在文件/u00/Kingbase/ES/V9/data/kingbase.conf 中,将系统配置参数 log_min_duration_statement 的值设置为 0:

```
log_min_duration_statement = 0        # 在日志文件中记录所有的 SQL 语句
```

修改完参数后还需要重新启动 KingbaseES 数据库。记录所有运行的 SQL 语句会产生大量的运行日志,一般情况下不需要这样配置。

注意:将 log_min_duration_statement 的值设置为 0,会记录所有的 SQL 语句及它们的执行时间,使用这个系统配置参数值的主要目的是进行性能分析与优化。

例 8.4 在运行日志文件中记录检查点操作。

基于例 8.1 的配置,在 kingbase.conf 中配置如下系统参数值:

```
log_checkpoints = on                  # 在日志文件中记录检查点
```

修改系统参数值后需要重新启动 KingbaseES 数据库。

例 8.5 在运行日志文件中记录超过死锁时间的锁等待事件。

基于例 8.1 的配置,在 kingbase.conf 中配置如下系统参数值:

```
log_lock_waits = on                   # 在日志文件中记录死锁信息
```

修改系统参数值后需要重新启动 KingbaseES 数据库。

8.2 数据库服务器的操作系统监控

由于服务器的内存、磁盘 I/O 和 CPU 这三者相互影响,因此任何一个过载,都会影响到其他两个。如果服务器内存过载,会引起内存和交换区之间产生额外的磁盘 I/O,在内存严重过载的情况下,系统一直在产生这些无用的磁盘 I/O,会引起服务器 CPU 产生磁盘 I/O 等待。因此,首先应该合理规划数据库服务器内存的分配,监控服务器内存的使用,确保 KingbaseES 数据库配置不会导致服务器内存过载,然后监控服务器的每个磁盘 I/O 的情况,最后监控 CPU。

在 KingbaseES 数据库服务器中,过量的磁盘 I/O 通常是由一些有性能问题的 SQL 语句所导致的,这需要 DBA 对这些有问题的 SQL 语句进行调优。如果数据库服务器不存在内存过载和磁盘 I/O 过载等问题,但是 CPU 的占比还比较高时,那就需要调整应用,来降低 CPU 的使用量或者为服务器增加更多的 CPU。

8.2.1 本节实验环境说明

KingbaseES 数据库用户不可以在生产环境进行 8.2 节的实战任务。使用在 1.4.5 节制作的虚拟机备份文件 dbserver-3DB-BestPractice.rar,将其解压缩在机械硬盘(hard disk drive,HDD)上并运行。在学习如何判断服务器的磁盘 I/O 过载时,虽然 SSD 可以承受很大的磁盘 I/O 压力，但是强行加大磁盘 I/O 负载，可能会损坏 SSD。

为了展示服务器内存、磁盘 I/O 和 CPU 的过载情况，可以使用内存测试工具 memtester 和负载压力测试工具 stress-ng。

memtester 是一个用于内存测试的开源工具，通过在内存中填充特定的模式和数据，检查读取的结果是否与预期一致，从而检测出内存中的错误和故障。

memtester 可以模拟不同类型的内存访问模式。

（1）**自动模式**：在整个可用内存范围内进行自动化的测试。

（2）**边界模式**：测试内存边界情况，如边界地址和边界对齐。

（3）**循环模式**：循环执行测试，连续多次对内存进行读写操作。

（4）**随机模式**：随机访问内存的不同地址，用以模拟内存真实的使用情况。

使用用户 root 下载并安装内存测试工具 memtester：

```
cd
wget https://pyropus.ca./software/memtester/old-versions/memtester-4.5.1.tar.gz --no-check-certificate
tar xzf memtester-4.5.1.tar.gz
cd memtester-4.5.1/
make
make install
```

stress-ng 是一个用于操作系统负载压力测试的开源工具，用于模拟 CPU、内存、硬盘、网络等多方面的负载，它提供了实时的结果输出和统计信息，帮助用户评估系统在不同条件下的稳定性和性能。

stress-ng 提供了丰富的压力测试模式。

（1）**CPU 负载模式**：用于测试 CPU 的计算能力和负载容量。

（2）**内存负载模式**：用于测试系统的内存使用和管理能力。

（3）**磁盘负载模式**：用于测试磁盘读写性能和文件系统的稳定性。

（4）**I/O 负载模式**：用于测试系统的输入/输出性能，包括文件操作、管道操作等。

（5）**网络负载模式**：用于测试网络带宽、连接数和网络协议的稳定性。

使用用户 root 下载并安装开源压力测试工具 stress-ng：

```
cd
wget http://ftp.ubuntu.com/ubuntu/ubuntu/pool/universe/s/stress-ng/stress-ng_0.13.12.orig.tar.xz
xz -dk stress-ng_0.13.12.orig.tar.xz
tar xf stress-ng_0.13.12.orig.tar
cd stress-ng-0.13.12/
make
make install
```

在 KingbaseES 数据库系统中，运行 TPC-H 测试脚本 tpchtest，仿真一个活动的 KingbaseES 数据库，让这个脚本运行 15min 以上，预热操作系统和数据库。

```
[kingbase@dbsvr ~]$ cd /home/kingbase/tpch_sql
[kingbase@dbsvr tpch_sql]$ sh tpchtest
#持续不断地运行 TPC-H 测试 SQL 语句……
```

8.2.2 监控服务器内存

操作系统提供了 top 和 vmstat 命令来监控服务器内存和 CPU 的使用情况。通过下面的例子来学习如何判断一个系统是否处于严重的内存过载状态。

例 8.6 观察处于严重内存过载的系统状态。

(1) 在 KingbaseES 数据库的服务器上,打开第 1 个终端,执行 top 命令查看当前系统的性能情况,尤其是关注以下两行关于系统内存的使用信息:

```
KiB Mem : 16247600 total,   9822336 free,    1626912 used,   4798352 buff/cache
KiB Swap: 67108860 total, 67108860 free,            0 used. 12234492 avail Mem
```

可以看到,当前系统的总内存为 16 247 600KB(约为 15.5GB),空闲内存为 9 822 336KB(不到 9.8GB),已使用内存为 1 626 912KB(约为 1.5GB),用于缓冲区/高速缓存的内存为 4 798 352KB(约为 4.5GB),还有 12 234 492KB(约为 11.6GB)的可用内存。

(2) 打开第 2 个终端,使用用户 root 执行 memtester 命令,消耗所有的可用内存:

```
[root@dbsvr ~]# memtester 12234492KB
memtester version 4.5.1 (64-bit)
Copyright (C) 2001-2020 Charles Cazabon.
Licensed under the GNU General Public License version 2 (only).

pagesize is 4096
pagesizemask is 0xfffffffffffff000
want 11947MB (12528119808 bytes)
got  11947MB (12528119808 bytes), trying mlock ...locked.
Loop 1:
  Stuck Address       : setting   9
#运行中……
```

(3) 回到第 1 个终端,观察 top 命令的输出中关于系统内存的使用信息:

```
KiB Mem : 16247600 total,    521300 free, 13519004 used,   2207296 buff/cache
KiB Swap: 67108860 total, 66704012 free,    404848 used.    418792 avail Mem
```

可以看到,当前系统的空闲内存为 521 300KB(约为 509MB),已经使用内存为 13 519 004KB(约为 12.8GB),用于缓冲区/高速缓存的内存为 2 207 296KB(约为 2.1GB),还有 418 792KB(约为 408MB)可用内存。此时,服务器的内存趋近满载的状态。

(4) 打开第 3 个终端,使用用户 root 执行 stress-ng 命令,产生 2 个内存分配进程,每个进程分配 2048MB 内存,持续 600s:

```
[root@dbsvr ~]# stress-ng --vm 2 --vm-bytes 2048m --vm-keep --timeout 600
```

(5) 打开第 4 个终端,使用 root 用户执行下面的 vmstat 命令:

```
[root@dbsvr ~]# vmstat 5 5
procs -----------memory---------- ---swap------- -----io---- -system---- ------cpu-----
 r  b   swpd   free   buff  cache   si   so    bi    bo   in   cs us sy id wa st
 9  3 2417380 170764      0 1720312  389  435  1458  2057  688  718 41  8 51  0  0
 5  4 2688484 152720      0 1741092 197099 90014 229878 93578 22959 22054 33 49  1 17  0
 4  3 2783204 162424      0 1926160 114967 36483 207768 109708 13000 13405 50 44  0  6  0
```

```
     6   3 2680804 200332          0 1873628 171205 32857 257123 37635    17008 18674 43 44    0 12   0
     6   3 2285284 173644          0 1914940 121826 47939 217780 69086    14976 14879 49 42    0  8   0
```

可以看到，swap 的 so 列值不为 0 且比较大，表示系统存在大量的换出操作（将内存中的页面写入交换区上，以便腾出内存），这是系统严重缺少内存的主要标志；swap 的 si 列的值，表示系统存在换入操作（将交换区中的页面读回内存），可以用来辅助判断系统是否缺乏内存，如果 so 列的值持续较大，且伴随着 si 列的值也比较大，那么说明系统当前内存过载。

（6）回到第 1 个终端，查看 top 命令关于系统内存的信息：

```
KiB Mem : 16247600 total,    154180 free, 14127012 used,   1966408 buff/cache
KiB Swap: 67108860 total, 64856804 free,   2252056 used.    38036 avail Mem
```

可以看到，即使系统工作在极度缺乏内存的状况下，也会维持一定大小的可用内存（此处为 38 036KB）。

（7）返回到运行 stress-ng 命令的第 3 个终端，按 Ctrl+C 组合键终止命令的运行。

（8）返回到运行 memtester 命令的第 2 个终端，按 Ctrl+C 组合键终止命令的运行。

8.2.3　监控服务器磁盘 I/O

操作系统提供了 iostat 命令来监控服务器磁盘 I/O 的使用情况。通过下面的例子来学习如何判断一个系统是否处于严重的磁盘 I/O 过载状态。

例 8.7　观察处于严重磁盘 I/O 过载的系统状态。

（1）打开第 1 个终端窗口，使用 root 用户执行 stress-ng 命令，产生 16 个随机读写的磁盘 I/O，每个磁盘 I/O 在目录/u00/temp 下写 10GB，持续 600s：

```
root@dbsvr ~]# mkdir -p /u00/temp
root@dbsvr ~]# stress-ng --hdd 16 --hdd-opts "wr-rnd" --hdd-bytes 10G --
              timeout 600 --temp-path /u00/temp
```

（2）打开第 2 个终端窗口，执行下面的命令：

```
[root@dbsvr ~]# iostat -xk -d /dev/sdb 5 5
Linux 3.10.0-1160.el7.x86_64 (dbsvr)    02/22/2024    _x86_64_    (4 CPU)
Device:          rrqm/s   wrqm/s     r/s     w/s    rkB/s    wkB/s avgrq-sz avgqu-sz   await r_await w_await  svctm  %util
sdb                0.07    12.67   14.88   62.06  2991.49  9391.00   321.85     3.40   44.05    1.38   54.28   0.50   3.85

Device:          rrqm/s   wrqm/s     r/s     w/s    rkB/s    wkB/s avgrq-sz avgqu-sz   await r_await w_await  svctm  %util
sdb                0.00   150.40    0.00 1187.80     0.00 132165.10   222.54   140.39  118.47    0.00  118.47   0.84 100.02

Device:          rrqm/s   wrqm/s     r/s     w/s    rkB/s    wkB/s avgrq-sz avgqu-sz   await r_await w_await  svctm  %util
sdb                0.00   170.20    0.00 1111.80     0.00 120860.20   217.41   141.17  119.68    0.00  119.68   0.90  99.90

Device:          rrqm/s   wrqm/s     r/s     w/s    rkB/s    wkB/s avgrq-sz avgqu-sz   await r_await w_await  svctm  %util
sdb                0.00   122.00    0.00 1255.60     0.00 148964.10   237.28   140.48  119.65    0.00  119.65   0.80  99.96

Device:          rrqm/s   wrqm/s     r/s     w/s    rkB/s    wkB/s avgrq-sz avgqu-sz   await r_await w_await  svctm  %util
sdb                0.00   188.40    0.00  907.20     0.00 143918.00   317.28   139.91  120.25    0.00  120.25   1.09  99.30
```

可以看到，磁盘/dev/sdb 的 **%util** 的值超过 85%（忽略第 1 个输出，该值是自开机以来的平均值），表示磁盘/dev/sdb 的服务能力已经接近饱和。**avgrq-sz** 的值是磁盘 I/O 请求队列的长度，该列有值的话，表示一直有磁盘 I/O 请求等待。磁盘设备/dev/sdb 当前基本处于过载的状态。

注意：这里的测试结果是基于 HDD 环境，如果是基于 SSD 环境，%util 的值会低于 90%。

（3）打开第 3 个终端窗口，执行下面的 vmstat 命令：

```
[root@dbsvr ~]# vmstat 5 5
procs --------memory---------- ---swap-- -----io---- -system-- ------cpu-----
 r  b   swpd    free    buff    cache   si   so    bi     bo     in    cs us sy id wa st
 6  6 794908 10723728       0 4459540 1471 1122  2617   7750    656   510  5  6 87  2  0
 4  6 794908 9692700        0 5489336    2    0     2 601182   8859  6334 22 55  2 21  0
 4  4 794908 9606892        0 5572096    0    0     0 527311   8727  6538 22 64  1 13  0
 6  6 794652 9723356        0 5457208   31    0    31 533857   8946  6604 23 58  1 17  0
 5  6 794652 9374772        0 5806444    0    0     0 554167   8774  6030 22 59  1 18  0
```

可以看到，在 procs 的 b 列，有不少进程长时间阻塞在磁盘 I/O 操作上，因此，可以判断系统目前处于磁盘 I/O 过载状态。

（4）当磁盘 I/O 过载的时候，常会伴随着大量的 CPU 消耗在磁盘 I/O 等待上。在命令 stress-ng 执行期间，打开第 4 个终端，执行 top 命令并查看 top 命令的输出：

```
%Cpu0  : 18.1 us, 55.7 sy,  0.0 ni,  3.4 id, 22.5 wa,  0.0 hi,  0.3 si,  0.0 st
%Cpu1  : 15.4 us, 61.1 sy,  0.0 ni,  3.4 id, 19.5 wa,  0.0 hi,  0.7 si,  0.0 st
%Cpu2  : 19.2 us, 38.1 sy,  0.0 ni,  5.8 id, 28.8 wa,  0.0 hi,  8.1 si,  0.0 st
%Cpu3  : 22.1 us, 49.3 sy,  0.0 ni,  1.7 id, 25.8 wa,  0.0 hi,  1.0 si,  0.0 st
```

可以看到，当前有大量的 CPU 在等待磁盘 I/O 完成操作。

8.2.4 监控服务器 CPU

通过下面的例子来学习如何判断一个系统是否处于严重的 CPU 过载状态。

例 8.8 观察处于严重 CPU 过载的系统状态。

（1）打开第 1 个终端，使用用户 root 执行 stress-ng 命令，产生 8 个 CPU 压力负载，持续 600s：

```
[root@dbsvr ~]# stress-ng --cpu 8 --cpu-method all   --timeout 600
```

（2）打开第 2 个终端，使用用户 root 执行 vmstat 命令：

```
[root@dbsvr ~]# vmstat 5 5
procs --------memory---------- ---swap-- -----io---- -system-- ------cpu-----
 r  b   swpd   free    buff    cache   si   so   bi    bo    in   cs us sy id wa st
 9  0  82176 3048332  12432 133128  3080 2426 3184 11807  881 1123  8  5 85  2  0
 8  0  82176 3047740  12432 133292    0    0    0  4008  481  100  0  0  0  0  0
 8  0  82176 3047740  12440 133292    0    0    2  4012  485  100  0  0  0  0  0
 8  0  82176 3047616  12440 133448    0    0    1  4031  493  100  0  0  0  0  0
10  0  82176 3047616  12448 133448    0    0    2  3596  463  100  0  0  0  0  0
```

可以看到，就绪队列中等待运行的进程/线程数（procs 中的 r 的值），如果长时间比系统 CPU 的核心数（测试环境为 4 核）大，说明系统处于缺乏 CPU 的状态。

（3）在命令 stress-ng 执行期间，打开第 3 个终端，执行 top 命令：

```
[root@dbsvr ~]# top
top - 17:53:28 up  1:07,  5 users,  load average: 7.62, 2.50, 1.66
```

```
Tasks: 374 total,   11 running, 363 sleeping,    0 stopped,    0 zombie
%Cpu0  :100.0 us,   0.0 sy,   0.0 ni,   0.0 id,   0.0 wa,   0.0 hi,   0.0 si,   0.0 st
%Cpu1  : 99.2 us,   0.8 sy,   0.0 ni,   0.0 id,   0.0 wa,   0.0 hi,   0.0 si,   0.0 st
%Cpu2  :100.0 us,   0.0 sy,   0.0 ni,   0.0 id,   0.0 wa,   0.0 hi,   0.0 si,   0.0 st
%Cpu3  :100.0 us,   0.0 sy,   0.0 ni,   0.0 id,   0.0 wa,   0.0 hi,   0.0 si,   0.0 st
KiB Mem : 16247592 total, 13417924 free,   1648596 used,   1181072 buff/cache
KiB Swap: 67108860 total, 67108860 free,         0 used. 13987864 avail Mem
    PID USER      PR  NI    VIRT    RES    SHR S  %CPU %MEM     TIME+ COMMAND
   5835 root      20   0   52204   4564   1320 R  50.8  0.0   0:47.41 stress-ng
   5834 root      20   0   52204   4564   1320 R  50.0  0.0   0:47.06 stress-ng
   5836 root      20   0   52204   4564   1320 R  50.0  0.0   0:47.31 stress-ng
   5837 root      20   0   52204   4564   1320 R  50.0  0.0   0:47.48 stress-ng
   5839 root      20   0   52204   4564   1320 R  50.0  0.0   0:47.37 stress-ng
```

可以看到，当前 CPU 已经没有空闲了（0.0 id），主要是 stress-ng 造成的 CPU 过载。可以根据下面的经验值来判断 CPU 的负载状态。

（1）CPU 轻载：CPU 的空闲百分比在 80% 左右。

（2）CPU 正常负载：CPU 的空闲百分比在 40% 左右。

（3）CPU 满载：CPU 的空闲百分比在 20% 左右。

（4）CPU 过载：CPU 的空闲百分比低于 10%。

8.2.5　监控服务器文件系统

每天 DBA 首先要做的事情，就是执行下面的 df 命令，查看 KingbaseES 数据库服务器的文件系统空间使用情况：

```
[root@dbsvr ~]# df -h
Filesystem                    Size  Used Avail Use% Mounted on
#省略了一些输出
/dev/mapper/centos-root       10G   104M  9.9G   2% /
/dev/mapper/centos-usr        20G   5.9G   15G  30% /usr
/dev/sda1                     10G   187M  9.9G   2% /boot
/dev/mapper/bakvg-baklv      800G    33M  800G   1% /dbbak
/dev/mapper/centos-tmp        10G    34M   10G   1% /tmp
/dev/mapper/centos-var        40G   2.0G   39G   5% /var
/dev/mapper/centos-home       40G    39M   40G   1% /home
/dev/mapper/centos-opt        40G   357M   40G   1% /opt
/dev/mapper/dbvg-u00lv       200G   5.7G  195G   3% /u00
/dev/mapper/dbvg-u03lv       200G    33M  200G   1% /u03
/dev/mapper/dbvg-u01lv       200G    33M  200G   1% /u01
/dev/mapper/dbvg-u02lv       200G   1.8G  199G   1% /u02
/dev/mapper/archvg-u04lv     800G   2.3G  798G   1% /u04
```

当发现某个文件系统的使用率（Use% 的值）超过了 80%，需要进一步确定造成文件系统高使用率的原因。在 KingbaseES 数据库服务器上，造成文件系统使用率比较高的常见原因如下。

（1）表膨胀导致文件系统的使用率异常升高；

（2）未及时清除无用的数据库备份，导致备份所在的文件系统使用率升高；

(3)未能及时释放 WAL 日志文件所占的空间,导致所在的文件系统使用率异常升高,造成这种情况的原因可能是 wal_keep_size 参数配置过大、归档配置不当(归档配置参数 archive_mode 和 archive_command)、频繁的大事务、主备集群的备机故障没能及时恢复等;

(4)业务正常进行中,业务数据的自然增长。

原因(1)导致的文件系统高使用率,处理方式请参看本书 7.4 节的内容。原因(2)导致的文件系统高使用率,请将过期的备份删除。原因(3)导致的文件系统高使用率,请尽快找到 WAL 日志文件不能及时归档的原因,并作出相应的处理。

下面的例 8.9,通过进行文件系统扩容,处理原因(4)导致的文件系统高使用率的问题。

例 8.9 处理因业务增长导致的文件系统磁盘空间高使用率的问题。

首先,执行下面的命令,在/u00/Kingbase/ES/V9 目录下,创建一个大小为 180GB 的文件(模仿 KingbaseES 数据库的数据存储量增长),然后,查看文件系统/u00 的磁盘空间使用率:

```
[root@dbsvr ~]# dd if=/dev/zero of=/u00/Kingbase/ES/V9/noUseBigFile bs=1G count=180
150+0 records in
150+0 records out
161061273600 bytes (161 GB) copied, 125.593 s, 1.3 GB/s
[root@dbsvr ~]# df -h /u00
Filesystem              Size  Used Avail Use% Mounted on
/dev/mapper/dbvg-u00lv  200G  188G   13G  94% /u00
```

可以看到,目前文件系统/u00 的存储空间使用率高达 94%,超过了警戒线 80%,只有 13GB 的可用空间。

执行下面的步骤可以为文件系统/u00 增加 200GB 的存储空间。

(1)查看所有的逻辑卷的情况:

```
[root@dbsvr ~]# lvs
  LV    VG    Attr       LSize   Pool Origin Data%  Meta%  Move Log Cpy%Sync Convert
  u04lv archvg -wi-ao---- 800.00g
  baklv bakvg  -wi-ao---- 800.00g
#省略了一些输出
  u00lv dbvg   -wi-ao---- 200.00g
  u01lv dbvg   -wi-ao---- 200.00g
  u02lv dbvg   -wi-ao---- 200.00g
  u03lv dbvg   -wi-ao---- 200.00g
```

可以看到,逻辑卷 u00lv 属于卷组 dbvg,大小是 200GB。

(2)查看卷组 dbvg 的详细情况:

```
[root@dbsvr ~]# vgdisplay dbvg
  --- Volume group ---
  VG Name               dbvg
#省略了一些输出
  VG Size               1.17 TiB
  PE Size               4.00 MiB
```

```
Total PE              307199
Alloc PE / Size       204800 / 800.00 GiB
Free  PE / Size       102399 / <400.00 GiB
VG UUID               KxlaRK-t20J-HwFZ-uWXl-fKsP-jicm-shD6Dt
```

可以看到，卷组 dbvg 现在还有 400GB 的空闲空间。

（3）执行下面的命令扩大逻辑卷 u00lv 的大小，扩大 200GB：

```
[root@dbsvr ~]# lvextend -L +200G dbvg/u00lv
  Size of logical volume dbvg/u00lv changed from 200.00 GiB (51200 extents) to 400.00 GiB (102400 extents).
  Logical volume dbvg/u00lv successfully resized.
```

（4）执行下面的命令扩展逻辑卷 u00lv 所在的文件系统：

```
[root@dbsvr ~]# xfs_growfs /u00
meta-data=/dev/mapper/dbvg-u00lv isize=512    agcount=4, agsize=13107200 blks
         =                       sectsz=512   attr=2, projid32bit=1
         =                       crc=1        finobt=0 spinodes=0
data     =                       bsize=4096   blocks=52428800, imaxpct=25
         =                       sunit=0      swidth=0 blks
naming   =version 2              bsize=4096   ascii-ci=0 ftype=1
log      =internal               bsize=4096   blocks=25600, version=2
         =                       sectsz=512   sunit=0 blks, lazy-count=1
realtime =none                   extsz=4096   blocks=0, rtextents=0
data blocks changed from 52428800 to 104857600
```

（5）执行下面的命令确认已经成功地扩大了逻辑卷 u00lv：

```
[root@dbsvr ~]# df -h /u00
Filesystem              Size  Used Avail Use% Mounted on
/dev/mapper/dbvg-u00lv  400G  188G  213G  47% /u00
```

可以看到，文件系统 /u00 扩大 200GB 后，存储空间使用率为 47%。

8.3 数据库的运行监控

8.3.1 监控会话

例 8.10 查看当前的会话数。

在业务高峰期间，可以执行下面的 SQL 语句统计当前 KingbaseES 数据库的会话连接总数：

```
[kingbase@dbsvr ~]$ ksql -d test -U system
system@test=# SELECT COUNT(*) AS total_sessions FROM sys_stat_activity;
 total_sessions
----------------
              9
(1 row)
```

如果业务高峰期间，上述查询得到的值与系统参数 max_connections 的值比较接近，

DBA 需要在合适的时机调大这个参数的值到一个安全的水平,并且需要监控操作系统内存一段时间,确保 KingbaseES 数据库服务器的内存不会因为增加了会话连接数而导致内存过载(参见 8.2.2 节)。

按数据库和用户统计业务高峰期间的会话连接总数,可以执行下面的 SQL 查询:

```
system@test=# SELECT datname, usename, COUNT(*) AS session_count
test-#          FROM sys_stat_activity
test-#          GROUP BY datname, usename
test-#          ORDER BY session_count DESC;
 datname | usename | session_count
---------+---------+---------------
         |         |       4
  test   | system  |       3
         | system  |       2
(3 rows)
```

其中:

(1) 有 4 个会话不关联特定的数据库用户和数据库,这是由后台进程或系统进程产生的会话;

(2) 有 3 个会话是数据库用户 system 连接访问数据库 test;

(3) 有 2 个会话是数据库用户 system 执行一些不需要特定数据库上下文的操作。

请不要退出这个会话,等待 10min,继续下面的例 8.11。

例 8.11 查看空闲超过 10min 的会话。

打开一个新的终端,执行下面的 ksql 命令和 SQL 语句:

```
[kingbase@dbsvr ~]$ ksql -d test -U system
system@test=# SELECT pid,now() - query_start AS duration,query,state
test-#          FROM sys_stat_activity
test-#          WHERE state = 'idle' AND now() - query_start > interval '10 minutes';
  pid  |         duration          |                     query                      | state
-------+---------------------------+------------------------------------------------+-------
  3659 | +000000000 00:11:06.065988000 | SELECT datname, usename, COUNT(*) AS session_count+| idle
       |                           | FROM sys_stat_activity                         +|
       |                           | GROUP BY datname, usename                      +|
       |                           | ORDER BY session_count DESC;                   |
(1 row)
```

所有的空闲时间超过 10min 的会话都是无用会话,可用使用 5.2.3 节的方法终止这些会话,还可以执行下面的 SQL 语句终止所有空闲时间超过 10min 的会话:

```
system@test=# SELECT sys_terminate_backend(pid)
test-#          FROM sys_stat_activity
test-#          WHERE state = 'idle' AND now() - query_start > INTERVAL '10 minutes';
 sys_terminate_backend
-----------------------
 t
(1 row)
```

8.3.2 监控长时间的活动事务

如果 KingbaseES 数据库上有持续时间很长的活动事务,将阻碍 KingbaseES 数据库系统通过事务 ID 冻结操作来回收事务 ID 的执行。如果这个事务持续时间足够长,而且数据库恰巧是一个高负载、高事务速率的环境,这有可能会导致系统中最新的事务和最老的事务年龄差$\geqslant 2^{31}$,最终导致系统发生事务 ID 的回卷问题。

作为预防,DBA 每天要检查 KingbaseES 数据库中有哪些正在运行的长事务。

例 8.12 监控持续时间很长的事务,并提交该事务。

(1) 在第 1 个终端,执行 ksql 命令并在 ksql 中开始一个事务:

```
[kingbase@dbsvr ~]$ ksql -d test -U system
system@test=# BEGIN;
system@test=# SELECT * from tpch.orders LIMIT 1;
#省略了所有的输出
system@test=#
```

(2) 请等待至少 1min,在第 2 个终端,执行下面的 ksql 命令和 SQL 语句,查看在哪个会话(对应 1 个后端服务进程)上,有持续时间超过 1min 的活动事务:

```
[kingbase@dbsvr ~]$ ksql -d test -U system
system@test=# SELECT pid,now() - xact_start AS duration,state
test-#          FROM sys_stat_activity
test-#          WHERE xact_start IS NOT NULL AND
test-#              now() - xact_start > interval '1 minutes' AND
test-#              state IN ('active', 'idle in transaction')
test-#          ORDER BY duration DESC;
  pid  |           duration            |        state
-------+-------------------------------+---------------------
 50506 | +000000000 00:14:30.175783000 | idle in transaction
(1 row)
system@test=#
```

可以看到,在 pid 为 50506 的后端服务进程的会话连接上,有一个持续时间超过 14min 的事务,该事务目前处于空闲状态(idle in transaction),但是没有结束。

(3) 回到第 1 个终端,结束 ksql 中的事务:

```
system@test=# END;
```

(4) 在第 2 个终端,再次执行(2)中的 SQL 语句,查看在哪个会话(对应 1 个后端服务进程)上有持续时间超过 1min 的活动事务:

```
system@test=# SELECT pid,now() - xact_start AS duration,state
test-#          FROM sys_stat_activity
test-#          WHERE xact_start IS NOT NULL AND
test-#              now() - xact_start > interval '1 minutes' AND
test-#              state IN ('active', 'idle in transaction')
test-#          ORDER BY duration DESC;
pid | duration | state
----+----------+-------
```

```
(0 rows)
system@test=#
```

可以看到，在后端服务进程的会话连接上已经没有任何持续时间超过 1min 的活动事务了。

还可以通过终止执行时间很长的事务所在的后端服务进程来终止这个长事务。知道该长事务的后端服务进程的 pid，可以执行下面的 SQL 语句来终止这个后端服务进程：

```
SELECT sys_terminate_backend(pid) FROM sys_stat_activity
```

还可以执行下面的 SQL 语句，终止所有持续时间超过 2h 的长事务所在的后端服务进程：

```
SELECT sys_terminate_backend(pid)
FROM sys_stat_activity
WHERE state IN ('active') AND now() - xact_start > INTERVAL '2 hours';
```

8.3.3 监控长时间运行的 SQL 语句

很多时候，DBA 需要查看当前 KingbaseES 数据库有哪些运行时间很长的 SQL 语句，可以通过配置运行日志（参见 8.1.2 节）来找到运行时间很长的 SQL 语句，或者使用 9.2 节提到的 KWR 工具。

下面示例使用 SQL 语句来发现当前系统中正在运行的，并且运行了很长时间的 SQL 语句。

（1）在第 1 个终端，执行 ksql 命令并在 ksql 中运行 SQL 语句，删除索引：

```
[kingbase@dbsvr ~]$ ksql -d test -U system
system@test=# SET search_path TO tpch;
system@test=# DROP INDEX lineitem_l_partkey_idx;
```

删除索引后，将导致接下来运行的 SQL 语句需要运行非常长的时间。

（2）在第 1 个终端，执行 ksql 命令并在 ksql 中运行 SQL 语句：

```
system@test=# \timing
system@test=# select sum(l_extendedprice) / 7.0 as avg_yearly
test-#          from lineitem,part
test-#          where p_partkey = l_partkey and p_brand = 'Brand#45' and
test-#                p_container = 'LG PKG' and l_quantity < (
test(#                                          select 0.2 * avg(l_quantity)
test(#                                          from lineitem
test(#                                          where l_partkey = p_partkey
test(#                                          );
#语句执行中……
```

（3）请等待至少 1min，在第 2 个终端，执行下面的 ksql 命令和 SQL 语句，查看当前 KingbaseES 数据库上有哪些 SQL 语句正在运行，并且运行时间超过了 1min：

```
[kingbase@dbsvr ~]$ ksql -d test -U system
system@test=# SELECT pid, now() - query_start AS duration,query,state
test-#         FROM sys_stat_activity
```

```
test-#           WHERE now() - query_start > interval '1 minutes' AND state = 'active';
  pid  |         duration           |                      query                                    | state
-------+----------------------------+---------------------------------------------------------------+-------
 13066 | +000000000 00:04:13.096289000 | select sum(l_extendedprice) / 7.0 as avg_yearly            +| active
       |                            | from lineitem,part                                           +|
       |                            | where p_partkey = l_partkey and p_brand = 'Brand#45' and     +|
       |                            |   p_container = 'LG PKG' and l_quantity < (                  +|
       |                            |       select 0.2 * avg(l_quantity)                           +|
       |                            |       from lineitem                                          +|
       |                            |       where l_partkey = p_partkey                            +|
       |                            |   );                                                          |
(1 row)
system@test=#
```

可以看到,在 pid 为 13066 的后端服务进程的会话连接上,正在运行的 SQL 查询语句,已经运行了 4 分 13 秒,并且还在继续运行(state 的值为 active)。

如果不想让这条语句继续运行,可用使用 5.2.3 节中例 5.12 的方法,取消会话中运行的 SQL 语句,或者使用例 5.13 的方法,直接终止运行 SQL 语句的会话连接。

8.3.4 监控锁

需要构建一个锁监控的测试环境,步骤如下。

(1) 执行下面的命令,创建一个测试表:

```
[kingbase@dbsvr ~]$ ksql -d test -U system -c "CREATE TABLE locktest(col1 int, col2 int);"
```

(2) 打开第 1 个终端,执行下面的 ksql 命令和 SQL 语句,开始一个事务:

```
[kingbase@dbsvr ~]$ ksql -d test -U system
system@test=# BEGIN;
system@test=# lock table locktest in Access Exclusive mode;
LOCK TABLE
system@test=#
```

(3) 打开第 2 个终端执行下面的 SQL 语句:

```
[kingbase@dbsvr ~]$ ksql -d test -U system
system@test=# BEGIN;
system@test=# lock table locktest in Access share mode;
#语句阻塞中……
```

(4) 打开第 3 个终端执行下面的 SQL 语句:

```
[kingbase@dbsvr ~]$ ksql -d test -U system
system@test=# BEGIN;
system@test=# lock table locktest in exclusive mode;
#语句阻塞中……
```

(5) 打开第 4 个终端执行下面的 SQL 语句:

```
[kingbase@dbsvr ~]$ ksql -d test -U system
system@test=# BEGIN;
system@test=# lock table locktest in Access Exclusive MODE;
#语句阻塞中……
```

现在创建好了测试环境,可以开始在 KingbaseES 数据库上完成锁监控任务了。

例 8.13 查看哪些进程没有获得锁,正在阻塞中。

在第 5 个终端上,执行下面的 ksql 命令和 SQL 语句:

```
[kingbase@dbsvr ~]$ ksql -d test -U system
system@test=# SELECT pid , locktype , relation::regclass, virtualxid, transactionid,
mode, granted
test-#        FROM sys_locks where granted is false
test-#        ORDER BY pid;
  pid  | locktype | relation | virtualxid | transactionid |        mode         | granted
-------+----------+----------+------------+---------------+---------------------+---------
 69362 | relation | locktest |            |               | AccessShareLock     | f
 69364 | relation | locktest |            |               | ExclusiveLock       | f
 69366 | relation | locktest |            |               | AccessExclusiveLock | f
(3 rows)
```

可以看到,当前有 3 个进程持有表级锁。

例 8.14 查看未获得锁的进程被谁阻塞。

在第 5 个终端上,执行下面的 ksql 命令和 SQL 语句:

```
system@test=# SELECT pid, sys_blocking_pids(pid) as blocked_by ,locktype ,
test-#              relation::regclass,virtualxid, transactionid, mode
test-#        FROM  sys_locks
test-#        WHERE granted is false
test-#        ORDER BY  pid;
  pid  |    blocked_by     | locktype | relation | virtualxid | transactionid |        mode
-------+-------------------+----------+----------+------------+---------------+---------------------
 69362 | {69360}           | relation | locktest |            |               | AccessShareLock
 69364 | {69360}           | relation | locktest |            |               | ExclusiveLock
 69366 | {69360,69362,69364}| relation | locktest |            |               | AccessExclusiveLock
(3 rows)
system@test=#
```

可以看到,pid 为 69362 和 pid 为 69364 的后端服务进程,都被 pid 为 69360 的后端服务进程阻塞,pid 为 69366 的后端服务进程被 pid 为 69300、69362 和 69364 的后端服务进程阻塞。因此,当前阻塞的源头是进程 39360 的后端服务进程。

例 8.15 查看未获得封锁的进程被阻塞的语句。

```
system@test=# SELECT pl.pid, unnest(sys_blocking_pids(pl.pid)) as blocked_by ,
test-#               locktype ,relation::regclass,virtualxid, transactionid, mode ,query
test-#        FROM  sys_locks pl ,sys_stat_activity ps
test-#        WHERE pl.pid =ps.pid AND granted is false
test-#        ORDER BY pid;
  pid  | blocked_by | locktype | relation | virtualxid | transactionid |        mode         |                  query
-------+------------+----------+----------+------------+---------------+---------------------+------------------------------------------
 69362 |   69360    | relation | locktest |            |               | AccessShareLock     | lock table locktest in Access share mode;
 69364 |   69360    | relation | locktest |            |               | ExclusiveLock       | lock table locktest in exclusive mode;
 69366 |   69360    | relation | locktest |            |               | AccessExclusiveLock | lock table locktest in Access Exclusive MODE;
 69366 |   69362    | relation | locktest |            |               | AccessExclusiveLock | lock table locktest in Access Exclusive MODE;
 69366 |   69364    | relation | locktest |            |               | AccessExclusiveLock | lock table locktest in Access Exclusive MODE;
(5 rows)
system@test=#
```

由于 pid 为 69366 的进程,被其他三个进程(pid 分别为 69360、69362 和 69364)阻塞,因此,pid 为 69366 的进程显示了三次正在执行等待的 SQL 语句"lock table locktest in Access Exclusive MODE;"。pid 为 69362 的后端服务进程被 pid 为 69360 的后端服务进程阻塞,

正在执行等待的 SQL 语句是"lock table locktest in Access share mode；"；pid 为 69364 的后端服务进程被 pid 为 69360 的后端服务进程阻塞，正在执行等待的 SQL 语句是"lock table locktest in exclusive mode；"。

DBA 可以按如下的方式处理锁等待。

（1）提醒阻塞源头的应用人员，尽快提交或回滚事务，释放持有的锁资源；

（2）如果阻塞源头是不重要的业务，并且找不到对业务负责的工作人员，可以取消正在执行的 SQL 语句或将这个会话终止（参见 5.2.3 节）。

8.3.5 监控 vacuum 操作

执行下面的 SQL 语句，查看当前连接的数据库中有哪些表正被自动清理工作进程执行 vacuum 操作及执行 vacuum 操作的进度：

```
system@test=# SELECT p.pid,p.datname,p.query,p.backend_type,a.phase,
test-#              a.heap_blks_scanned / a.heap_blks_total::float * 100 AS "% scanned",
test-#              a.heap_blks_vacuumed / a.heap_blks_total::float * 100 AS "% vacuumed",
test-#              sys_size_pretty(sys_table_size(a.relid)) AS "table size",
test-#              sys_size_pretty(sys_indexes_size(a.relid)) AS "indexes size",
test-#              sys_get_userbyid(c.relowner) AS owner
test-#         FROM sys_stat_activity p JOIN sys_stat_progress_vacuum a ON a.pid = p.pid
test-#                                  JOIN sys_class c ON c.oid = a.relid
test-#        WHERE p.query LIKE 'autovacuum%';
 pid | datname | query | backend_type | phase | % scanned | % vacuumed | table size | indexes size | owner
-----+---------+-------+--------------+-------+-----------+------------+------------+--------------+------
(0 rows)
system@test=#
```

由于目前没有 vacuum 操作在运行中，因此当前查询不到任何记录。

取消正在进行的自动清理，可以执行下面的 SQL 语句：

```
SELECT  sys_cancel_backend(< pid >)
FROM    sys_stat_progress_vacuum
WHERE   relid = '<schema_name>.<table_name>' :: regclass;
```

其中的 pid 值来自 sys_stat_activity.pid 的查询结果。

例 8.16 取消 pid 为 286964 的自动清理进程的操作。

执行下面的 SQL 语句取消 pid 为 286964 的自动清理进程的操作：

```
SELECT  sys_cancel_backend(286964)
FROM    sys_stat_progress_vacuum
WHERE   relid = 'tpch.orders' :: regclass;
```

取消自动清理操作后，执行下面的 SQL 语句，发出手动清理冻结操作：

```
VACUUM ( TRUNCATE off,
         INDEX_CLEANUP false,
         VERBOSE, FREEZE ) < schema_name >.< table_name >;
```

例 8.17 发出手动清理冻结操作。

执行下面的 SQL 语句手动发起对表 tpch.orders 的清理冻结操作：

```
system@test=# VACUUM (TRUNCATE off,INDEX_CLEANUP false,VERBOSE,FREEZE ) tpch.
```

```
orders;
INFO:  aggressively vacuuming "tpch.orders"
INFO:  "orders": found 0 removable, 161 nonremovable row versions in 3 out of 27681 pages
DETAIL:  0 dead row versions cannot be removed yet, oldest xmin: 2517
There were 0 unused item identifiers.
Skipped 0 pages due to buffer pins, 27678 frozenTup pages.
0 pages are entirely empty.
CPU: user: 0.00 s, system: 0.00 s, elapsed: 0.00 s.
VACUUM
system@test=#
```

如果手动清理操作时间很长,执行下面的 SQL 语句监视清理操作的进度:

```
SELECT *, relid :: regclass,
       heap_blks_scanned / heap_blks_total :: float * 100 "% scanned",
       heap_blks_vacuumed / heap_blks_total :: float * 100 "% vacuumed"
FROM   sys_stat_progress_vacuum;
```

8.3.6 监控事务 ID 回卷风险

执行下面的 SQL 语句按数据库查看事务回卷的风险:

```
system@test=# SELECT  datname,age(datfrozenxid),
test-#               (age(datfrozenxid)::numeric/1000000000 * 100)::numeric(8,2) as "% WRAPAROUND RISK"
test-#         FROM sys_database ORDER BY 2 DESC;
  datname  | age  | % wraparound risk
-----------+------+-------------------
 test      | 1081 |         0.00
 kingbase  | 1081 |         0.00
 template1 | 1081 |         0.00
 template0 | 1081 |         0.00
 security  | 1081 |         0.00
(5 rows)
system@test=#
```

执行下面的 SQL 语句查看 test 数据库 tpch 模式下所有表的事务回卷风险:

```
system@test=# SELECT n.nspname AS schema_name,c.relname AS table_name,
test-#               age(c.relfrozenxid) AS txid_age,
test-#               setting::bigint AS autovacuum_freeze_max_age,
test-#               (setting::bigint - age(c.relfrozenxid)) AS age_left_before_vacuum
test-#         FROM  sys_class c JOIN sys_namespace n ON c.relnamespace = n.oid JOIN
test-#               sys_database d ON d.datname = 'test' CROSS JOIN sys_settings s
test-#         WHERE n.nspname = 'tpch'   AND         -- 指定模式名
test-#               c.relkind = 'r'      AND         -- 只显式表
test-#               s.name = 'autovacuum_freeze_max_age'
test-#         ORDER BY age_left_before_vacuum;
 schema_name | table_name | txid_age | autovacuum_freeze_max_age | age_left_before_vacuum
-------------+------------+----------+---------------------------+------------------------
 tpch        | nation     |   672    |        200000000          |       199999328
 tpch        | customer   |   671    |        200000000          |       199999329
```

```
--省略了一些输出
(8 rows)
system@test=#
```

执行下面的 SQL 语句查看事务回卷的风险：

```
system@test=# WITH max_age AS (
test(#            SELECT 2000000000 as max_old_xid,setting AS autovacuum_freeze_max_age
test(#            FROM sys_catalog.sys_settings
test(#            WHERE name = 'autovacuum_freeze_max_age' ),
test-#     per_database_stats AS (
test(#            SELECT datname,m.max_old_xid::int,m.autovacuum_freeze_max_age::int,
test(#                 age(d.datfrozenxid) AS oldest_current_xid
test(#            FROM sys_catalog.sys_database d JOIN max_age m ON (true)
test(#            WHERE d.datallowconn )
test-#    SELECT max(oldest_current_xid) AS oldest_current_xid,
test-#          max(ROUND(100 * (oldest_current_xid/max_old_xid::float)))
test-#            AS percent_towards_wraparound,
test-#          max(ROUND(100 * (oldest_current_xid/autovacuum_freeze_max_age::float)))
test-#            AS percent_towards_emergency_autovac
test-#    FROM per_database_stats;
 oldest_current_xid | percent_towards_wraparound | percent_towards_emergency_autovac
--------------------+----------------------------+-----------------------------------
               1206 |                          0 |                                 0
(1 row)
system@test=#
```

如果监控到有事务 ID 回卷的风险，首先，按照 8.3.2 节找到持续很长时间的活动事务，并对持续时间很长的活动事务采取合理的处理方式（提交或回滚），然后，按照 7.5.2 节手工冻结事务号。

8.4　接手一个生产数据库

接手一个正在运行的 KingbaseES 生产数据库是一项综合任务，以下是一些关键步骤和建议，帮助新任 DBA 有效地接手和管理新的 KingbaseES 数据库。

（1）获取 KingbaseES 数据库服务器操作系统的账号及密码，并进行验证。

（2）获取 KingbaseES 数据库用户的账号及密码，并进行验证。

（3）通过前任 DBA，了解 KingbaseES 数据库服务器或 KingbaseES 主备集群的系统物理拓扑结构、存储子系统、网络等情况。

（4）获取 KingbaseES 数据库备份和恢复的相关运维文档，通过将生产数据库的物理备份恢复到备份测试机上进行验证。

（5）如果 KingbaseES 数据库是主备集群，还需要接收集群管理运维的相关文档，包括集群启动、关闭、日常运维和故障处理等文档。

（6）获取 KingbaseES 数据库集簇的信息、用户表空间信息、用户数据库信息（包括模式）。

（7）通过前任 DBA，了解 KingbaseES 生产数据库中的用户数据库。

（8）从前任 DBA 那里获取与应用相关的日常问题处理文档及处理问题时的联系人（联

系人姓名、电话、电子邮件、联系人所在部门等信息）。

（9）与前任 DBA 共同运维 KingbaseES 生产数据库 1 周以上的时间，这期间重点观察 KingbaseES 生产数据库在业务高峰期间的性能及日常问题处理。

注意：对于在接手过程中发现的问题，不要急于处理，应当与业务负责人、技术负责人、开发人员及 DBA 组中的其他技术人员，经过充分讨论和测试之后，再采取行动。

第 9 章 数据库性能问题诊断工具

KingbaseES 数据库与 Oracle 数据库的特性高度兼容,其在数据库性能问题诊断方面,提供了 KWR、KSH 和 KDDM 这 3 个性能问题诊断工具,它们的功能特性和操作方式与 Oracle 数据库类似。

性能问题诊断工具 KWR、KSH 和 KDDM 英文全称如下。

(1) **KWR**:KingbaseES auto workload repertories(KingbaseES 自动负载信息库)。

(2) **KSH**:KingbaseES session history(KingbaseES 会话历史)。

(3) **KDDM**:KingbaseES auto database diagnostic monitor(KingbaseES 自动数据库诊断监控)。

本章将详细介绍 KingbaseES 数据库的 KWR、KSH 和 KDDM 这 3 个性能问题诊断工具的配置和使用方法。

本章的实验环境基于 TPC-H 测试,使用 1.4.5 节制作的虚拟机备份文件 dbserver-3DB-BestPractice.rar,解压缩并运行后,执行下面的 shell 脚本,来仿真一个活动的生产数据库:

```
[kingbase@dbsvr ~]$ cd /home/kingbase/tpch_sql/
[kingbase@dbsvr tpch_sql]$ sh tpchtest
#TPC-H 的测试 SQL 语句在持续不断地运行中……
```

在这个仿真活动的生产数据库上,使用 KWR、KSH 和 KDDM 这 3 个性能问题诊断工具来诊断 KingbaseES 数据库的性能问题。

9.1 性能问题诊断工具概述

KWR、KSH 和 KDDM 这 3 个性能问题诊断工具,是通过扩展插件 sys_kwr 来实现的。扩展插件 sys_kwr 依赖于内置扩展插件 sys_stat_statements。内置扩展插件 sys_stat_statements 负责收集 SQL 语句的执行信息,在 KingbaseES 数据库集簇的所有数据库上,都会自动安装该插件。

KingbaseES 数据库 KWR、KSH 和 KDDM 工作原理如图 9-1 所示,主要包括如下 4 个阶段。

(1) 数据信息的产生,主要来源有以下 3 方面。

第9章 数据库性能问题诊断工具

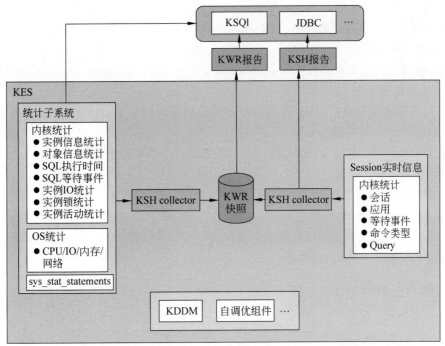

图 9-1 KingbaseES 数据库 KWR、KSH 和 KDDM 工作原理

① 统计信息子系统会收集并存储 KingbaseES 数据库实例在运行过程中不断产生的统计数据，如对某个表的访问次数、数据页的内存命中次数、某个等待事件发生的次数和总时间、SQL 语句的解析时间等。

② 扩展插件 sys_stat_statements 收集关于 SQL 语句的执行信息。

③ 采集自操作系统的统计信息。

（2）扩展插件 sys_kwr 负责启动 kwr collector、ksh collector 和 ksh write 等后台进程，根据用户的配置采集统计信息并存储快照信息。

（3）基于保存的快照生成 KWR、KSH 和 KDDM 报告。

（4）通过阅读和解析这些报告对数据库的性能问题进行多维度的诊断分析。

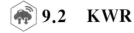 ## 9.2 KWR

KWR 通过周期性自动地记录性能统计相关的快照，分析出 KingbaseES 数据库操作系统运行环境、KingbaseES 数据库时间组成、等待事件和 TOP SQL 等性能指标，为数据库性能调优提供指导。

9.2.1 KWR 的使用场景

KWR 的主要使用场景如下。

（1）**性能监控与调优**：KWR 收集有关数据库活动的详细信息，如 SQL 查询、等待事件、计算资源使用情况等，帮助 DBA 识别和解决性能问题。

（2）**历史数据分析**：KWR 存储历史性能数据，允许进行时间段内的比较分析，帮助

DBA 了解数据库性能随时间发生的变化。

（3）**瓶颈诊断**：通过分析 KWR 报告，可以识别数据库的瓶颈，如缓存命中率低、磁盘 I/O 等待时间长、CPU 使用率高等。

（4）**SQL 优化**：KWR 提供了有关最频繁和最耗时 SQL 语句的信息，这有助于 DBA 优化查询性能和索引策略。

（5）**系统资源使用分析**：KWR 收集了有关系统资源（如 CPU、内存、磁盘 I/O）使用情况的数据，可以用来做容量规划和资源分配。

（6）**事故后分析**：在数据库出现性能问题或故障后，KWR 报告可以帮助 DBA 分析事故原因，以预防未来的问题。

（7）**趋势分析和预测**：利用 KWR 存储的历史数据，可以分析数据库的使用趋势，并对将来的性能和资源需求做出预测。

（8）**比较和基准测试**：系统升级前后或修改配置前后，可以使用 KWR 数据来对比性能的变化，确保改动带来了预期的效果。

9.2.2 配置 KWR

首先，用户 kingbase 使用 vi 编辑器，编辑 KingbaseES 数据库的启动参数文件 kingbase.conf，设置如下参数。

（1）配置启用 KWR 系统必须加载的库（默认已经满足要求，可以忽略）：

```
shared_preload_libraries = 'liboracle_parser,sys_kwr,sys_stat_statements'
```

（2）配置统计信息子系统采集的数据信息：

```
track_counts = on              #参数值为 on(默认值),表示启用统计数据库活动
track_sql = on                 #参数值为 on,表示统计 SQL 时间、SQL 等待事件、SQL 磁盘 I/O
                               #参数 track_sql 默认值为 off
track_instance = on            #参数值为 on,表示统计实例级磁盘 I/O、锁、关键活动
                               #参数 track_instance 默认值为 off
track_io_timing = on           #参数值为 on,表示统计磁盘 I/O 耗时
                               #参数 track_io_timing 默认值为 off
track_wait_timing = on         #参数值为 on(默认值),表示统计累积式等待事件的时间
track_functions = 'all'        #参数值为 all,表示统计用户自定义函数使用情况
                               #参数 track_functions 默认值为 off
```

建议用户配置以上系统参数，否则 KWR 报告中会缺失部分内容。

（3）配置内置扩展插件 sys_stat_statements 应该采集的数据信息：

```
sys_stat_statements.track = 'top'  #控制哪些语句可以被该模块跟踪,建议配置为 top
                                   #参数值为 top,表示跟踪顶级(通过客户端发出的)语句
                                   #参数值为 all,表示跟踪嵌套的语句
                                   #参数值为 none,表示禁用语句状态收集
```

（4）配置扩展插件 sys_kwr 在生成快照时应该采集的数据信息、生成和管理方式及设置 KWR 报表的信息：

```
sys_kwr.track_objects = on   #KWR 1.4 新增参数,控制统计数据库对象使用情况
                             #默认值为 on,建议开启
```

```
sys_kwr.track_os = on          #KWR 1.4新增参数,控制统计系统数据,默认值为 on,建议开启
sys_kwr.history_days = 8       #设置快照保留日期,最少 1 天,最多 1000 天
                               #默认值为 8,表示只保存 8 天的 KWR 采集信息
                               #在每次创建新的快照时,会删除过期的快照
sys_kwr.enable = on            #设置是否自动生成快照,建议开启
sys_kwr.interval = 60          #设置自动生成快照的频率
                               #默认值为 60,表示 60min 自动采样一次
sys_kwr.topn = 20              #设置显示 kwr 报告中排名前 n 条的信息
                               #最少为 10,最多为 100
                               #默认值为 20,表示只显示 20 条 SQL 语句
sys_kwr.language = 'chinese'   #KWR 1.3 新增参数,设置 KWR 报告、KWR diff 报告使用语言
                               #默认为中文(chinese 或 chn),可选为英文(english 或 eng)
```

修改完上面的启动参数后,需要执行 sys_reload_conf( ) 函数重新加载参数文件,让 KWR 开始工作:

```
[kingbase@dbsvr ~]$ ksql -d kingbase -U system
system@kingbase=# SELECT sys_reload_conf();
```

至此,完成了 KWR 的配置:每小时自动采样 1 个 KWR 快照,快照只保存 8 天。

一旦 DBA 配置 KWR 并开启自动快照功能,KWR 会自动在 KingbaseES 数据库集簇的 kingbase 数据库上安装扩展插件 sys_kwr:

```
system@kingbase=# SELECT * FROM sys_extension WHERE extname IN ('sys_kwr','sys_stat_statements');
  oid  |      extname       | extowner | extnamespace | extrelocatable | extversion | extconfig | extcondition
-------+--------------------+----------+--------------+----------------+------------+-----------+--------------
 16836 | sys_kwr            |       10 |         2200 | f              | 1.7        |           |
 14092 | sys_stat_statements|       10 |         2200 | t              | 1.10       |           |
(2 rows)
```

如果 DBA 想通过 kingbase 数据库以外的其他数据库(如 test)来管理 KWR、KSH 和 KDDM,那么首先需要打开一个新的终端,使用 ksql 命令连接到 test 数据库,执行 SQL 语句"CREATE EXTENSION sys_kwr;",安装扩展插件 sys_kwr:

```
[kingbase@dbsvr ~]$ ksql -d test -U system
system@test=# SELECT * FROM sys_extension WHERE extname IN ('sys_kwr','sys_stat_statements');
  oid  |      extname       | extowner | extnamespace | extrelocatable | extversion | extconfig | extcondition
-------+--------------------+----------+--------------+----------------+------------+-----------+--------------
 14092 | sys_stat_statements|       10 |         2200 | t              | 1.10       |           |
(1 row)
system@test=# CREATE EXTENSION sys_kwr;
system@test=# SELECT * FROM sys_extension WHERE extname IN ('sys_kwr','sys_stat_statements');
  oid  |      extname       | extowner | extnamespace | extrelocatable | extversion | extconfig | extcondition
-------+--------------------+----------+--------------+----------------+------------+-----------+--------------
 19849 | sys_kwr            |       10 |         2200 | f              | 1.7        |           |
 14092 | sys_stat_statements|       10 |         2200 | t              | 1.10       |           |
(2 rows)
```

然后,DBA 就可以连接到 test 数据库来管理 KWR、KSH 和 KDDM。建议只使用默认的 kingbase 数据库来管理 KWR、KSH 和 KDDM,执行下面的 SQL 语句删除 test 库上的扩展插件 sys_kwr:

```
system@test=# DROP EXTENSION sys_kwr;
```

```
system@test=# \q
[kingbase@dbsvr ~]$
```

9.2.3 创建 KWR 快照

KWR 快照有如下两种生成方式。

（1）自动生成；

（2）使用 SQL 语句手动生成。

在 KingbaseES 生产数据库上，配置自动生成 KWR 快照。在 9.2.2 节，通过配置以下参数实现 KWR 快照的自动生成：

```
sys_kwr.enable = on            #启用 KWR 快照自动生成
sys_kwr.interval = 60          #设置每间隔 60min 自动生成一个 KWR 快照
sys_kwr.history_days = 8       #只保存 8 天的 KWR 采集信息，自动删除过期的旧快照
```

修改完上面的参数后，重新启动 KingbaseES 数据库，开启自动生成 KWR 快照的功能，并且会马上创建一个 KWR 快照（如 snap_id=1）：

```
[kingbase@dbsvr ~]$ ksql -d kingbase -U system
system@kingbase=#  SELECT * FROM perf.kwr_snapshots;
 snap_id |      snap_time       | sess_count | snap_version
---------+----------------------+------------+--------------
       1 | 2024-02-24 11:55:09+08|          0 | 1.7
(1 rows)
```

后台进程 kwr collector 每分钟检查最后一次快照时间跟当前时间的间隔是否大于 sys_kwr.interval，如果大于则立刻创建新的快照并更新最后快照时间。

数据库用户或 DBA 想立即生成一个 KWR 快照，此时可以执行 SQL 语句"SELECT perf.create_snapshot();"，手动创建一个 KWR 快照：

```
system@kingbase=# SELECT perf.create_snapshot();
 create_snapshot
-----------------
               2
(1 row)
```

执行完 perf.create_snapshot() 函数后，将创建 1 个新的 KWR 快照，同时在表 perf.kwr_snapshots 中添加一行数据，记录新创建 KWR 快照的信息。

为了继续下面的实战学习，执行下面的 SQL 语句再创建 4 个 KWR 快照：

```
system@kingbase=# SELECT perf.create_snapshot();
system@kingbase=# SELECT perf.create_snapshot();
system@kingbase=# SELECT perf.create_snapshot();
system@kingbase=# SELECT perf.create_snapshot();
```

9.2.4 查看 KWR 快照

执行 SQL 语句查看当前 KingbaseES 数据库记录的 KWR 快照信息：

```
system@kingbase=# SELECT * FROM perf.kwr_snapshots;
 snap_id |       snap_time        | sess_count | snap_version
---------+------------------------+------------+--------------
       1 | 2024-02-24 11:55:09+08 |          0 | 1.7
       2 | 2024-02-24 11:59:38+08 |          1 | 1.7
       3 | 2024-02-24 12:03:26+08 |          1 | 1.7
       4 | 2024-02-24 12:03:27+08 |          1 | 1.7
       5 | 2024-02-24 12:03:27+08 |          1 | 1.7
       6 | 2024-02-24 12:03:27+08 |          1 | 1.7
(6 rows)
```

可以看到，当前 KingbaseES 数据库有 6 个 KWR 快照。

9.2.5 删除 KWR 快照

一般情况下不需要手工管理已经生成的快照，KWR 会在每次创建新的快照时，自动删除已经超过 sys_kwr.history_days 的快照数据，避免因为快照数据过多引起性能问题。

执行下面的 SQL 语句手工删除指定范围的 KWR 快照（如 1～3 号 KWR 快照）：

```
system@kingbase=# SELECT * from perf.drop_snapshots(1,3);
 drop_snapshots
----------------

(1 row)
system@kingbase=# SELECT * FROM perf.kwr_snapshots;
 snap_id |       snap_time        | sess_count | snap_version
---------+------------------------+------------+--------------
       4 | 2024-02-24 12:03:27+08 |          1 | 1.7
       5 | 2024-02-24 12:03:27+08 |          1 | 1.7
       6 | 2024-02-24 12:03:27+08 |          1 | 1.7
(3 rows)
```

可以看到，现在还剩下 snap_id 范围为 4～6 的快照。

执行下面的 SQL 语句删除 4 号 KWR 快照：

```
system@kingbase=# SELECT * from perf.drop_snapshots(4,4);
 drop_snapshots
----------------

(1 row)
system@kingbase=# SELECT * FROM perf.kwr_snapshots;
 snap_id |       snap_time        | sess_count | snap_version
---------+------------------------+------------+--------------
       5 | 2024-02-24 12:03:27+08 |          1 | 1.7
       6 | 2024-02-24 12:03:27+08 |          1 | 1.7
(2 rows)
```

也就是说，把要删除的 KWR 快照的起始值设置为一样（示例中为 4）就可以删除指定 snap_id 的快照。

执行下面的 SQL 语句删除 5～6 号 KWR 快照：

```
system@kingbase=# SELECT * from perf.drop_snapshots(5,6);
NOTICE:  The last snapshot 6 is always kept for baseline
 drop_snapshots
----------------

(1 row)
system@kingbase=# SELECT * FROM perf.kwr_snapshots;
 snap_id |      snap_time       | sess_count | snap_version
---------+----------------------+------------+--------------
       6 | 2024-02-03 08:34:23+08|          1 | 1.7
(1 row)
system@kingbase=#
```

可以看到,执行 perf.drop_snapshots(5,6) 函数,只删除了 5 号快照,无法使用 perf.drop_snapshots(5,6) 函数来删除 KWR 的最后一个快照(6 号快照),系统将保留最后一个 KWR 快照作为性能分析的基准点。

DBA 想删除所有的 KWR 快照,可以执行 perf.reset_snapshots() 函数:

```
system@kingbase=# SELECT * from perf.reset_snapshots();
 reset_snapshots
-----------------

(1 row)
system@kingbase=# SELECT * FROM perf.kwr_snapshots;
 snap_id | snap_time | sess_count | snap_version
---------+-----------+------------+--------------
(0 rows)
system@kingbase=#
```

可以看到,执行 perf.reset_snapshots() 函数,将删除所有的 KWR 快照(包括最后一个快照)。

不过稍等一会后,系统会自动再创建一个 KWR 快照(7 号快照),作为性能分析的基准点:

```
system@kingbase=# SELECT * FROM perf.kwr_snapshots;
 snap_id |      snap_time       | sess_count | snap_version
---------+----------------------+------------+--------------
       7 | 2024-02-24 12:09:54+08|          1 | 1.7
(1 row)
system@kingbase=#
```

为了继续下面的实战学习,执行下面的 SQL 语句再创建 3 个 KWR 快照:

```
system@kingbase=# SELECT perf.create_snapshot();
system@kingbase=# SELECT perf.create_snapshot();
system@kingbase=# SELECT perf.create_snapshot();
system@kingbase=# SELECT * FROM perf.kwr_snapshots;
 snap_id |      snap_time       | sess_count | snap_version
---------+----------------------+------------+--------------
       7 | 2024-02-24 12:09:54+08|          1 | 1.7
```

```
        8 | 2024-02-24 12:10:28+08 |         1 | 1.7
        9 | 2024-02-24 12:10:28+08 |         1 | 1.7
       10 | 2024-02-24 12:10:28+08 |         1 | 1.7
(4 rows)
system@kingbase=#
```

9.2.6 生成 KWR 报告

KWR 报告是基于快照生成的,报告中的数据信息反映的是两个快照之间系统的运行状况。KWR 报告有两种形式:TEXT 和 HTML,推荐使用 HTML 格式,因为它更便于阅读。

执行 SQL 语句基于 KWR 快照 7 和 8 生成 TEXT 格式的 KWR 报告:

```
system@kingbase=# SELECT * FROM perf.kwr_report(7,8,'text');
```

执行 SQL 语句基于 KWR 快照 7 和 8 生成 HTML 格式的 KWR 报告:

```
system@kingbase=# SELECT * FROM perf.kwr_report(7,8,'html');
```

生成的 TEXT 格式和 HTML 格式的 KWR 报告,保存在 KingbaseES 数据库 data 目录的 sys_log 子目录(/u00/Kingbase/ES/V9/data/sys_log)下,可以使用浏览器查看它们。

如果想将生成的 KWR 报告保存到指定的磁盘目录下,用 kwr_report_to_file()函数,第 4 个参数指定文件全路径:

```
system@kingbase=# SELECT * FROM perf.kwr_report_to_file(7,8,'html', '/home/
kingbase/kwr_7_8.html');
 kwr_report_to_file
--------------------
 t
(1 row)
system@kingbase=# \! ls -l /home/kingbase/kwr_7_8.html
-rw-------. 1 kingbase dba 237168 Feb 24 12:11 /home/kingbase/kwr_7_8.html
system@kingbase=#
```

9.2.7 生成 KWR 运行期对比报告 KWR DIFF

KWR 报告是 DIFF 报告的基础。在数据库运行过程中,如在业务的高峰期和低谷期或者在参数调整之后,数据库的性能指标都会发生变化,分析引起这种变化的原因对于分析数据库性能及进行优化调整是非常必要的。

DIFF 报告的作用就是分析两个 KWR 报告的差异,找出性能变化的原因及分析性能优化的效果。创建 DIFF 报告的函数如下:

```
perf.kwr_diff_report_to_file({snap_1}, {snap_2}, {snap_3}, {snap_4}, {file_
path})
```

其中的参数如下。

(1) **snap_1**:生成第一份 KWR 报告的起始快照 ID。

(2) **snap_2**:生成第一份 KWR 报告的结束快照 ID。

（3）**snap_3**：生成第二份 KWR 报告的起始快照 ID。

（4）**snap_4**：生成第二份 KWR 报告的结束快照 ID。

（5）**file_path**：报告生成的位置，如/home/kingbase/kwr_diff_report.html。

执行下面的 SQL 语句生成一个 KWR 快照 7~8、KWR 快照 9~10 的 KWR DIFF 报告，将报告保存为/home/kingbase/kwr_diff_rpt.html：

```
system@kingbase=# SELECT *
kingbase-#        FROM perf.kwr_diff_report_to_file(7,8,9,10,'/home/kingbase/kwr_diff_rpt.html');
 kwr_diff_report_to_file
-------------------------
 t
(1 row)
system@kingbase=# \! ls -l /home/kingbase/kwr_diff_rpt.html
-rw-------. 1 kingbase dba 250200 Feb 24 12:11 /home/kingbase/kwr_diff_rpt.html
system@kingbase=#
```

9.3 KSH

系统视图 sys_stat_activity 记录每个会话的当前信息，对于某些信息需要获取其累计值，用以判断系统的运行瓶颈，如会话的等待事件时间。因为获取了累计值，就可以知道系统到底在等待什么，所以 KingbaseES 数据库引入了 KSH，DBA 可以进行会话历史的分析，针对报告呈现的性能瓶颈进行优化。

开启 KSH 功能后，后台进程 ksh collector 以每秒采样的方式收集会话数据，并将采集的数据放入内存的 Ringbuf 队列中，采集的数据主要包括会话、应用、等待事件、命令类型、QueryId 等。当 Ringbuf 队列使用的内存缓冲区充满度达到 80% 时，后台进程 ksh write 把其中的部分数据写入磁盘文件中。

KSH 聚焦于会话累计的信息，用于当系统出现异常时，观察当前或历史某个时刻系统在执行和运行了什么任务。KSH 可以看作是对 KWR 的补充，当开启 KSH 功能时，KWR 报告中包含 KSH 的相关内容。

KSH 也可以单独配置生成自己的报告，当系统出现异常时，单独使用 KSH 报告诊断系统可能存在的问题及其原因。

9.3.1 KSH 的使用场景

KSH 的主要使用场景如下。

（1）**实时性能监控**：KSH 提供了有关当前活动会话的实时数据，可以帮助 DBA 快速识别和解决正在发生的性能问题。

（2）**短期性能问题分析**：对于那些可能不会出现在 KWR 报告中的短暂问题，KSH 可以提供详细的信息来帮助诊断。

（3）**等待事件分析**：KSH 收集关于数据库等待事件的数据，这有助于识别性能瓶颈，如锁争用、磁盘 I/O 瓶颈或网络问题等。

（4）**SQL 语句性能诊断**：KSH 记录了执行中的 SQL 语句信息，可以用来识别和优化性能不佳的查询。

（5）**会话和用户活动监控**：KSH 提供关于特定会话或用户活动的详细信息，有助于理解用户行为对性能的影响。

（6）**历史性能数据分析**：虽然 KSH 主要用于实时监控，但它也存储了一段时间内的历史数据，可以用来分析过去的性能问题。

（7）**事故后分析**：在出现性能问题后，KSH 可以提供关键信息来帮助确定问题发生的原因。

（8）**优化资源分配**：通过分析 KSH 数据，可以确定哪些资源（如 CPU、内存）最为紧张，从而做出更有效的资源调配和优化决策。

（9）**验证性能调优措施**：在进行性能调优后，可以使用 KSH 数据来验证调优措施的效果。

9.3.2 配置 KSH

KSH 功能也是通过扩展插件 sys_kwr 来实现的。

要配置 KSH，用户 kingbase 使用 vi 编辑器，编辑 KingbaseES 数据库的启动参数文件 kingbase.conf，设置如下参数。

（1）配置启用 KWR 系统必须加载的库（默认已经满足要求，可以忽略）：

```
shared_preload_libraries = 'liboracle_parser,sys_kwr,sys_stat_statements'
```

（2）配置统计信息子系统采集的数据信息：

```
track_activities = on       #参数值为 on(默认值)，表示跟踪活动会话的等待事件、SQL、状态等
```

（3）配置内置扩展插件 sys_stat_statements 的参数：

```
sys_stat_statements.track = 'top'  #在启用 KWR 时已经配置了该参数
                                   #用于控制哪些语句可以被该模块跟踪，建议配置为 top
```

（4）配置扩展插件 sys_ksh 的参数，启动 KSH 功能：

```
sys_ksh.history_days = 8            #在启用 KSH 时已经配置了该参数
sys_ksh.language = 'chinese'        #在启用 KSH 时已经配置了该参数

sys_ksh.collect_ksh = on            #参数值为 on,表示启动 KSH 功能,默认是关闭的
                                    #打开该参数,因为每秒采集一次数据,会占用系统的资源
sys_ksh.ringbuf_size = 1000000      #设置 sys_ksh ringbuf 缓冲区的大小,以及在内存中
                                    #可以存储多少条数据
                                    #默认是 100 000 条
```

修改完上面的启动参数后，需要执行 sys_reload_conf()函数重新加载参数文件，让 KSH 开始工作：

```
system@kingbase=# SELECT sys_reload_conf();
```

9.3.3　查看 KSH 数据

要查看 KSH 生成的数据，需要等待一段时间，KSH 会自动收集一些数据。

执行下面的 SQL 语句可以查看保存于内存 Ringbuf 的 KSH 数据：

```
system@kingbase=# SELECT * FROM perf.session_history;
--省略了输出
```

可以通过视图 perf.ksh_history 查看保存于数据库的历史数据：

```
system@kingbase=# SELECT * FROM perf.ksh_history;
--省略了输出
```

9.3.4　生成 KSH 报告

开启 KSH 功能后，后台进程 ksh collector 就会自动每秒采集信息，使用 perf.ksh_report()函数生成 KSH 报告。KSH 报告也有两种形式：TEXT 和 HTML，可以生成在线报告或把报告输出到文件中。

基于指定 KWR 快照区间生成 KSH 报告的函数如下：

```
SELECT * FROM perf.ksh_report_by_snapshots({start_snapid},{end_snapid},{slot_width}, {format});
SELECT * FROM perf.ksh_report_to_file_by_snapshots({start_snapid}, {end_snapid}, {file_path}, {format}, {slot_width});
```

其中的参数如下。

(1) **start_snapid**：起始快照号。

(2) **end_snapid**：结束快照号。

(3) **file_path**：输出的文件路径。

(4) **format**：报告生成格式，可选择 HTML 和 TEXT 两种格式，默认为 TEXT 格式。

(5) **slot_width**：报告最小区间，输入 0 时系统自动计算合适的宽度，默认值为 0。

例 9.1　基于指定 KWR 快照区间生成 KSH 报告。

执行下面的 SQL 语句手工创建一个 KWR 快照：

```
system@kingbase=# SELECT perf.create_snapshot();
create_snapshot
-----------------
             11
(1 row)
system@kingbase=# SELECT * FROM perf.kwr_snapshots;
 snap_id |        snap_time        | sess_count | snap_version
---------+-------------------------+------------+--------------
       7 | 2024-02-24 12:09:54+08  |          1 | 1.7
       8 | 2024-02-24 12:10:28+08  |          1 | 1.7
       9 | 2024-02-24 12:10:28+08  |          1 | 1.7
      10 | 2024-02-24 12:10:28+08  |          1 | 1.7
      11 | 2024-02-24 12:21:14+08  |          2 | 1.7
(5 rows)
```

等待 10min 后,执行下面的 SQL 语句手工创建一个 KWR 快照:

```
system@kingbase=# SELECT perf.create_snapshot();
create_snapshot
-----------------
             12
(1 row)
```

执行下面的 SQL 语句查看当前的 KWR 快照:

```
system@kingbase=# SELECT * FROM perf.kwr_snapshots;
 snap_id |       snap_time        | sess_count | snap_version
---------+------------------------+------------+--------------
       7 | 2024-02-24 12:09:54+08 |          1 | 1.7
       8 | 2024-02-24 12:10:28+08 |          1 | 1.7
       9 | 2024-02-24 12:10:28+08 |          1 | 1.7
      10 | 2024-02-24 12:10:28+08 |          1 | 1.7
      11 | 2024-02-24 12:21:14+08 |          2 | 1.7
      12 | 2024-02-24 12:32:33+08 |          2 | 1.7
(6 rows)
```

执行下面的 SQL 语句根据 KWR 快照 11~12 来生成 HTML 文件格式的 KSH 报告:

```
system@kingbase=# SELECT * FROM perf.ksh_report_by_snapshots(11,12,0,'html');
system@kingbase=# \! ls -l /u00/Kingbase/ES/V9/data/sys_log/*KSH.html
-rw-------. 1 kingbase dba 65257 Feb 24 12:45 /u00/Kingbase/ES/V9/data/sys_log/
20240224_124505_KSH.html
system@kingbase=#
```

默认生成的 KSH 报告位于目录 /u00/Kingbase/ES/V9/data/sys_log 下,文件名带有生成时间。

基于时间点生成 KSH 报告的函数如下:

```
SELECT * FROM perf.ksh_report({start_ts}, {duration}, {slot_width}, {format});
SELECT * FROM perf.ksh_report_to_file({start_ts},{duration},{slot_width},{file
_path},{format});
```

其中的参数如下。

(1) **start_ts**:报告开始时间,默认值为当前时间-15min。

(2) **duration**:报告时长,默认值为 15min,最大不超过 60min。

(3) **slot_width**:报告最小区间,输入 0 时系统自动计算合适的宽度,默认值为 0。

(4) **format**:报告生成格式,可选择 HTML 和 TEXT 两种格式,默认为 TEXT 格式。

(5) **file_path**:输出的文件路径。

例 9.2 基于时间点生成 KSH 报告。

执行下面的 SQL 语句生成 TEXT 文件格式的 KSH 报告:

```
system@kingbase=# SELECT * FROM perf.ksh_report('2024-02-24 12:21:14+08', 5, 0,
'text');
system@kingbase=# \! ls -l /u00/Kingbase/ES/V9/data/sys_log/*KSH.text
-rw-------. 1 kingbase dba 33853 Feb 24 12:47 /u00/Kingbase/ES/V9/data/sys_log/
```

```
20240224_124700_KSH.text
system@kingbase=#
```

使用用户 kingbase，执行下面的命令和 SQL 语句生成 HTML 文件格式的 KSH 报告：

```
system@kingbase=# SELECT * FROM perf.ksh_report('2024-02-24 12:21:14+08', 5, 0,
'html');
system@kingbase=# \! ls -l /u00/Kingbase/ES/V9/data/sys_log/*KSH.html
-rw-------. 1 kingbase dba 65257 Feb 24 12:45 /u00/Kingbase/ES/V9/data/sys_log/
20240224_124505_KSH.html
-rw-------. 1 kingbase dba 57347 Feb 24 12:48 /u00/Kingbase/ES/V9/data/sys_log/
20240224_124814_KSH.html
system@kingbase=#
```

默认生成的 KSH 报告位于目录 /u00/Kingbase/ES/V9/data/sys_log 下，文件名带有报告的生成时间。

9.4　KDDM

KingbaseES 自动数据库诊断监控（KDDM）是基于 KWR 快照采集的性能指标和数据库时间（DB time），自动分析等待事件、磁盘 I/O、网络、内存和 SQL 执行时间等，通过生成 KDDM 报告，给出一系列性能优化建议，帮助数据库用户实现数据库性能的快速调优。KDDM 也是通过扩展插件 sys_kwr 来实现的。

9.4.1　KDDM 的使用场景

KDDM 的主要使用场景如下。

（1）**自动性能分析**：KDDM 自动分析 KWR 数据，识别潜在的性能问题和瓶颈，减少人工分析所需的时间和专业知识。

（2）**问题诊断和建议**：KDDM 不仅识别问题，还提供具体的改进建议，如 SQL 优化、资源配置调整、系统设置更改等。

（3）**SQL 性能调优**：通过分析 SQL 执行计划和性能统计数据，KDDM 可以优化长时间运行或资源密集型的 SQL 查询。

（4）**资源利用率优化**：KDDM 分析 CPU、内存、磁盘 I/O 等资源的使用情况，识别过度使用资源或不足使用资源的区域，并给出优化建议。

（5）**等待事件和锁争用分析**：KDDM 能识别导致数据库性能降低的等待事件和锁争用问题，并提供解决方案。

（6）**趋势分析和预测**：利用历史数据，KDDM 可以预测未来可能出现的性能问题，从而提前采取预防措施。

（7）**系统变更前后的比较分析**：在进行系统升级、配置调整或优化措施后，可以使用 KDDM 来比较变更前后的性能差异，确保改动带来了正面效果。

（8）**容量规划**：KDDM 的分析结果可以用于数据库的容量规划，如扩展硬件资源或调整配置。

（9）**事故后分析**：当数据库出现严重性能问题或故障时，KDDM 可以快速定位问题原因并提供解决方案。

9.4.2 生成 KDDM 用户报告

生成 KWR 快照后,就可以继续生成 KDDM 报告。目前 KDDM 报告仅支持 TEXT 格式,不支持 HTML 格式输出。

执行下面的 SQL 语句,根据 KWR 快照 11～12 生成一个 KDDM 用户报告,报告将显示在终端上,并在目录/u00/Kingbase/ES/V9/data/sys_log 中生成一个 TEXT 格式的 KDDM 报告:

```
system@kingbase=# SELECT perf.kddm_report(11,12);
system@kingbase=# \! ls -l /u00/Kingbase/ES/V9/data/sys_log/kddm_11_12*
-rw-------. 1 kingbase dba 48247 Feb  3 14:51 /u00/Kingbase/ES/V9/data/sys_log/kddm_11_12_20240203_145131.txt
system@kingbase=#
```

执行下面的 SQL 语句在指定的位置生成 KDDM 报告:

```
system@kingbase=# SELECT perf.kddm_report_to_file(11, 12, '/home/kingbase/kddmreport.txt');
system@kingbase=# \! ls -l /home/kingbase/kddmreport.txt
-rw-------. 1 kingbase dba 48247 Feb  3 14:54 /home/kingbase/kddmreport.txt
system@kingbase=#
```

执行下面的 SQL 语句为指定 KWR 快照范围区间的某个指定 Query Id 生成 KDDM 报告:

```
system@kingbase=# \! grep "Query Id" /home/kingbase/kddmreport.txt |tail -1
Query Id:3140549141248054114
system@kingbase=# SELECT * FROM perf.kddm_sql_report(11, 12, 3140549141248054114);
--忽略了报告输出
```

这个功能主要用来分析耗时较多的 SQL 语句占用 CPU、磁盘 I/O 资源的使用情况。

9.4.3 获取数据库配置参数的建议值

KDDM 提供系统配置参数建议功能,根据数据库服务器的硬件情况和用户指定的业务类型,显示建议结果。

使用 kingbase 用户执行 perf.kddm_guc_advisor()函数获取 KingbaseES 数据库配置参数的建议值:

```
system@kingbase=# SELECT * FROM perf.kddm_guc_advisor();
             kddm_guc_advisor
---------------------------------------------------
                                                   +
 基础信息:                                          +
 --------                                          +
                                                   +
   最大连接数:300,自动获取                          +
```

```
        CPU 核心数:4,自动获取                                            +
        总物理内存:15 GB,自动获取                                         +
        应用类型:oltp,使用默认值                                          +
                                                                      +
     建议参数列表:                                                       +
     ------------                                                     +
                                                                      +
       连接数相关:                                                      +
         max_connections = 300                                         +
                                                                      +
       内存相关:                                                        +
         shared_buffers = 3967 MB                                      +
         effective_cache_size = 12 GB                                  +
         wal_buffers = 16 MB                                           +
     --省略了一些输出
(1 row)
```

9.5 性能诊断工具 KWR 实战

需要执行下面的步骤打造一个 KingbaseES 数据库性能问题诊断的实战环境。

(1) 执行下面的 SQL 语句查看当前的 KWR 快照信息:

```
system@kingbase=#  SELECT * FROM perf.kwr_snapshots;
 snap_id |       snap_time        | sess_count | snap_version
---------+------------------------+------------+--------------
       7 | 2024-02-24 12:09:54+08 |          1 | 1.7
       8 | 2024-02-24 12:10:28+08 |          1 | 1.7
       9 | 2024-02-24 12:10:28+08 |          1 | 1.7
      10 | 2024-02-24 12:10:28+08 |          1 | 1.7
      11 | 2024-02-24 12:21:14+08 |          2 | 1.7
      12 | 2024-02-24 12:32:33+08 |          2 | 1.7
(6 rows)
```

(2) 继续运行 TPC-H 的测试脚本 sh tpchtest,打开另外一个终端执行命令和 SQL 语句:

```
[kingbase@dbsvr ~]$ #因为删除索引后,第 17、第 20 条语句运行很慢,先禁止运行这 2 条语句
[kingbase@dbsvr ~]$ cd tpch_sql/
[kingbase@dbsvr tpch_sql]$ mkdir 1
[kingbase@dbsvr tpch_sql]$ mv stream_0_17.sql 1
[kingbase@dbsvr tpch_sql]$ mv stream_0_20.sql 1
[kingbase@dbsvr ~]$ #删除索引后,运行第 17 条语句,将会运行很慢,通过 KWR 来发现该条语句
[kingbase@dbsvr ~]$ ksql -d test -U system
system@test=# SET search_path TO tpch;
system@test=# DROP INDEX lineitem_l_partkey_idx;
system@test=# \timing
system@test=# select sum(l_extendedprice) / 7.0 as avg_yearly
test-#          from lineitem,part
test-#          where p_partkey = l_partkey and p_brand = 'Brand#45' and
```

```
test-#                  p_container = 'LG PKG' and l_quantity < (  select 0.2 * avg(l_quantity)
test(#                                                             from lineitem
test(#                                                             where l_partkey = p_partkey);
--语句执行中……
```

这条语句需要执行比较长的时间，请耐心等待执行完成：

```
   avg_yearly
---------------------
 288975.151428571429
(1 row)
```

（3）执行下面的 SQL 语句查看 KWR 的快照信息：

```
system@kingbase=# SELECT * FROM perf.kwr_snapshots;
 snap_id |      snap_time       | sess_count | snap_version
---------+----------------------+------------+--------------
       7 | 2024-02-24 12:09:54+08 |          1 | 1.7
       8 | 2024-02-24 12:10:28+08 |          1 | 1.7
       9 | 2024-02-24 12:10:28+08 |          1 | 1.7
      10 | 2024-02-24 12:10:28+08 |          1 | 1.7
      11 | 2024-02-24 12:21:14+08 |          2 | 1.7
      12 | 2024-02-24 12:32:33+08 |          2 | 1.7
      13 | 2024-02-24 13:33:25+08 |          3 | 1.7
      14 | 2024-02-24 14:33:39+08 |          3 | 1.7
(8 rows)
```

执行下面的 SQL 语句基于 KWR 快照 12～14 生成 HTML 格式的 KWR 报告：

```
system@kingbase=# SELECT * from perf.kwr_report(12,14,'html');
#按小写字母 q
system@kingbase=# \! ls -l /u00/Kingbase/ES/V9/data/sys_log/kwr_12_14_*
-rw-------. 1 kingbase dba 360690 Feb 24 14:41 /u00/Kingbase/ES/V9/data/sys_log/
kwr_12_14_20240224_144111.html
system@kingbase=#
```

可以看到，生成 HTML 格式的 KWR 报告，保存在目录/u00/Kingbase/ES/V9/data/sys_log 中，名为 kwr_12_14_20240224_144111.html。

使用用户 kingbase 登录 CentOS 的 Gnome 图形界面，打开一个终端，执行下面的命令：

```
[kingbase@dbsvr ~]$ google-chrome
```

打开一个 google chrome 浏览器，在 chrome 浏览器的 URL 窗口输入 file:///u00/Kingbase/ES/V9/data/sys_log，将显示日志目录下的文件列表，如图 9-2 所示。选中刚刚生成的 KWR 报告文件（示例是 kwr_12_14_20240224_144111.html 的 KWR 文件），查看 KWR 报告，如图 9-3 所示。

解读 KWR 报告，可以参考 KingbaseES 数据库的官方手册，或者访问以下的 URL：

https://help.kingbase.com.cn/v9/perfor/performance-optimization/performance-optimization-6.html#id29

通过解读 KWR 报告，可以很容易地发现 KingbaseES 数据库在运行中的一些问题，如

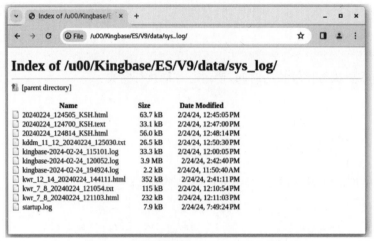

图 9-2 使用浏览器查看 KWR 报告目录

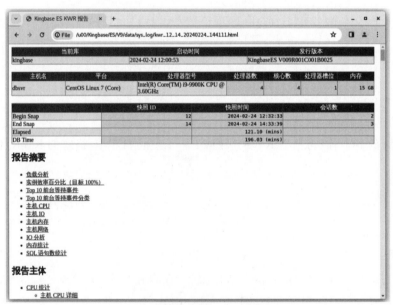

图 9-3 使用浏览器查看 KWR 报告文件

按执行时间排序的 SQL 语句，如图 9-4 所示。

可以发现 QueryID 为-4989494339632256003 的 SQL 语句只执行了 1 次，耗时 4626.94s，该 SQL 语句为：

```
select sum(l_extendedprice) / 7.0 as avg_yearly
from lineitem, part
where p_partkey = l_partkey and p_brand = 'Brand#45' and
        p_container = 'LG PKG' and
        l_quantity < ( select 0.2 * avg(l_quantity) from lineitem where l_partkey = p_partkey )
```

需要 DBA 或开发人员来优化这条语句。

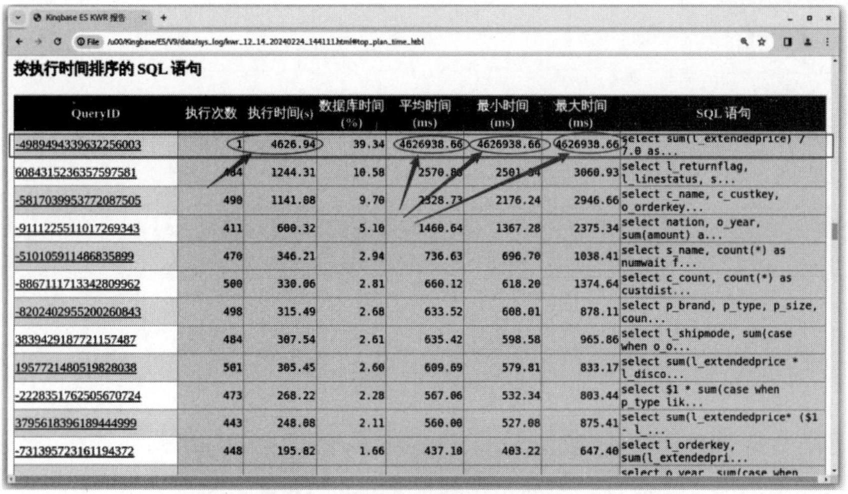

第 9 章 数据库性能问题诊断工具

图 9-4 KWR 报告的按执行时间排序的 SQL 语句

第 10 章

SQL 语句执行计划

提高数据库查询性能的关键因素之一是生成要执行的 SQL 语句并选择一个高效的执行计划。由于数据库优化器自身的局限性及其他原因（如缺乏统计数据、索引等），为 SQL 语句选择的执行计划，并不一定是最好的。因此，需要开发人员或 DBA 能够读懂 SQL 语句的执行计划，对有问题的语句分析其产生原因并进行优化。

本章的实验环境可以使用 1.4.5 节制作的虚拟机备份文件 dbserver-3DB-BestPractice.rar，解压缩并运行。基于 TPC-H 数据集构建一些实例，引导读者学会阅读 SQL 语句的执行计划，并掌握初步优化 SQL 语句的技能。

10.1　SQL 语句的执行过程

SQL 语句是高度的非过程化编程语言，它描述了需要数据库完成的任务，但没有描述数据库应当如何完成需要的功能。例如，SQL 语句 SELECT C2 FROM TBL，该语句要求从表 TBL 中查找出 C2 列的所有值，但该 SQL 语句并没有告诉数据库如何去完成这项任务、如何查找表 TBL 在磁盘的位置、以何种方式获取 C2 列的值等。在实际执行时，需要通过基于过程的语言（如 C 语言）来描述数据库应当如何完成 SQL 语言指定的功能。

SQL 语句的执行过程如图 10-1 所示。

当客户端程序发出 SQL 语句，数据库后端服务进程接收到 SQL 语句后，调用解析器（parser）进行语法分析和词法分析，生成一棵语法树。语法树被转交给分析器（analyzer），进行语义分析，生成查询树。查询树被转交给查询重写器，基于规则进行查询重写，生成了一棵重写后的查询树。重写后的查询树转交给优化器（optimizer），生成一个高度优化的执行计划。执行计划转交给执行器（executor），产生最终结果，由数据库后端服务进程将查询结果转交给客户端程序。

关系系统中 SQL 语句的特点是非过程化，用户只需提出"干什么"，不必指出"怎么干"。同一个 SQL 语句会有多种执行方法，不同执行方法的性能可能会相差几个数量级。因此，查询优化器在 SQL 语句的处理过程中发挥重要的作用，它对 SQL 语句的关系代数表达式进行变换，并基于系统的统计信息选出尽可能最优的执行方式交给系统的执行器去执行。SQL 语句的具体执行方式是一个个具体操作算子组成的执行树，称为执行计划。

KingbaseES 数据库采用基于代价的查询优化方法。基于数据库中大量数据进行统计

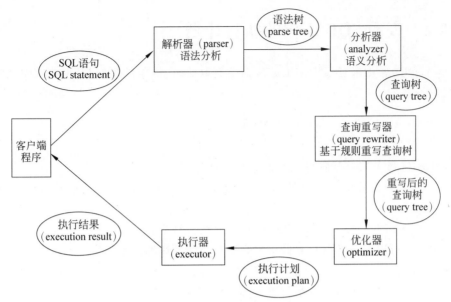

图 10-1　SQL 语句的执行过程

（或抽样统计）得到的统计结果和系统中的各种参数（如缓冲区大小、数据分布、存取路径等），优化器可以准确地估计出各种查询路径所需的资源消耗和时间开销，权衡利弊之后选出一个最适合的查询计划。

KingbaseES 数据库对于每种操作要考虑两类代价：磁盘 I/O 代价和 CPU 代价。顾名思义，磁盘 I/O 代价是访问磁盘的开销，包括磁盘读取和磁盘写入，它与操作中使用的表和索引的大小、页面读取和写入的成本代价有关。CPU 代价通常表示数据库表中的数据、表达式的计算等处理对 CPU 的开销。

在进行操作代价估算的过程中，需要估算操作结果集的大小，如利用统计信息来估算所选操作结果集的大小，从而估算出整个操作的代价。统计信息包括表的行数、列的唯一值数量、数据分布等，统计信息是进行代价估算的基础。基于这些成本估算，KingbaseES 数据库的优化器可以比较不同执行计划的成本代价，并选择成本代价最低的执行计划作为最优计划。

KingbaseES 数据库的执行器采用流水线形式进行操作结果的传递，因此，KingbaseES 数据库总代价分成两部分：启动代价（startup cost）和运行代价（run cost）。

启动代价是操作节点阻塞执行时的时间，在这段时间内，该操作节点不会向上层传递任何结果。运行代价是操作节点从输出第一条结果开始，到整个节点操作完毕的时间。以外部归并排序算法为例，读取数据文件、建立有序组和写临时文件的过程是阻塞过程，这部分时间代价记做启动代价。在读临时文件的过程中，可以一边归并一边输出排序结果，这部分时间代价记做运行代价。通常会记录每个操作的启动代价和总代价（total cost），即启动代价和运行代价的总和。

本章重点讨论如何查看和分析 SQL 执行计划，本章的例子继续使用 TPC-H 数据集。

10.2 查看 SQL 语句的执行计划

KingbaseES 数据库通过 EXPLAIN 命令获取 SQL 语句的执行计划。EXPLAIN 命令的格式如下：

```
EXPLAIN [ ( option [, ...] ) ] statement
EXPLAIN [ ANALYZE ] [ VERBOSE ] statement
```

EXPLAIN 命令主要有两种执行方式。

(1) 使用 EXPLAIN SQL 获取执行计划，输出将只显示 SQL 语句的执行计划，SQL 语句并不真正执行。

(2) 使用 EXPLAIN ANALYZE SQL 获取执行计划，除了显示 SQL 语句的执行计划外，还会实际执行该 SQL 语句，并显示实际运行时间等详细信息。

KingbaseES 数据库的 SQL 语句执行计划是由完成该 SQL 语句的多个操作组成的操作树，每个操作是操作树中的一个节点，针对每一个操作节点，主要包括以下描述信息。

(1) 操作的名称，如各种扫描操作、连接操作、排序、聚集等。

(2) 操作的对象，如表、索引等。

(3) 对该操作的代价估算。

下面是一个获取 SQL 语句的执行计划的例子：

```
[kingbase@dbsvr ~]$ ksql -d test -U system
system@test=# SET search_path = 'tpch';
system@test=# EXPLAIN select * from orders;
                          QUERY PLAN
-----------------------------------------------------------------
 Seq Scan on orders  (cost=0.00..42681.00 rows=1500000 width=111)
(1 row)
```

这是一个最简单的单表查询语句，该语句的执行计划只有一个操作节点，说明该 SQL 语句的执行只包含一个全表顺序扫描（seq scan）操作，该操作的操作对象是表 orders，括号中的信息，是对该操作的代价估算，主要包括以下信息。

(1) cost：显示查询计划的代价估算值。cost 有两个值：一个值表示获取该操作第 1 条结果的代价，即启动代价；另一个值是该操作执行完成的总代价。

(2) rows：表示估计的结果集行数。

(3) width：表示估计的结果集中每个元组的宽度。

使用 EXPLAIN ANALYZE 执行同样的 SQL 语句，结果如下：

```
system@test=# EXPLAIN ANALYZE select * from orders;
                          QUERY PLAN
-----------------------------------------------------------------
 Seq Scan on orders  (cost=0.00..42681.00 rows=1500000 width=111) (actual time=
0.012..252.283 rows=1500000 loops=1)
 Planning Time: 0.192 ms
 Execution Time: 280.494 ms
```

(3 rows)

可以看出执行结果中不仅包含了 SQL 语句的执行计划,还有真正执行该 SQL 语句的实际开销。

在本例中,SQL 语句依然采用的是全表顺序扫描的执行方式,对该操作的代价估算有两个内容:第一个括号中的信息是系统估算的代价;第二个括号中的信息是系统真正执行的结果,包括以下信息。

(1) actual time:显示该操作真正执行的代价值,包括启动代价和总代价。

(2) rows:表示该操作返回的结果集行数。

(3) loops:表示该操作循环执行的次数。

10.3 阅读 SQL 语句的执行计划

本节通过一些具体的实例来介绍如何阅读 SQL 语句的执行计划。

10.3.1 单表查询

单表查询是最简单的 SQL 语句,它的执行方式通常有以下 4 种。

(1) 全表顺序扫描(seq scan):该操作扫描表中的每个数据页面,并逐行检查每个元组是否满足过滤条件。

(2) 索引扫描(index scan):首先根据索引键查找满足条件的索引项,然后再根据索引项指向数据表中元组的指针,找到相关的元组。

(3) 只索引扫描(index only scan):从索引中直接获取所需的列,无须访问表的行数据。

(4) 位图扫描(bitmap scan):该执行方式包括两个步骤。

① 位图索引扫描(bitmap index scan):首先使用索引扫描,找到满足索引键的所有索引项。

② 位图堆扫描(bitmap heap scan):把所获得的索引项指向数据表中元组的指针,进行排序,然后根据指针在表中找到指定的元组。

下面的 SQL 语句查询订单总额大于 50 000 的订单信息:

```
system@test=# EXPLAIN select * from orders where o_totalprice>50000;
                    QUERY PLAN
-----------------------------------------------------------------
 Seq Scan on orders  (cost=0.00..46431.00 rows=1291060 width=111)
   Filter: (o_totalprice > '50000'::numeric)
(2 rows)
```

该 SQL 语句有 WHERE 条件,因此执行计划中对全表顺序扫描操作增加了过滤(filter),括号中(的内容)就是对结果集的过滤条件。

该 SQL 语句的执行计划中估算出结果集大小是 1 291 060,选择率是 1 291 060/1 500 000 = 86%,即表中有 86% 的元组满足条件,即使在 o_totalprice 列上有索引,优化器也会采用全表顺序扫描的方式执行该 SQL 语句。

下面的 SQL 语句查询订单总额等于 5 000 000 的订单信息：

```
system@test=# EXPLAIN select * from orders where o_totalprice>5000000;
                               QUERY PLAN
-------------------------------------------------------------------------
 Index Scan using idx_orders_o_totalprice on orders  (cost=0.43..8.44 rows=1 width=111)
   Index Cond: (o_totalprice > '5000000'::numeric)
(2 rows)
```

该 SQL 语句的执行计划中估算出结果集大小是 1，WHERE 条件的选择率非常低，优化器也会采用索引扫描的方式执行该 SQL 语句，使用 o_totalprice 列上的索引 idx_orders_o_totalprice。索引键的选择操作使用 index cond 表示，其后面括号中的信息是索引的选择条件。

下面的 SQL 语句查询订单总额大于 500 000 的订单信息：

```
system@test=# EXPLAIN select * from orders where o_totalprice>500000;
                               QUERY PLAN
-------------------------------------------------------------------------
 Bitmap Heap Scan on orders  (cost=101.02..11385.05 rows=4206 width=111)
   Recheck Cond: (o_totalprice > '500000'::numeric)
   ->  Bitmap Index Scan on idx_orders_o_totalprice  (cost=0.00..99.97 rows=4206 width=0)
         Index Cond: (o_totalprice > '500000'::numeric)
(4 rows)
```

该 SQL 语句的执行计划中估算出结果集大小是 4206 条，WHERE 条件的选择率不大不小，使用索引扫描未必是很好的选择，这里优化器选择的是使用位图扫描的执行方式。该执行方式分为两个步骤：位图索引扫描和位图堆扫描。在执行计划中，位图索引扫描操作使用->进行了一个缩进，说明了这两个操作执行的先后顺序，缩进的操作先执行。

下面的 SQL 语句查询订单总额大于 500 000 或订单号小于 100 的订单信息：

```
system@test=# EXPLAIN select * from orders where o_totalprice>500000 or o_orderkey<100;
                               QUERY PLAN
-------------------------------------------------------------------------
 Bitmap Heap Scan on orders  (cost=106.70..11447.73 rows=4230 width=111)
   Recheck Cond: ((o_totalprice > '500000'::numeric) OR (o_orderkey < 100))
   ->  BitmapOr  (cost=106.70..106.70 rows=4230 width=0)
         ->  Bitmap Index Scan on idx_orders_o_totalprice  (cost=0.00..99.97 rows=4206 width=0)
               Index Cond: (o_totalprice > '500000'::numeric)
         ->  Bitmap Index Scan on orders_pkey  (cost=0.00..4.61 rows=24 width=0)
               Index Cond: (o_orderkey < 100)
(7 rows)
```

该 SQL 语句的 WHERE 条件中，使用 OR 连接两个条件，并且在这两个条件引用的列上都有索引，这时就可以分别使用两个索引查找满足条件的索引项，然后对索引项中的指针进行 BitmapOr 操作，找出所有满足条件的索引项，根据其指针再到表中找到满足条件的元组。

执行计划中估算满足 o_totalprice>500 000 条件的元组是 4206 条,满足 o_orderkey<100 条件的元组是 24 条,进行 BitmapOr 操作后的元组条数是 4230 条。

下面的 SQL 语句查询订单总额大于 500 000 并且订单号小于 100 的订单信息:

```
system@test=# EXPLAIN select o_orderkey from orders where o_totalprice>500000
and o_orderkey<100;
                              QUERY PLAN
-----------------------------------------------------------------------
 Index Scan using orders_pkey on orders  (cost=0.43..8.91 rows=1 width=4)
   Index Cond: (o_orderkey < 100)
   Filter: (o_totalprice > '500000'::numeric)
(3 rows)
```

该 SQL 语句的 WHERE 条件中,使用 AND 连接两个条件,因为对满足 o_orderkey<100 条件的元组的估算只有 24 条,所以直接使用在该列上的索引 orders_pkey 进行索引扫描,找到满足条件的索引项,然后根据索引项中的指针到表中查找元组,在 o_totalprice 列上的条件充当过滤条件来找出最终满足两个条件的元组。

如果数据库应用经常一起访问 o_orderkey,o_totalprice 这两列,则可以在这两列上创建复合索引,然后看下面 SQL 语句的执行计划:

```
system@test=# CREATE INDEX i_orderkey_totalprice ON orders(o_orderkey, o_
totalprice);
system@test=# ANALYZE;
system@test=# EXPLAIN select o_orderkey from orders where o_totalprice>500000
and o_orderkey<100;
                              QUERY PLAN
-----------------------------------------------------------------------
 Index Only Scan using i_orderkey_totalprice on orders  (cost=0.43..8.67 rows=1 width=4)
   Index Cond: ((o_orderkey < 100) AND (o_totalprice > '500000'::numeric))
(2 rows)
```

可以看到同样的 SQL 语句,这次使用的是在复合索引 i_orderkey_totalprice 上的索引扫描,并且是只索引扫描,因为索引中已经包含了该 SQL 语句执行需要的所有信息(WHERE 条件和目标列中引用的表列),只需要查询索引就可以了,不需要再去访问表。

10.3.2 多表连接查询

如果 SQL 语句中访问多张表,则需要进行表之间的连接(join)操作。

KingbaseES 中的连接操作有以下 3 种执行方式。

(1) 嵌套循环连接(nested loop):对外表中的每一行元组,扫描一遍内表,构造新元组,如果满足条件,则作为结果元组输出。

(2) 归并连接(merge join):通常用于等值连接操作,如果进行连接的两个表在连接属性上已经排序,则使用归并连接算法进行连接操作具有比较好的性能。

(3) 哈希连接(hash join)：通常用于等值连接操作，对其中的一个表创建哈希表，对另外一个表的元组进行探测，满足条件则输出结果元组。

下面的 SQL 语句查询订单号小于 10 000 的订单信息：

```
system@test=# EXPLAIN select * from orders,lineitem
test=#                  where o_orderkey=l_orderkey and o_orderkey<10000;
                                QUERY PLAN
---------------------------------------------------------------------------
 Nested Loop  (cost=0.86..20201.79 rows=9342 width=240)
   ->  Index Scan using orders_pkey on orders  (cost=0.43..116.29 rows=2335 width=111)
         Index Cond: (o_orderkey < 10000)
   ->  Index Scan using i_l_orderkey on lineitem  (cost=0.43..8.43 rows=17 width=129)
         Index Cond: (l_orderkey = orders.o_orderkey)
(5 rows)
```

可以看出，该 SQL 语句的执行计划有 3 个节点，两个表都使用的是索引扫描，然后使用嵌套循环连接算法进行连接操作。首先使用表 orders 上的 orderskey 索引查找订单号小于 10 000 的元组，对每个满足条件的元组，到表 lineitem 查找具有相同订单号的订单明细信息，使用表 lineitem 上的 i_l_orderkey 索引。

使用 EXPLAIN ANALYZE 执行下面的 SQL 语句，可以看到实际执行的情况：

```
system@test=# EXPLAIN ANALYZE select * from orders ,lineitem
test-#                       where o_orderkey = l_orderkey and o_orderkey<10000;
                                QUERY PLAN
---------------------------------------------------------------------------
 Nested Loop  (cost=0.86..20201.79 rows=9342 width=240) (actual time=0.030..5.901 rows=9965 loops=1)
   ->  Index Scan using orders_pkey on orders  (cost=0.43..116.29 rows=2335 width=111) (actual time=0.019..0.340 rows=2503 loops=1)
         Index Cond: (o_orderkey < 10000)
   ->  Index Scan using i_l_orderkey on lineitem  (cost=0.43..8.43 rows=17 width=129) (actual time=0.001..0.001 rows=4 loops=2503)
         Index Cond: (l_orderkey = orders.o_orderkey)
 Planning Time: 0.320 ms
 Execution Time: 6.127 ms
(7 rows)
```

从该执行计划可以看出，表 orders 中有 2503 个订单号小于 10 000 的订单，对于每个满足条件的订单都需要到表 lineitem 上执行一次索引扫描，可以看到其 loops＝2503，即该操作执行了 2503 遍，这是嵌套循环连接算法的特点。

下面的 SQL 语句查询所有订单的订单明细信息：

```
system@test=# EXPLAIN select * from orders ,lineitem where o_orderkey = l_orderkey;
                                QUERY PLAN
---------------------------------------------------------------------------
 Merge Join  (cost=3.68..426608.41 rows=6001134 width=240)
```

```
     Merge Cond: (orders.o_orderkey = lineitem.l_orderkey)
     ->  Index Scan using orders_pkey on orders   (cost=0.43..66736.43 rows=1500000 width=111)
     ->  Index Scan using i_l_orderkey on lineitem   (cost=0.43..281120.44 rows=6001134 width=
 129)
(4 rows)
```

表orders和表lineitem上分别有150万和600多万条元组,如果使用嵌套循环连接算法,则在表lineitem上需要执行150万次索引扫描,显然性能有问题,正好在两个表上的连接属性上都有索引,查询优化器选择了归并连接算法,在两个表上的索引扫描形成两个有序的输入流,进行归并连接操作,输出结果元组。

如果表上没有索引,两个大表的连接怎么操作合适呢?删除表lineitem上的索引,看看相同SQL语句的执行计划:

```
system@test=# DROP INDEX i_l_orderkey;
system@test=# EXPLAIN select * from orders,lineitem where o_orderkey = l_
orderkey;
                          QUERY PLAN
---------------------------------------------------------------------
 Hash Join   (cost=86334.00..546261.36 rows=6001134 width=240)
   Hash Cond: (lineitem.l_orderkey = orders.o_orderkey)
   ->  Seq Scan on lineitem   (cost=0.00..184851.34 rows=6001134 width=129)
   ->  Hash   (cost=42681.00..42681.00 rows=1500000 width=111)
         ->  Seq Scan on orders   (cost=0.00..42681.00 rows=1500000 width=111)
(5 rows)
```

该SQL语句采用的是哈希连接算法,执行计划中有4个操作节点。首先扫描表orders,创建哈希表,然后扫描表lineitem,对其中的每个元组探测哈希表,输出满足连接条件的结果元组。

如果是多个表进行连接,首先选择其中的两个表进行连接,将生成的结果集与第三个表进行连接,以此类推,生成最后的结果集。

下面的SQL语句用于查询中国顾客的订单信息,禁用并行查询后的执行计划如下:

```
system@test=# SET max_parallel_workers_per_gather = 0;
system@test=# EXPLAIN   select * from orders,lineitem,customer,nation
test-#               where o_orderkey = l_orderkey   and o_custkey = c_custkey and
test-#                     c_nationkey = n_nationkey and n_name = 'chinese';
                     QUERY PLAN
---------------------------------------------------------------------
 Nested Loop   (cost=4053.92..129581.56 rows=240060 width=508)
   ->  Hash Join   (cost=4053.48..52959.48 rows=60000 width=379)
         Hash Cond: (orders.o_custkey = customer.c_custkey)
         ->  Seq Scan on orders   (cost=0.00..42681.00 rows=1500000 width=111)
         ->  Hash   (cost=3978.48..3978.48 rows=6000 width=268)
               ->  Nested Loop   (cost=114.92..3978.48 rows=6000 width=268)
                     ->  Seq Scan on nation   (cost=0.00..1.31 rows=1 width=109)
                           Filter: (n_name = 'chinese'::bpchar)
                     ->  Bitmap Heap Scan on customer   (cost=114.92..3917.17 rows=6000 width=159)
                           Recheck Cond: (c_nationkey = nation.n_nationkey)
```

```
                        -> Bitmap Index Scan on i_c_nationkey  (cost=0.00..113.42 rows=6000
width=0)
                              Index Cond: (c_nationkey = nation.n_nationkey)
    -> Index Scan using lineitem_pkey on lineitem  (cost=0.43..1.11 rows=17 width=129)
           Index Cond: (l_orderkey = orders.o_orderkey)
(14 rows)
```

该 SQL 语句是 4 张表的连接查询，根据其执行计划可以构建出如图 10-2 所示的操作树。可以看到该 SQL 语句的执行方式如下。

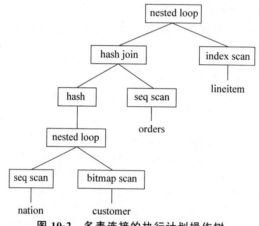

图 10-2 多表连接的执行计划操作树

（1）采用 Nestedloop 连接算法，连接表 nation 和表 customer：

① 顺序扫描（Seq Scan）表 nation，使用过滤条件 n_name = 'chinese' 来查找符合条件的行（满足条件的只有 1 行）。

② 以位图索引扫描（Bitmap Index Scan）的方式，扫描索引 i_c_nationkey，确定表 customer 中有哪些行的 c_nationkey 值，与①中找到的行的 nation.n_nationkey 值相匹配。

③ 根据②位图索引扫描的结果，以位图堆扫描（Bitmap Heap Scan）的方式，扫描表 customer，查找表 customer 中关联的行。

④ 使用嵌套循环连接算法，将表 nation 和表 customer 的结果组合在一起，并准备与表 orders 进行连接。

（2）然后采用 Hashjoin 连接算法，将（1）返回的结果集与 orders 表进行连接：

① 基于步骤（1）表 nation 和表 customer 之间嵌套循环生成的连接结果集，创建哈希表。

② 全表顺序扫描 orders 表，探测上一步生成的哈希表，对于满足哈希连接条件的元组，生成结果集（满足条件的订单）。

（3）最后采用 Nestedloop 连接算法，将（2）返回的结果集，与表 lineitem 进行连接：对步骤（2）返回的每个元组，根据连接属性值，索引扫描 lineitem 表，找到满足连接条件的元组，生成结果元组输出。

SQL 语句的执行方式跟 FROM 子句中表的顺序无关，查询优化器会根据估算的代价选择合适的表连接顺序和连接算法。

10.3.3 分组聚集查询

KingbaseES 数据库中的分组聚集操作有两种执行方式。

(1) 先排序,然后对有序的输入流进行分组聚焦计算。

(2) 使用哈希表进行分组,然后再进行聚焦计算。

下面的 SQL 语句查询所有订单总额累计最高的 5 名顾客信息:

```
system@test=# EXPLAIN select c_custkey, c_name, sum(o_totalprice) as amount
test-#                from orders , customer
test-#                where o_custkey = c_custkey
test-#                group by c_custkey
test-#                order by amount
test-#                limit 5;
                              QUERY PLAN
-----------------------------------------------------------------------
 Limit  (cost=268202.23..268202.25 rows=5 width=55)
   ->  Sort  (cost=268202.23..268577.23 rows=150000 width=55)
         Sort Key: (sum(orders.o_totalprice))
         ->  GroupAggregate  (cost=222193.65..265710.79 rows=150000 width=55)
               Group Key: customer.c_custkey
               ->  Merge Join  (cost=222193.65..256335.79 rows=1500000 width=31)
                     Merge Cond: (customer.c_custkey = orders.o_custkey)
                     ->  Index Scan using customer_pkey on customer  (cost=0.42..7519.43 rows
                                                                     =150000 width=23)
                     ->  Materialize  (cost=222192.48..229692.48 rows=1500000 width=12)
                           ->  Sort  (cost=222192.48..225942.48 rows=1500000 width=12)
                                 Sort Key: orders.o_custkey
                                 ->  Seq Scan on orders  (cost=0.00..42681.00 rows=1500000
                                                          width=12)
(12 rows)
```

在这个执行计划中,出现了 3 个比较陌生的操作节点。

(1) Sort 操作:根据排序键进行排序。

(2) Materialize 物化操作:把中间结果集写到外存文件中。

(3) GroupAggregate 分组聚集操作:根据分组键进行聚集操作,如计算总和、平均值等。

从执行计划中可以看出该 SQL 语句的执行方式如下。

(1) 表 customer 和表 orders 采用归并连接算法进行连接操作,归并连接算法需要进行连接的两个表在连接属性上是有序的。

① 表 customer 在连接属性上有索引,可以直接索引扫描。

② 表 orders 在连接属性上没有索引,采用全表顺序扫描,然后在连接属性上进行排序,获得有序的输入流。

(2) 进行聚集计算。

(3) 进行排序,输出前 5 名顾客的信息。

如果在表 orders 的连接属性 o_custkey 上创建索引，则该 SQL 语句的执行计划中在表 orders 上直接使用索引：

```
system@test=# CREATE INDEX i_o_custkey ON orders(o_custkey);
system@test=# EXPLAIN select c_custkey, c_name, sum(o_totalprice) as amount
test-#              from orders , customer
test-#              where o_custkey = c_custkey
test-#              group by c_custkey
test-#              order by amount desc
test-#              limit 5;
                                QUERY PLAN
------------------------------------------------------------------------
 Limit  (cost=188306.57..188306.58 rows=5 width=55)
   ->  Sort  (cost=188306.57..188681.57 rows=150000 width=55)
         Sort Key: (sum(orders.o_totalprice)) DESC
         ->  GroupAggregate  (cost=1.87..185815.13 rows=150000 width=55)
               Group Key: customer.c_custkey
               ->  Merge Join  (cost=1.87..176440.13 rows=1500000 width=31)
                     Merge Cond: (orders.o_custkey = customer.c_custkey)
                     ->  Index Scan using i_o_custkey on orders  (cost=0.43..
                            149795.80 rows=1500000 width=12)
                     ->  Index Scan using customer_pkey on customer  (cost=0.42..
                            7519.43 rows=150000 width=23)
(9 rows)
```

10.3.4 子查询

KingbaseES 数据库中对子查询的处理有两种执行方式。
（1）把子查询转换成连接操作。
（2）把子查询单独处理成子计划。
下面的 SQL 语句查询顾客 Customer#000000005 的订单信息：

```
system@test=# EXPLAIN select * from orders where o_custkey = (
test-#         select c_custkey from customer where c_name = 'Customer#000000005');
                                QUERY PLAN
------------------------------------------------------------------------
 Bitmap Heap Scan on orders  (cost=5477.56..5544.51 rows=17 width=111)
   Recheck Cond: (o_custkey = $0)
   InitPlan 1 (returns $0)
     ->  Seq Scan on customer  (cost=0.00..5473.00 rows=1 width=4)
           Filter: ((c_name)::text = 'Customer#000000005'::text)
   ->  Bitmap Index Scan on i_o_custkey  (cost=0.00..4.56 rows=17 width=0)
         Index Cond: (o_custkey = $0)
(7 rows)
```

表 orders 中只有顾客号，没有顾客名，如果需要根据顾客名查询订单信息，可以先到表 customer 中根据顾客名查出该顾客的顾客号，然后在表 orders 中查出该顾客的订单信息。显然该 SQL 语句是一个不相关的子查询，子查询中查出顾客 Customer#000000005 的顾客号，与父查询无关，因此，在执行计划中，该子查询单独处理成一个子计划 InitPlan，其查询结果$0作为父查询条件中的参数。

上面的SQL语句可以改写成语义相同的连接查询,其执行计划如下:

```
system@test=# EXPLAIN select * from orders,customer
test-#                 where o_custkey = c_custkey and c_name = 'Customer#000000005';
                                    QUERY PLAN
-------------------------------------------------------------------------------
 Nested Loop  (cost=4.56..5544.68 rows=10 width=270)
   ->  Seq Scan on customer  (cost=0.00..5473.00 rows=1 width=159)
         Filter: ((c_name)::text = 'Customer#000000005'::text)
   ->  Bitmap Heap Scan on orders  (cost=4.56..71.51 rows=17 width=111)
         Recheck Cond: (o_custkey = customer.c_custkey)
         ->  Bitmap Index Scan on i_o_custkey  (cost=0.00..4.56 rows=17 width=0)
               Index Cond: (o_custkey = customer.c_custkey)
(7 rows)
```

下面的SQL语句用于查询订单总额大于650万的顾客姓名:

```
system@test=# EXPLAIN select c_custkey, c_name
test-#                 from customer
test-#                 where 6500000 <
test(#                    (select sum(o_totalprice) from orders where o_custkey = c_custkey);
                                    QUERY PLAN
-------------------------------------------------------------------------------
 Seq Scan on customer  (cost=0.00..10739904.41 rows=50000 width=23)
   Filter: ('6500000'::numeric < (SubPlan 1))
   SubPlan 1
     ->  Aggregate  (cost=71.55..71.56 rows=1 width=32)
           ->  Bitmap Heap Scan on orders  (cost=4.56..71.51 rows=17 width=8)
                 Recheck Cond: (o_custkey = customer.c_custkey)
                 ->  Bitmap Index Scan on i_o_custkey  (cost=0.00..4.56 rows=17
                                                              width=0)
                       Index Cond: (o_custkey = customer.c_custkey)
(8 rows)
```

这是一个相关子查询的例子,对于父查询中的每个顾客,都到子查询中根据该顾客的顾客号在表orders中查询该顾客的订单总额返回给父查询,父查询再判断是否满足条件,如果满足条件,该顾客信息作为结果元组输出。这类相关子查询,在KingbaseES数据库中处理为SubPlan。

上面的SQL语句可以改写成语义相同的连接查询,其执行计划如下:

```
system@test=# EXPLAIN select c_custkey, c_name,sum(o_totalprice) as amount
test-#                 from customer,orders
test-#                 where o_custkey = c_custkey
test-#                 group by c_custkey
test-#                 having sum(o_totalprice) > 6500000;
                                    QUERY PLAN
-------------------------------------------------------------------------------
 GroupAggregate  (cost=1.87..190315.13 rows=50000 width=55)
```

```
          Group Key: customer.c_custkey
          Filter: (sum(orders.o_totalprice) > '6500000'::numeric)
       -> Merge Join  (cost=1.87..176440.13 rows=1500000 width=31)
            Merge Cond: (customer.c_custkey = orders.o_custkey)
         -> Index Scan using customer_pkey on customer  (cost=0.42..7519.43 rows
                                                           =150000 width=23)
         -> Index Scan using i_o_custkey on orders  (cost=0.43..149795.80 rows=
                                                       1500000 width=12)
(7 rows)
```

这两种 SQL 语句的写法具有相同的语义,但执行方式不同,因此,SQL 语句的性能也可能不同。由此可以选择一个性能较高的 SQL 语句的写法。

10.4 影响 SQL 语句的执行计划

影响 SQL 语句的执行计划是通过改变查询的执行方式,使查询能够更高效地执行,以提高查询性能和响应时间。

10.4.1 更新数据库的统计信息

当遇到 SQL 语句执行慢时,首先想到的是系统的统计信息是否及时更新了。查询优化器使用数据库的统计信息来评估执行计划的成本。如果统计信息不准确或过时,可能会导致执行计划选择不当。通过更新统计信息,可以提供更准确的信息给查询优化器,以得到更优的执行计划。

KingbaseES 数据库提供了以下两种收集统计信息的机制。

(1) **自动收集统计信息**:KingbaseES 数据库的自动统计分析(autovacuum)是默认启用的。通过自动统计分析,系统会定期收集表和索引的统计信息。在 kingbase.conf 配置文件中,可以配置和调整 autovacuum 的相关参数。

(2) **手动收集统计信息**:执行 ANALYZE 命令,可同时执行清理和统计信息收集,这条命令会收集整个数据库的统计信息。如果要收集某个表的统计数据,可以执行命令 ANALYZE table_name。

下面的 SQL 语句收集当前连接的数据库 test 的统计信息:

```
system@test=# ANALYZE;
```

如果只想收集单个表(如表 orders)的统计信息,可以执行下面的 SQL 语句:

```
system@test=# ANALYZE orders;
ANALYZE
system@test=# select nspname, relname, relpages, reltuples from sys_class, sys
_namespace
test-#          where relnamespace = sys_namespace.oid and nspname='tpch' and
relname='orders';
 nspname | relname | relpages | reltuples
---------+---------+----------+-----------
 tpch    | orders  |    27681 |    1.5e+06
(1 row)
```

在系统表 sys_class 中可以查询到表中元组数和该表所占用的操作系统文件的页面数。

10.4.2　创建合适的索引

合适的索引可以加速查询操作。通过创建、删除或修改索引，可以改变执行计划中的访问路径，以便更高效地利用索引进行数据检索。

下面是索引影响 SQL 语句执行计划的实战示例。这条语句是 TPC-H 测试中的一条 SQL 语句：

```
system@test=# DROP INDEX lineitem_l_partkey_idx;
--说明:删除该索引将导致接下来的 SQL 语句需要运行很长的时间
system@test=# \timing
system@test=# select sum(l_extendedprice) / 7.0 as avg_yearly
test-#          from lineitem,part
test-#          where p_partkey = l_partkey and p_brand = 'Brand#45' and
                  p_container = 'LG PKG'
test-#              and l_quantity < ( select 0.2 * avg(l_quantity)
test(#                                   from lineitem
test(#                                   where l_partkey = p_partkey );
    avg_yearly
---------------------
 288975.151428571429
(1 row)
Time: 2971403.114 ms (49:31.403)
```

执行下面的 SQL 语句查看该 SQL 语句的执行计划：

```
system@test=# EXPLAIN select sum(l_extendedprice) / 7.0 as avg_yearly
test-#     from lineitem,part
test-#     where p_partkey = l_partkey and p_brand = 'Brand#45' and
test-#         p_container = 'LG PKG'and l_quantity < ( select 0.2 * avg(l_quantity)
test(#                                      from lineitem
test(#                                      where l_partkey = p_partkey );
                            QUERY PLAN
-----------------------------------------------------------------
 Aggregate  (cost=2205463.41..2205463.42 rows=1 width=32)
   ->  Hash Join  (cost=6388.40..2205458.21 rows=2080 width=8)
         Hash Cond: (lineitem.l_partkey = part.p_partkey)
         Join Filter: (lineitem.l_quantity < (SubPlan 1))
         ->  Seq Scan on lineitem  (cost=0.00..184845.72 rows=6000572 width=17)
         ->  Hash  (cost=6385.80..6385.80 rows=208 width=4)
               ->  Gather  (cost=1000.00..6385.80 rows=208 width=4)
                     Workers Planned: 2
                     ->  Parallel Seq Scan on part  (cost=0.00..5365.00 rows=87 width=4)
                           Filter: ((p_brand = 'Brand#45'::bpchar) AND (p_container = 'LG
                                                                       PKG'::bpchar))
         SubPlan 1
           ->  Aggregate  (cost=199847.23..199847.25 rows=1 width=32)
                 ->  Seq Scan on lineitem lineitem_1  (cost=0.00..199847.15 rows=32 width=5)
                       Filter: (l_partkey = part.p_partkey)
```

```
(14 rows)
Time: 1.549 ms
```

观察这个执行计划,可以发现如下信息。

(1) 对表 lineitem 和表 part 的扫描都是全表顺序扫描。

(2) 执行计划中每个节点的执行代价都比较高。

因此有理由怀疑该 SQL 语句是因为缺少了一些索引,而导致 SQL 语句运行速度慢。

在上面的 SQL 语句执行计划中,有如下情形。

(1) lineitem.l_partkey = part.p_partkey(连接条件)。

(2) lineitem.l_quantity < (SubPlan 1)(连接过滤条件)。

连接条件和过滤条件所涉及的列,一般来说,是需要创建索引的候选列,很有可能是缺失了这几个索引。

现在可以创建这些索引,并重新进行测试来验证如下两个事实。

(1) 创建这些索引对上面的 SQL 语句的性能有巨大的帮助。

(2) 创建这些索引对其他的 SQL 语句的性能没有副作用,不会导致性能变差。

首先基于连接条件 lineitem.l_partkey = part.p_partkey,创建两个索引:

```
system@test=# CREATE INDEX lineitem_l_partkey_idx  ON lineitem(l_partkey);
system@test=# CREATE INDEX partsupp_ps_partkey_idx ON partsupp(ps_partkey);
system@test=# ANALYZE;
```

然后执行下面的 SQL 语句,查看该 SQL 语句的执行情况:

```
system@test=# EXPLAIN ANALYZE
test-#       select sum(l_extendedprice) / 7.0 as avg_yearly
test-#       from lineitem,part
test-#       where p_partkey = l_partkey and p_brand = 'Brand#45' and p_container = 'LG PKG'
test-#       and l_quantity < ( select 0.2 * avg(l_quantity)
test(#                          from    lineitem
test(#                          where   l_partkey = p_partkey);
                        QUERY PLAN
-----------------------------------------------------------------
 Aggregate  (cost=208315.46..208315.48 rows=1 width=32) (actual time=1102.767..1102.825 rows=1 loops=1)
   ->  Hash Join  (cost=6388.40..208310.26 rows=2080 width=8) (actual time=17.763..1102.606 rows=482 loops=1)
         Hash Cond: (lineitem.l_partkey = part.p_partkey)
         Join Filter: (lineitem.l_quantity < (SubPlan 1))
         Rows Removed by Join Filter: 5169
         ->  Seq Scan on lineitem  (cost=0.00..184852.15 rows=6001215 width=17) (actual time=0.004..502.598 rows=6001215 loops=1)
         ->  Hash  (cost=6385.80..6385.80 rows=208 width=4) (actual time=17.292..17.346 rows=189 loops=1)
               Buckets: 1024  Batches: 1  Memory Usage: 15kB
               ->  Gather  (cost=1000.00..6385.80 rows=208 width=4) (actual time=0.454..17.286 rows=189 loops=1)
                     Workers Planned: 2
```

```
                    Workers Launched: 2
                    -> Parallel Seq Scan on part  (cost=0.00..5365.00 rows=87 width=4)
                                      (actual time=0.175..12.450 rows=63 loops=3)
                          Filter: ((p_brand = 'Brand#45'::bpchar) AND
                                   (p_container = 'LG PKG'::bpchar))
                          Rows Removed by Filter: 66604
            SubPlan 1
              -> Aggregate  (cost=131.63..131.64 rows=1 width=32) (actual time=0.045..0.045
                                                      rows=1 loops=5651)
                    -> Bitmap Heap Scan on lineitem lineitem_1  (cost=4.68..131.54 rows=32
                                                      width=5)
                                      (actual time=0.010..0.040 rows=31 loops=5651)
                          Recheck Cond: (l_partkey = part.p_partkey)
                          Heap Blocks: exact=175056
                          -> Bitmap Index Scan on lineitem_l_partkey_idx  (cost=0.00..4.67 rows
                                                      =32 width=0)
                                      (actual time=0.007..0.007 rows=31 loops=5651)
                                Index Cond: (l_partkey = part.p_partkey)
 Planning Time: 0.357 ms
 Execution Time: 1102.862 ms
(23 rows)
Time: 1103.766 ms (00:01.104)
```

可以看到，增加了这两个索引后，SQL 语句的执行时间由原来的 49 分 32 秒减少到 1.393 秒，这是一个非常惊人的性能改善。

继续基于连接的过滤条件 lineitem.l_quantity ＜（SubPlan 1），创建 1 个索引：

```
system@test=# CREATE INDEX lineitem_l_quantity_idx ON lineitem(l_quantity);
system@test=# ANALYZE;
```

创建完索引后，重新测试该条 SQL 语句：

```
system@test=# EXPLAIN ANALYZE
test-#           select sum(l_extendedprice) / 7.0 as avg_yearly
test-#           from lineitem,part
test-#           where p_partkey = l_partkey and p_brand = 'Brand#45' and
                 p_container = 'LG PKG'
test-#              and l_quantity < ( select 0.2 * avg(l_quantity)
test(#                                 from    lineitem
test(#                                 where   l_partkey = p_partkey);
                                      QUERY PLAN
-----------------------------------------------------------------
 Aggregate  (cost=208316.19..208316.20 rows=1 width=32) (actual time=1539.794..1539.835
rows=1 loops=1)
   -> Hash Join  (cost=6388.18..208311.04 rows=2060 width=8) (actual time=42.038..1539.614
rows=482 loops=1)
         Hash Cond: (lineitem.l_partkey = part.p_partkey)
         Join Filter: (lineitem.l_quantity < (SubPlan 1))
```

```
                    Rows Removed by Join Filter: 5169
            ->  Seq Scan on lineitem  (cost=0.00..184853.15 rows=6001215 width=17) (actual time
                                                        =0.005..538.916 rows=6001215 loops=1)
            ->  Hash  (cost=6385.60..6385.60 rows=206 width=4) (actual time=38.418..38.458
                                                        rows=189 loops=1)
                  Buckets: 1024  Batches: 1  Memory Usage: 15kB
                  ->  Gather  (cost=1000.00..6385.60 rows=206 width=4) (actual time=0.219..38.
                                                        384 rows=189 loops=1)
                        Workers Planned: 2
                        Workers Launched: 2
                        ->  Parallel Seq Scan on part  (cost=0.00..5365.00 rows=86 width=4)
                                                (actual time=0.205..24.001 rows=63 loops=3)
                              Filter: ((p_brand = 'Brand#45'::bpchar) AND (p_container = 'LG PKG'::
                                                        bpchar))
                              Rows Removed by Filter: 66604
         SubPlan 1
           ->  Aggregate  (cost=131.63..131.64 rows=1 width=32) (actual time=0.066..0.066
                                                        rows=1 loops=5651)
                 ->  Bitmap Heap Scan on lineitem lineitem_1  (cost=4.68..131.54 rows=32
                                                        width=5)
                                                (actual time=0.012..0.061 rows=31 loops=5651)
                       Recheck Cond: (l_partkey = part.p_partkey)
                       Heap Blocks: exact=175056
                       ->  Bitmap Index Scan on lineitem_l_partkey_idx  (cost=0.00..4.67 rows
                                                        =32 width=0)
                                                (actual time=0.006..0.006 rows=31 loops=5651)
                             Index Cond: (l_partkey = part.p_partkey)
 Planning Time: 0.248 ms
 Execution Time: 1539.867 ms
(23 rows)
Time: 1541.397 ms (00:01.541)
```

可以看到,增加新的索引后,性能并没有大的改善(执行多次后得出的结论),甚至有时性能会有所降低,需要删除该索引:

```
system@test=# DROP INDEX lineitem_l_quantity_idx;
```

至此,完成了该条 SQL 语句的优化。

10.4.3 影响执行计划的配置参数

KingbaseES 数据库提供了一系列的配置参数来控制在生成执行计划时采用的执行算法,这些配置参数都是用户级参数,在对 SQL 语句进行调优时,可以使用这些参数来调整 SQL 语句的执行计划,并验证其性能,从而找出合适的执行方法。

对于表扫描的执行算法,可以使用下面的参数进行控制:

```
enable_seqscan = on          #参数值为 on 时,在执行计划中使用顺序扫描(sequential scan)
enable_indexscan = on        #参数值为 on 时,在执行计划中使用索引扫描(index scan)
enable_indexonlyscan = on    #参数值为 on 时,在执行计划中使用仅索引扫描(index-only scan)
enable_bitmapscan = on       #参数值为 on 时,在执行计划中使用位图扫描(bitmap scan)
```

下面的 SQL 语句查询订单总额大于 500 000 的订单信息：

```
system@test=# EXPLAIN select * from orders where o_totalprice>500000;
                                    QUERY PLAN
-----------------------------------------------------------------------
 Bitmap Heap Scan on orders   (cost=101.98..11631.08 rows=4330 width=111)
   Recheck Cond: (o_totalprice > '500000'::numeric)
   ->  Bitmap Index Scan on idx_orders_o_totalprice  (cost=0.00..100.90 rows=
                                                     4330 width=0)
         Index Cond: (o_totalprice > '500000'::numeric)
(4 rows)
```

如果位图扫描这种执行算法性能不是太好，可以尝试关闭参数 enable_bitmapscan，则 SQL 语句的执行计划就会发生变化，如下所示：

```
system@test=# SET enable_bitmapscan = off;       --关闭参数 enable_bitmapscan
system@test=# EXPLAIN select * from orders where o_totalprice>500000;
                                    QUERY PLAN
-----------------------------------------------------------------------
 Index Scan using idx_orders_o_totalprice on orders   (cost=0.43..16207.87 rows=
                                                      4330 width=111)
   Index Cond: (o_totalprice > '500000'::numeric)
(2 rows)
system@test=# SET enable_bitmapscan = on;        --重新打开参数 enable_bitmapscan
```

对于表连接操作，可以使用下面的参数进行控制：

```
#enable_hashjoin = on    #参数值为 on 时,在执行计划中使用哈希连接(hash join)
#enable_nestloop = on    #参数值为 on 时,在执行计划中使用嵌套循环连接(nested loop join)
#enable_mergejoin = on   #参数值为 on 时,在执行计划中使用合并连接(merge join)
```

下面的 SQL 语句查询订单号小于 10 000 的订单信息：

```
system@test=# SET max_parallel_workers_per_gather = 0;
system@test=# EXPLAIN select * from orders,lineitem
test-#              where o_orderkey = l_orderkey and o_orderkey<10000;
                                    QUERY PLAN
-----------------------------------------------------------------------
 Nested Loop   (cost=0.86..76403.88 rows=9386 width=240)
   ->  Index Scan using orders_pkey on orders   (cost=0.43..116.48 rows=2346 width=111)
         Index Cond: (o_orderkey < 10000)
   ->  Index Scan using lineitem_pkey on lineitem   (cost=0.43..32.35 rows=17 width=129)
         Index Cond: (l_orderkey = orders.o_orderkey)
(5 rows)
```

关闭参数 enable_nestloop 后，SQL 语句的执行计划如下所示：

```
system@test=# SET enable_nestloop = off;
system@test=# EXPLAIN select * from orders,lineitem
test-#              where o_orderkey = l_orderkey and o_orderkey<10000;
                                    QUERY PLAN
-----------------------------------------------------------------------
 Hash Join   (cost=145.81..200752.95 rows=9386 width=240)
```

```
            Hash Cond: (lineitem.l_orderkey = orders.o_orderkey)
            ->  Seq Scan on lineitem   (cost=0.00..184853.54 rows=6001354 width=129)
            ->  Hash  (cost=116.48..116.48 rows=2346 width=111)
                  ->  Index Scan using orders_pkey on orders   (cost=0.43..116.48 rows=2346
                                                                        width=111)
                        Index Cond: (o_orderkey < 10000)
(6 rows)
system@test=# set enable_nestloop = on;
```

该 SQL 语句的连接算法变成了哈希连接算法，可以实际执行下该 SQL 语句，看看性能是否有变化。通过这种方式，可以尝试并对比结果从而得到哪种执行算法的效率更高，给 SQL 语句调优提供了一种手段。

同理，对于聚集操作的方法，也可以使用下面的参数来控制：

```
enable_hashagg = on
```

10.4.4　使用查询提示 SQL hint

在某些情况下，在 SQL 语句中使用查询提示（SQL hint）来指定强制使用特定的算法，从而改变执行计划。这可以用于处理某些特殊情况下优化器无法得到最佳执行计划的场景。

查询优化器通过在目标 SQL 语句 SELECT 之后给出的特殊形式的注释来读取 hint 注释。注释的形式以字符序列/＊＋开头，以＊/结尾，如/＊＋SeqScan(tablename)＊/。

SQL hint 的功能由系统参数 enable_hint 控制，该参数默认为 off。如果希望使用 SQL hint 功能，在系统配置文件 kingbase.conf 中添加如下参数：

```
enable_hint = on
```

该参数是语句级参数，也可以直接使用 SET 语句设置。

下面的例子执行一个没有 SQL hint 的语句：

```
system@test=# SHOW enable_hint;
 enable_hint
--------------
 off
(1 row)
system@test=# SET enable_hint = on;
SET
system@test=# SHOW enable_hint;
 enable_hint
--------------
 on
(1 row)
system@test=# EXPLAIN select * from orders where o_totalprice>500000;
                                 QUERY PLAN
-------------------------------------------------------------------------------
 Bitmap Heap Scan on orders   (cost=101.98..11631.08 rows=4330 width=111)
   Recheck Cond: (o_totalprice > '500000'::numeric)
   ->  Bitmap Index Scan on idx_orders_o_totalprince   (cost=0.00..100.90 rows=
```

```
                                                       4330 width=0)
        Index Cond: (o_totalprice > '500000'::numeric)
(4 rows)
```

可以看到,该语句的执行计划中,使用位图扫描方法扫描表 orders。

使用 SQL hint 提示 /*+IndexScan(orders)*/,强制语句扫描表 orders 时使用索引扫描方法:

```
system@test=# EXPLAIN select /*+IndexScan(orders)*/ * from orders where o_totalprice>500000;
                              QUERY PLAN
---------------------------------------------------------------------
 Index Scan using idx_orders_o_totalprice on orders  (cost=0.43..16207.87 rows=
                                                       4330 width=111)
   Index Cond: (o_totalprice > '500000'::numeric)
(2 rows)
system@test=#
```

可以看到,执行计划确实在扫描表 orders 的时候,使用了索引扫描方法。

下面是使用 SQL hint 改变连接算法的例子:

```
system@test=# EXPLAIN select * from orders ,lineitem
test-#                where o_orderkey = l_orderkey and o_orderkey<10000;
                              QUERY PLAN
---------------------------------------------------------------------
 Nested Loop  (cost=0.86..76403.88 rows=9386 width=240)
   ->  Index Scan using orders_pkey on orders  (cost=0.43..116.48 rows=2346 width
                                                 =111)
         Index Cond: (o_orderkey < 10000)
   ->  Index Scan using lineitem_pkey on lineitem  (cost=0.43..32.35 rows=17
                                                     width=129)
         Index Cond: (l_orderkey = orders.o_orderkey)
(5 rows)
```

可以看到,该语句的执行计划中,使用嵌套循环连接方法进行两个表的连接操作。

使用 SQL hint 提示 /*+hashjoin(orders lineitem)*/,强制 SQL 语句使用哈希连接算法,如下所示:

```
system@test=# EXPLAIN select/*+hashjoin(orders lineitem)*/ * from orders
,lineitem
test-#                where o_orderkey = l_orderkey and o_orderkey<10000;
                              QUERY PLAN
---------------------------------------------------------------------
 Hash Join  (cost=145.81..200752.95 rows=9386 width=240)
   Hash Cond: (lineitem.l_orderkey = orders.o_orderkey)
   ->  Seq Scan on lineitem  (cost=0.00..184853.54 rows=6001354 width=129)
   ->  Hash  (cost=116.48..116.48 rows=2346 width=111)
         ->  Index Scan using orders_pkey on orders  (cost=0.43..116.48 rows=2346
                                                       width=111)
```

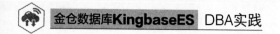

```
         Index Cond: (o_orderkey < 10000)
(6 rows)
```

可以看到,执行计划确实在表 orders 和表 lineitem 连接操作时使用了哈希连接算法。

KingbaseES 数据库的 SQL hint 的功能还在一直完善中,有关该功能的最新信息,可以参考官方网站。

第 11 章

物理数据库备份与恢复

本章首先介绍 KingbaseES 数据库备份与恢复的基本概念，然后聚焦于 KingbaseES 数据库的物理备份和恢复，重点介绍了 KingbaseES 数据库脱机冷备份与恢复和基于 sys_rman 的备份与恢复（单机环境）。

本章的实验环境可以使用 1.4.5 节制作的虚拟机备份文件 dbserver-3DB-BestPractice.rar。

 ## 11.1 数据库备份与恢复的基本概念

11.1.1 逻辑备份与物理备份

KingbaseES 数据库的**逻辑备份**使用 sys_dump、exp 等工具，导出 KingbaseES 数据库中的数据库对象（如用户、表空间、数据库、模式、表、索引、视图、存储过程等）。KingbaseES 数据库的**逻辑恢复**使用 sys_restore、imp 等工具，将 KingbaseES 数据库的逻辑备份导入 KingbaseES 数据库。

KingbaseES 数据库的逻辑备份和恢复具有以下特点。

(1) 备份特定的数据库对象和数据：逻辑备份关注表、视图、存储过程等对象及其数据。

(2) 更好的灵活性：允许用户只恢复数据库的特定部分，而不是整个数据库。

(3) 依赖于数据库结构：逻辑备份需要了解数据库的结构和模式。

(4) 移植性更强：逻辑备份的数据可以用于不同版本或不同类型的数据库系统。

(5) 速度慢：逻辑备份和恢复过程可能比物理备份慢，尤其是对于大型数据库。

KingbaseES 数据库的**物理备份**将备份 KingbaseES 数据库集簇的所有物理文件，包括数据文件、控制文件、WAL 日志文件和归档日志文件等，适用于备份数据规模比较大的 KingbaseES 数据库。

KingbaseES 数据库的物理备份和恢复具有以下特点。

(1) 备份数据库的所有物理文件：包括数据文件、控制文件和日志文件。

(2) 快速恢复：物理备份可以用于快速恢复整个数据库，因为物理备份包含了所有必要的文件。

(3) 不依赖于数据库结构：物理备份不关心数据的逻辑结构，只关心文件的物理内容。

(4) 灾难恢复：在严重故障如硬件损坏的情况下，物理备份是恢复数据库的关键。

（5）可能需要数据库停机：为确保数据一致性，某些物理备份可能需要在数据库停机时进行。

总的来说，物理备份和逻辑备份各有优势，KingbaseES 数据库用户和 DBA 通常需要结合这两种备份方法，为用户数据提供更为全面的保护。

11.1.2 冷备份和热备份

冷备份（cold backup）也称**离线备份**（offline backup），是关闭 KingbaseES 数据库后，备份整个 KingbaseES 数据库集簇。理解 KingbaseES 数据库集簇、数据库集簇的物理结构这两个基本概念，是掌握 KingbaseES 数据库集簇冷备份的关键。由于备份的时候需要关闭 KingbaseES 数据库服务器，因此对于需要 7×24 小时运行的用户生产数据库来说，不适合使用冷备份，更适合采用热备份。

热备份（hot backup）也称**联机备份**（online backup），是在 KingbaseES 数据库正在运行的时候，对 KingbaseES 数据库集簇进行备份。KingbaseES 数据库进行联机备份时，系统会持续修改这些物理文件，因此数据库的联机备份不是数据一致的备份。

使用 KingbaseES 数据库的联机数据库热备份来恢复一个生产数据库，需要有两个过程：复原和恢复。**复原**（restore）指的是将联机热备份中的 KingbaseES 数据库集簇的数据文件、控制文件和日志文件复制回 KingbaseES 数据库服务器的过程。**恢复**（recovery）指的是将归档日志文件和 WAL 日志文件应用到复原后的数据库文件，直到将 KingbaseES 数据库恢复到故障点、DBA 指定的时间点或事务 ID，将 KingbaseES 数据库恢复到一致性的状态。

因此在物理备份时最好执行 1 次切换 WAL 日志的操作，然后再备份自上次以来未备份系统生成的归档日志文件，如图 11-1 所示。

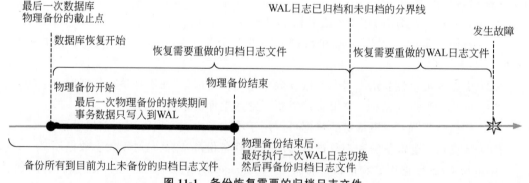

图 11-1　备份恢复需要的归档日志文件

如果在一个归档日志文件系列中，丢失了其中一个归档日志文件，会有什么问题？该怎么办呢？如图 11-2 所示，丢失了归档日志 k 后，从之前的物理备份开始，无论如何也无法恢复到日志 k 之后的数据库状态了。当然 DBA 也无法从现有的数据库状态出发，来恢复数据库过去的状态。因此丢失归档日志文件 k 之后的处理方法就是：DBA 需要马上进行一次数据库的物理备份。

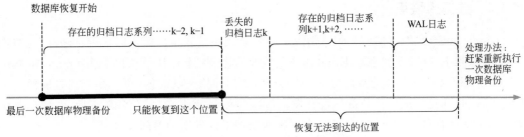

图 11-2　丢失归档日志文件的影响

11.1.3　全量备份、差异备份和增量备份

对于 KingbaseES 数据库联机热备份来说，可用如下 3 种方式来备份数据库。

（1）**全量备份**（full backup）：不依赖于其他的备份，因为全量备份已经备份了 KingbaseES 数据库集簇的所有文件和数据，可以用于独立恢复整个数据库集簇。

（2）**差异备份**（differential backup）：依赖于上一次的全量备份，差异备份将备份上一次全量备份后发生变化的数据，如图 11-3 所示。

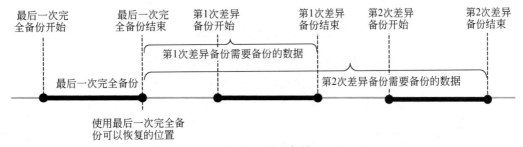

图 11-3　差异备份

（3）**增量备份**（incremental backup）：同样依赖于上一次的全量备份，增量备份将备份自从上一次全量备份或最后一次增量备份之后所有发生变化的数据，如图 11-4 所示。

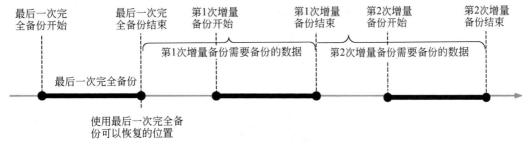

图 11-4　增量备份

当前，KingbaseES 数据库的增量备份支持两种粒度方式。

① 文件粒度：当某个数据文件的一个数据块发生变化后，增量备份将复制整个数据文件。

② 块粒度：当某个数据文件的一个数据块发生变化后，块增量备份只复制变化的数据块。

11.1.4 数据库恢复

KingbaseES 数据库的恢复默认情况下是恢复到系统发生故障点的最后时刻,实现数据不丢失,这也称为完全恢复。KingbaseES 数据库提供基于时间点的恢复(point-in-time recovery,PITR)。PITR 是指进行 KingbaseES 数据库恢复的时候,可以恢复到之前的一个时间点,该时间点后面的操作好像没有发生,也称为不完全恢复。如图 11-5 所示,使用备份文件和系统运行日志把 KingbaseES 数据库恢复到错误发生的时间点 Tc。

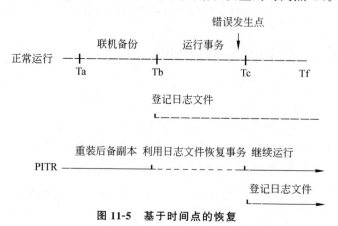

图 11-5 基于时间点的恢复

系统的日志逻辑是随着时间连续增长的,很容易标识日志记录的先后顺序,但是使用 PITR 技术把 KingbaseES 数据库恢复到指定时间点后,系统继续运行时,产生的日志与原来系统中时间点 Tc 之后的日志已经不是线性关系了,为了区分这两段日志,KingbaseES 数据库引入了时间线(timeline)的概念。

时间线标识了 KingbaseES 数据库运行的时间轴,KingbaseES 数据库系统总是在一个时间线上运行,如果对 KingbaseES 数据库系统执行了 PITR,KingbaseES 数据库系统恢复后会生成新的时间线,KingbaseES 数据库系统在新的时间线上运行,日志会在新的时间线上生成。如图 11-6 所示,KingbaseES 数据库系统在时间线 1 上运行时,在 T1 时刻出现误操作,使用备份把 KingbaseES 数据库系统恢复到 T1 时间点,生成时间线 2,在时间线 2 上继续运行。

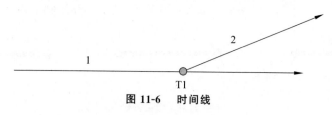

图 11-6 时间线

WAL 日志文件名 000000010000000000000001 中前 8 位 00000001 表示的是 KingbaseES 数据库的时间线。数据库初始化后,默认时间线是 1,随着数据库系统的运行,新时间线会在以下两种情况下产生。

(1) PITR 恢复后执行了 promote 操作 select sys_wal_replay_resume()。

(2) 备节点升为主节点。

每次创建一个新的时间线，KingbaseES 数据库都会在目录/u00/Kingbase/ES/V9/data/sys_wal/下创建一个时间线历史文件，文件名后缀为.history，里面的内容是由原时间线 history 文件的内容和追加一条当前时间线切换记录组成。假设 KingbaseES 数据库恢复启动后，切换到新的时间线 ID 为 2，那么文件名就是 00000002.history，该文件记录了当前时间线是在什么时间、从哪个时间线由于什么原因分出来的，时间线 history 文件可能含有多行记录。

执行下面的 ksql 命令和 SQL 语句可以查看当前的时间线：

```
[kingbase@dbsvr ~]$ ksql -d test -U system -c "select timeline_id from sys_control_checkpoint();"
 timeline_id
-------------
           1
(1 row)
```

11.2 为生产系统引入备份恢复测试机

在 KingbaseES 数据库生产环境中，准备一台备份恢复测试机，这可以带来如下好处。

(1) 通过在备份恢复测试机上恢复生产数据库上的 KignbaseES 数据库物理备份，可以验证备份的有效性。

(2) 当生产数据库发生一些逻辑错误（例如，在更新某个表行某个列值的时候，由于没有使用 WHERE 条件，造成该列的值都被更新为一样的值）的时候，通过在备份恢复测试机上，将数据库恢复到发生逻辑错误之前的时刻，可以恢复误操作表的数据，然后将该表从备份恢复测试机上导回生产数据库。

(3) 当生产数据库发生严重硬件问题的时候，备份恢复测试机可以作为临时服务器，保证生产持续进行。

为了方便备份恢复测试机和 KingbaseES 数据库生产服务器之间传送数据库备份文件，一个可行的方案如下。

(1) 为 KingbaseES 生产数据库服务器配置 rsync 服务，方便 KingbaseES 数据库服务器与备份恢复测试机之间的数据同步。

(2) 在 KingbaseES 生产数据库服务器和备份恢复测试机之间配置无密码 ssh 访问。

11.2.1 准备数据库备份恢复测试机

使用 1.4.5 节制作的虚拟机备份文件 dbtest-2DB.rar，解压缩并运行后，使用用户 kingbase，执行下面的命令，创建用于备份的目录：

```
[kingbase@dbtest ~]$ mkdir -p /dbbak/pbak/backupdb/archivelog
[kingbase@dbtest ~]$ mkdir -p /dbbak/pbak/hot.old/dbcluster/
[kingbase@dbtest ~]$ mkdir -p /dbbak/pbak/hot/dbcluster/
[kingbase@dbtest ~]$ mkdir -p /dbbak/pbak/cold
```

11.2.2 为数据库服务器配置 rsync 服务

执行下面的步骤为 KingbaseES 生产数据库服务器配置 rsync 服务。

(1) 以用户 root 的身份,使用 vi 编辑器编辑 rsyncd 服务的配置文件 /etc/rsyncd.conf,将如下内容添加到文件的末尾:

```
wid = kingbase
gid = dba
use chroot = no
address = 192.168.100.22
port=873
log file = /var/log/rsyncd.log
pid file = /var/run/rsyncd.pid
hosts allow = 192.168.100.19/24
[kingbaseESarchivelog]
path =/u04/Kingbase/ES/V9/archivelog
comment = KingbaseES DBMS archivelog Directory
read only = yes
dont compress = *.gz *.bz2 *.tgz *.zip *.rar *.z
auth users = kingbase
secrets file = /etc/rsyncd_users.db
```

(2) 执行下面的命令为 rsyncd 创建用户密码文件:

```
[root@dbsvr ~]# echo "kingbase:kingbase123" >/etc/rsyncd_users.db
[root@dbsvr ~]# chmod 600 /etc/rsyncd_users.db
```

(3) 执行下面的命令配置开机自动启动 rsyncd 服务,并启动 rsyncd 服务:

```
[root@dbsvr ~]# systemctl enable rsyncd.service    #设置 rsync 开机启动
[root@dbsvr ~]# systemctl start rsyncd.service     #启动 rsync 服务
```

(4) 执行下面的命令可以查看 rsyncd 服务的运行状态:

```
[root@dbsvr ~]# systemctl status rsyncd.service
   rsyncd.service - fast remote file copy program daemon
   Loaded: loaded (/usr/lib/systemd/system/rsyncd.service; enabled; vendor preset: disabled)
   Active: active (running) since Sat 2024-02-10 19:44:49 CST; 15s ago
 Main PID: 68369 (rsync)
    Tasks: 1
   CGroup: /system.slice/rsyncd.service
           └─68369 /usr/bin/rsync --daemon --no-detach

Feb 10 19:44:49 dbsvr systemd[1]: Started fast remote file copy program daemon.
Feb 10 19:44:49 dbsvr rsyncd[68369]: Unknown Parameter encountered: "id"
Feb 10 19:44:49 dbsvr rsyncd[68369]: IGNORING unknown parameter "id"
Feb 10 19:44:49 dbsvr rsyncd[68369]: params.c:Parameter() - Ignoring badly formed line in config file: port 873
Feb 10 19:44:49 dbsvr rsyncd[68369]: rsyncd version 3.1.2 starting, listening on port 873
```

```
[root@dbsvr ~]#
```

11.2.3 配置主机间的无密码 ssh

为了运维的方便，一般会在 KingbaseES 生产数据库服务器和备份恢复测试机之间配置 ssh 无密码访问，下面是具体的配置步骤。

(1) 在生产数据库服务器上，以用户 kingbase 的身份，执行命令 ssh-keygen，生成公钥和私钥：

```
[kingbase@dbsvr ~]$ ssh-keygen
Generating public/private rsa key pair.
Enter file in which to save the key (/home/kingbase/.ssh/id_rsa): #按回车键
Created directory '/home/kingbase/.ssh'.
Enter passphrase (empty for no passphrase):              #按回车键
Enter same passphrase again:                             #按回车键
#省略了许多输出
[kingbase@dbsvr ~]$
```

(2) 在生产数据库服务器上，执行命令 ssh-copy-id，将公钥复制到备份恢复测试机的 ~/.ssh/authorized_keys 文件中：

```
[kingbase@dbsvr ~]$ ssh-copy-id -i /home/kingbase/.ssh/id_rsa.pub kingbase@192.168.100.19
/usr/bin/ssh-copy-id: INFO: Source of key(s) to be installed: "/home/kingbase/.ssh/id_rsa.pub"
The authenticity of host '192.168.100.19 (192.168.100.19)' can't be established.
ECDSA key fingerprint is SHA256:r/lTnKpUArFJQBZzwmqXwfpKT281ru8+7/4Y2N1kr9E.
ECDSA key fingerprint is MD5:ff:d6:e4:64:fb:3a:2b:46:e6:44:d0:5a:68:01:5f:ac.
Are you sure you want to continue connecting (yes/no)? yes
/usr/bin/ssh-copy-id: INFO: attempting to log in with the new key(s), to filter
out any that are already installed
/usr/bin/ssh-copy-id: INFO: 1 key(s) remain to be installed -- if you are prompted
now it is to install the new keys
kingbase@192.168.100.19's password: #输入 kingbase 用户的密码 kingbase123
#省略了一些输出
[kingbase@dbsvr ~]$
```

(3) 在生产数据库服务器上，执行下面的 ssh 命令，无密码登录到备份恢复测试机上：

```
[kingbase@dbsvr ~]$ ssh 192.168.100.19
Last login: Sat Feb 10 18:16:27 2024
[kingbase@dbtest ~]$
```

(4) 在备份恢复测试机上，以用户 kingbase 的身份，执行命令 ssh-keygen，为备份恢复测试机生成公钥和私钥：

```
[kingbase@dbtest ~]$ ssh-keygen
Generating public/private rsa key pair.
Enter file in which to save the key (/home/kingbase/.ssh/id_rsa):  #按回车键
Enter passphrase (empty for no passphrase):                        #按回车键
Enter same passphrase again:                                       #按回车键
```

```
#省略了一些输出
[kingbase@dbtest ~]$
```

(5) 在备份恢复测试机上，执行命令 ssh-copy-id，将公钥复制到 KingbaseES 生产数据库服务器上的 /home/kingbase/.ssh/authorized_keys 文件中：

```
[kingbase@dbtest ~]$ ssh-copy-id -i /home/kingbase/.ssh/id_rsa.pub kingbase@192.168.100.22
/usr/bin/ssh-copy-id: INFO: Source of key(s) to be installed: "/home/kingbase/.ssh/id_rsa.pub"
The authenticity of host '192.168.100.22 (192.168.100.22)' can't be established.
ECDSA key fingerprint is SHA256:r/lTnKpUArFJQBZzwmqXwfpKT281ru8+7/4Y2N1kr9E.
ECDSA key fingerprint is MD5:ff:d6:e4:64:fb:3a:2b:46:e6:44:d0:5a:68:01:5f:ac.
Are you sure you want to continue connecting (yes/no)? yes
/usr/bin/ssh-copy-id: INFO: attempting to log in with the new key(s), to filter out any that are already installed
/usr/bin/ssh-copy-id: INFO: 1 key(s) remain to be installed -- if you are prompted now it is to install the new keys
kingbase@192.168.100.22's password: #输入 kingbase 用户的密码 kingbase123
#省略了一些输出
[kingbase@dbtest ~]$
```

(6) 在备份恢复测试机上，执行下面的 ssh 命令验证 KingbaseES 生产服务器和备份恢复测试机之间是否已经配置好了 ssh 无密码登录：

```
[kingbase@dbtest ~]$ ssh 192.168.100.22
Last login: Sat Feb 10 19:11:09 2024
[kingbase@dbsvr ~]$ ssh dbtest
The authenticity of host 'dbtest (192.168.100.19)' can't be established.
ECDSA key fingerprint is SHA256:r/lTnKpUArFJQBZzwmqXwfpKT281ru8+7/4Y2N1kr9E.
ECDSA key fingerprint is MD5:ff:d6:e4:64:fb:3a:2b:46:e6:44:d0:5a:68:01:5f:ac.
Are you sure you want to continue connecting (yes/no)? yes
Warning: Permanently added 'dbtest' (ECDSA) to the list of known hosts.
Last login: Sat Feb 10 19:38:34 2024 from dbsvr
[kingbase@dbtest ~]$ ssh dbsvr
The authenticity of host 'dbsvr (192.168.100.22)' can't be established.
ECDSA key fingerprint is SHA256:r/lTnKpUArFJQBZzwmqXwfpKT281ru8+7/4Y2N1kr9E.
ECDSA key fingerprint is MD5:ff:d6:e4:64:fb:3a:2b:46:e6:44:d0:5a:68:01:5f:ac.
Are you sure you want to continue connecting (yes/no)? yes
Warning: Permanently added 'dbsvr' (ECDSA) to the list of known hosts.
Last login: Sat Feb 10 19:41:02 2024 from dbtest
[kingbase@dbsvr ~]$
```

11.3 数据库脱机冷备份与恢复

11.3.1 数据库脱机冷备份

物理数据库脱机冷备份，需要备份 KingbaseES 数据库集簇的所有目录和文件。要成功进行 KingbaseES 数据库的脱机冷备份，关键是要获取数据库集簇的物理结构信息。

1. 获取数据库集簇的物理结构信息

（1）执行下面的命令和 SQL 语句查看 KingbaseES 数据库数据目录的位置信息：

```
[kingbase@dbsvr ~]$ ksql -d test -U system
system@test=# SELECT name, setting FROM sys_settings WHERE name = 'data_directory';
      name       |        setting
-----------------+------------------------
 data_directory | /u00/Kingbase/ES/V9/data
(1 row)
```

可以看到，KingbaseES 数据库的数据目录位于目录 /u00/Kingbase/ES/V9/data 下。

（2）执行下面的 SQL 语句查看 KingbaseES 数据库非系统默认表空间及其文件目录的位置信息：

```
system@test=# SELECT spcname,sys_tablespace_location(oid)
test-#          FROM sys_tablespace
test-#          WHERE spcname NOT IN ('sys_default','sys_global','sysaudit');
 spcname |     sys_tablespace_location
---------+----------------------------------------
 user_ts | /u02/Kingbase/ES/V9/data/userts/user_ts
 temp_ts | /u03/Kingbase/ES/V9/data/userts/temp_ts
(2 rows)
system@test=#
```

可以看到，当前有两个非系统默认的表空间，它们位于操作系统目录 /u02 和 /u03 下。

（3）执行下面的 SQL 语句查看 KingbaseES 数据库控制文件多元化信息：

```
system@test=# SELECT name, setting FROM sys_settings WHERE setting LIKE '%control%';
      name        |                  setting
------------------+------------------------------------------------
 control_file_copy| /u00/Kingbase/ES/V9/data/global/sys_control;
                  | /u01/Kingbase/ES/V9/data/global/sys_control;
                  | /u02/Kingbase/ES/V9/data/global/sys_control
(1 row)
system@test=#
```

可以看到，目前有 3 个控制文件复制，它们位于操作系统目录 /u00、/u01 和 /u02 下。

（4）执行下面的命令和 SQL 语句查看 KingbaseES 当前的 WAL 日志信息：

```
system@test=# SELECT sys_ls_waldir();
                    sys_ls_waldir
----------------------------------------------------------
 (00000001000000000000002C,67108864,"2024-03-02 22:07:47+08")
 (00000001000000000000002D,67108864,"2024-03-02 22:10:25+08")
 (00000001000000000000001F,67108864,"2024-03-03 19:39:14+08")
 (000000010000000000000020,67108864,"2024-03-02 22:07:26+08")
 (000000010000000000000021,67108864,"2024-03-02 22:07:28+08")
 (000000010000000000000022,67108864,"2024-03-02 22:07:30+08")
```

```
(0000000100000000000000023,67108864,"2024-03-02 22:07:32+08")
(0000000100000000000000024,67108864,"2024-03-02 22:07:33+08")
(0000000100000000000000025,67108864,"2024-03-02 22:07:35+08")
(0000000100000000000000026,67108864,"2024-03-02 22:07:36+08")
(0000000100000000000000027,67108864,"2024-03-02 22:07:42+08")
(0000000100000000000000028,67108864,"2024-03-02 22:07:42+08")
(0000000100000000000000029,67108864,"2024-03-02 22:07:43+08")
(000000010000000000000002A,67108864,"2024-03-02 22:07:45+08")
(000000010000000000000002B,67108864,"2024-03-02 22:07:46+08")
(15 rows)
system@test=# \!   ls -l /u00/Kingbase/ES/V9/data/sys_wal
total 983040
-rw-------. 1 kingbase dba 67108864 Mar  3 19:39 00000001000000000000001F
-rw-------. 1 kingbase dba 67108864 Mar  2 22:07 000000010000000000000020
-rw-------. 1 kingbase dba 67108864 Mar  2 22:07 000000010000000000000021
-rw-------. 1 kingbase dba 67108864 Mar  2 22:07 000000010000000000000022
-rw-------. 1 kingbase dba 67108864 Mar  2 22:07 000000010000000000000023
-rw-------. 1 kingbase dba 67108864 Mar  2 22:07 000000010000000000000024
-rw-------. 1 kingbase dba 67108864 Mar  2 22:07 000000010000000000000025
-rw-------. 1 kingbase dba 67108864 Mar  2 22:07 000000010000000000000026
-rw-------. 1 kingbase dba 67108864 Mar  2 22:07 000000010000000000000027
-rw-------. 1 kingbase dba 67108864 Mar  2 22:07 000000010000000000000028
-rw-------. 1 kingbase dba 67108864 Mar  2 22:07 000000010000000000000029
-rw-------. 1 kingbase dba 67108864 Mar  2 22:07 00000001000000000000002A
-rw-------. 1 kingbase dba 67108864 Mar  2 22:07 00000001000000000000002B
-rw-------. 1 kingbase dba 67108864 Mar  2 22:07 00000001000000000000002C
-rw-------. 1 kingbase dba 67108864 Mar  2 22:10 00000001000000000000002D
drwx------. 2 kingbase dba       43 Mar  3 19:39 archive_status
system@test=#
```

（5）执行下面的命令和 SQL 语句查看 KingbaseES 当前的归档日志目录信息：

```
system@test=# SHOW archive_command;
          archive_command
----------------------------------------------
 cp %p /u04/Kingbase/ES/V9/data/archivelog/%f
(1 row)
system@test=#
```

可以看到，归档日志位于目录/u04/Kingbase/ES/V9/data/archivelog 下。

（6）执行下面的 SQL 语句查看 KingbaseES 数据库的启动参数文件，客户端访问控制文件和客户端身份映射文件的信息：

```
system@test=# SELECT name, setting
test-#         FROM sys_settings
test-#         WHERE name in ('config_file','hba_file','ident_file');
    name     |              setting
-------------+----------------------------------------
 config_file | /u00/Kingbase/ES/V9/data/kingbase.conf
 hba_file    | /u00/Kingbase/ES/V9/data/sys_hba.conf
 ident_file  | /u00/Kingbase/ES/V9/data/sys_ident.conf
```

```
(3 rows)
system@test=#
```

综上可知,KingbaseES 数据库集簇的物理结构如下。

(1) 数据目录位于/u00/Kingbase/ES/V9/data 目录下。

(2) 控制文件位于以下 3 个目录。

① /u00/Kingbase/ES/V9/data/global/sys_control。

② /u01/Kingbase/ES/V9/data/global/sys_control。

③ /u02/Kingbase/ES/V9/data/global/sys_control。

(3) WAL 日志位于/u00/Kingbase/ES/V9/data/sys_wal 目录下。

(4) 在以下目录创建了用户表空间。

① /u02/Kingbase/ES/V9/data/userts/user_ts。

② /u03/Kingbase/ES/V9/data/userts/temp_ts。

此外,KingbaseES 数据库的归档日志位于/u04/Kingbase/ES/V9/data/archivelog 目录下。

2. 物理数据库脱机冷备份

如果 KingbaseES 数据库没有按照"附录 B:安装 KingbaseES 单机数据库的最佳实践"来进行安装部署,需要在关闭 KingbaseES 数据库服务器后,根据获取的 KingbaseES 数据库集簇的物理结构信息,备份所有必要的目录和文件,可能需要多条操作系统命令来完成备份。

如果按照附录 B 部署的 KingbaseES 单机生产数据库,其物理数据库脱机冷备份相当简单:

```
[kingbase@dbsvr ~]$ ksql -d test -U kingbase -c "drop table if exists dbbackupinfo;"
[kingbase@dbsvr ~]$ ksql -d test -U kingbase \
        -c "create table dbbackupinfo(backuptype varchar(40),backupdate date);"
[kingbase@dbsvr ~]$ ksql -d test -U kingbase -c "insert into dbbackupinfo values ('Cold',now());"
[kingbase@dbsvr ~]$ sys_ctl stop
[kingbase@dbsvr ~]$ cd /
[kingbase@dbsvr /]$ tar cf /dbbak/pbak/cold/dbclusterbackup.tar u0?/Kingbase/ES/V9/data/*
[kingbase@dbsvr /]$ scp /dbbak/pbak/cold/dbclusterbackup.tar 192.168.100.19:/dbbak/pbak/cold
[kingbase@dbsvr /]$ sys_ctl start
```

物理数据库脱机冷备份包含以下 5 个步骤:记录本次备份的信息、关闭 KingbaseES 数据库、执行 KingbaseES 数据库物理脱机冷备份(仅需一条 tar 命令)、将数据库物理脱机冷备份传送到备份恢复测试机、重新启动 KingbaseES 数据库。

可以将以上的命令保存到 coldbk 的 shell 脚本文件中,之后每次进行物理脱机冷备份,只需要执行 coldbk 脚本就可以了。直接运行脚本而不是命令,可以让现场的 DBA 避免出现误操作。

11.3.2 物理数据库脱机冷备份恢复

物理数据库脱机冷备份的恢复也相当简单，下面是在**备份恢复测试机 dbtest** 上的恢复过程：

```
[kingbase@dbtest ~]$ sys_ctl stop
[kingbase@dbtest ~]$ rm -rf /u0?/Kingbase/ES/V9/data/*
[kingbase@dbtest ~]$ cd /
[kingbase@dbtest /]$ tar xf /dbbak/pbak/cold/dbclusterbackup.tar
[kingbase@dbtest /]$ sys_ctl start
[kingbase@dbtest /]$ ksql -d test -U kingbase -c "SELECT * FROM dbbackupinfo;"
 backuptype |      backupdate
------------+---------------------
 Cold       | 2024-02-11 17:22:01
(1 row)
[kingbase@dbtest /]$ ksql -d test -U kingbase -c "DROP TABLE dbbackupinfo;"
[kingbase@dbtest /]$
```

物理数据库脱机冷备份恢复包含以下 5 个步骤：关闭备份测试机上的 KignbaseES 数据库、删除备份测试机上的 KingbaseES 数据库集簇、在备份测试机上使用物理数据库脱机冷备份恢复 KingbaseES 数据库集簇（仅需一条 tar 命令）、在备份测试机上重新启动 KingbaseES 数据库、验证已经恢复了脱机物理冷备份。

为了安全起见，请将上述命令保存到 restoreColdbackup 的 shell 脚本文件中，之后每次在备份恢复测试机上恢复物理脱机冷备份时，只需要执行 restoreColdbackup 脚本就可以了。直接运行脚本而不是命令，可以让现场的 DBA 避免出现误操作。

11.4 数据库联机热备与恢复

KingbaseES 数据库使用备份恢复工具 sys_rman 实现数据库联机热备与恢复。

11.4.1 sys_rman 备份恢复工具简介

KingbaseES 数据库用户一般使用 sys_rman 备份恢复工具来备份生产数据库，sys_rman 备份工具有以下 4 个主要组件。

（1）**sys_backup.conf**：初始化脚本 sys_backup.sh 使用的配置文件。

① 模板文件位于目录 $KINGBASE_HOME/Server/share 下。

② 执行脚本 sys_backup.sh 时，会按照如下顺序读取 sys_backup.conf 配置文件，首先读取配置文件 $KINGBASE_HOME/Server/bin/sys_backup.conf；如果不存在配置文件 $KINGBASE_HOME/Server/bin/sys_backup.conf，则继续读取配置文件 $KINGBASE_HOME/Server/share/sys_backup.conf。

（2）**sys_backup.sh**：用于协助管理员完成 sys_rman 初始配置的脚本文件。

① 脚本 sys_backup.sh 位于 $KINGBASE_HOME/Server/bin/目录下。

② 只能在 REPO 节点上执行脚本 sys_backup.sh。

③ 脚本 sys_backup.sh 完成 sys_rman 的初始化，并自动生成 sys_rman.conf 文件。

（3）**sys_rman.conf**：sys_rman 运行时使用的配置文件。

① 由初始化脚本 sys_backup.sh 自动生成。

② 位于存储库（REPO）目录下（如/home/kingbase/kbbr_repo）。

③ 不建议 DBA 人为修改配置文件 sys_rman.conf。

（4）**sys_rman**：执行备份还原核心二进制可执行文件；位于 KingbaseES 安装目录的./bin 目录下；可以在数据库节点和备份服务器上执行 sys_rman，会读取 sys_rman.conf 配置文件信息。

sys_rman 的备份存储库 REPO 保存了 KingbaseES 数据库备份的元数据和备份集，可以位于要备份的 KingbaseES 数据库节点内部（REPO 内部部署模式，如图 11-7 所示），也可以是位于另外一台 KingbaseES 数据库服务器外部（REPO 外部部署模式，如图 11-8 所示）。

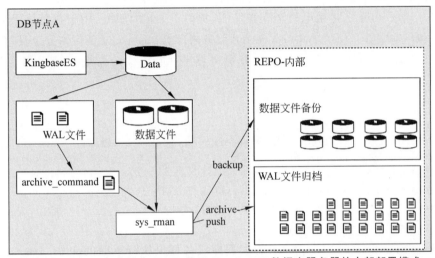

图 11-7　sys_rman 存储库 REPO 位于 KingbaseES 数据库服务器的内部部署模式

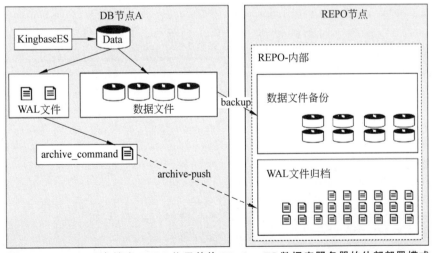

图 11-8　sys_rman 存储库 REPO 位于其他 KingbaseES 数据库服务器的外部部署模式

基于备份数据的安全考虑，推荐使用 REPO 外部部署模式。用户现场的 KingbaseES 生产数据库，一般采用主备集群的方式部署，将在第 15 章给出 REPO 外部部署模式的实战实例。本章的单机 KingbaseES 数据库 sys_rman 备份恢复实战实例采用 REPO 内部部署的模式。

11.4.2 配置 sys_rman

将在单机 KingbaseES 数据库服务器上，配置 sys_rman 使用**存储库 REPO 内部部署模式**，进行数据库备份。假设 sys_rman 使用存储库 REPO 位于目录/dbbak/pbak/kbbr_repo，需要确保在 KingbaseES 数据库服务器上，使用用户 kingbase 可以创建这个目录：

```
[kingbase@dbsvr ~]$ mkdir -p /dbbak/pbak/kbbr_repo
[kingbase@dbsvr ~]$ rmdir /dbbak/pbak/kbbr_repo
```

创建完成后需要删除这个目录，否则在执行初始化脚本的时候会报错。

sys_securecmd 服务是由电科金仓开发的，类似于 ssh 的服务，用于在可信服务器之间执行命令和传输文件。使用 sys_securecmd 服务来部署 sys_rman 备份恢复工具。

使用用户 root，在 KingbaseES 数据库服务器上执行下面的命令，布署 sys_securecmd 服务：

```
[root@dbsvr ~]# cd /u00/Kingbase/ES/V9/kingbase/Server/bin
[root@dbsvr bin]# ./sys_HAscmdd.sh init
successfully initialized the sys_securecmdd, please use "./sys_HAscmdd.sh start" to start the sys_securecmdd
[root@dbsvr bin]# ./sys_HAscmdd.sh start
Created symlink from /etc/systemd/system/multi-user.target.wants/securecmdd.service to /etc/systemd/system/securecmdd.service.
[root@dbsvr bin]#
```

使用用户 kingbase，在 KingbaseES 数据库服务器上执行下面的命令，为 sys_rman 准备配置文件 $KINGBASE_HOME/Server/bin/sys_backup.conf：

```
[kingbase@dbsvr ~]$ cp /u00/Kingbase/ES/V9/kingbase/Server/share/sys_backup.conf \
                      /u00/Kingbase/ES/V9/kingbase/Server/bin
[kingbase@dbsvr ~]$ vi /u00/Kingbase/ES/V9/kingbase/Server/bin/sys_backup.conf
```

修改如下的行：

```
_target_db_style="single"
_one_db_ip="192.168.100.22"
_repo_ip="192.168.100.22"
_stanza_name="kingbase"
_os_user_name="kingbase"
_repo_path="/dbbak/pbak/kbbr_repo"
_single_data_dir="/u00/Kingbase/ES/V9/data"
_single_bin_dir="/u00/Kingbase/ES/V9/kingbase/Server/bin"
_single_db_user="system"
_single_db_port="54321"
# on means sys_securecmd, off means normal ssh
_use_scmd=on
```

如果想修改 **sys_rman** 自动备份策略，可以修改如下的行：

```
# count of keep, over the count FULL-backup will be remove(设置保存数据库完全备份的
份数)
_repo_retention_full_count=5
# count of days, interval to do FULL-backup(设置间隔几天做一次数据库的完全备份)
_crond_full_days=7
# count of days, interval to do DIFF-backup(设置间隔几天做一次数据库的差异备份)
_crond_diff_days=0
# count of days, interval to do INCR-backup(设置间隔几天做一次数据库的增量备份)
_crond_incr_days=1
# HOUR to do the FULL-backup(设置做数据库完全备份的小时数)
_crond_full_hour=2
# HOUR to do the DIFF-backup(设置做数据库差异备份的小时数)
_crond_diff_hour=3
# HOUR to do the INCR-backup(设置做数据库增量备份的小时数)
_crond_incr_hour=4
```

如果想限制 sys_rman 备份时使用的磁盘 I/O 带宽,可以修改如下的行:

```
# band witdh limit, fixed in Mb/s, default 0 means no limit
# (设置磁盘 I/O 带宽的最大限制单位为 Mb/s,如果值为 0,表示不限制备份使用的磁盘 I/O 带宽)
_band_width=0
```

使用用户 kingbase,在 KingbaseES 数据库服务器上执行脚本 sys_backup.sh,生成 sys_rman 的配置文件 sys_rman.conf:

```
[kingbase@dbsvr ~]$ /u00/Kingbase/ES/V9/kingbase/Server/bin/sys_backup.sh init
# pre-condition: check the non-archived WAL files
# generate single sys_rman.conf...DONE
# update single archive_command with sys_rman.archive-push...DONE
# create stanza and check...(maybe 60+ seconds)
# create stanza and check...DONE
# initial first full backup...(maybe several minutes)
# initial first full backup...DONE
# Initial sys_rman OK.
'sys_backup.sh start' should be executed when need back-rest feature.
[kingbase@dbsvr ~]$
```

执行完脚本 sys_backup.sh,将创建目录/dbbak/pbak/kbbr_repo/kbbr_repo/,并在该目录下生成 sys_rman 的配置文件 sys_rman.conf,然后执行一次全量备份。

如果想配置 sys_rman 按照参数文件 sys_backup.conf 中设置的备份计划定时自动进行备份,可以使用用户 kingbase 执行下面的脚本命令:

```
[kingbase@dbsvr ~]$ /u00/Kingbase/ES/V9/kingbase/Server/bin/sys_backup.
sh start
# pre-condition: check the non-archived WAL files
Enable some sys_rman in crontab-daemon
no crontab for kingbase
Set full-backup in 7 days
Set incr-backup in 1 days
0 2 */7 * * /u01/Kingbase/ES/V9/Server/bin/sys_rman --config=/home/
kingbase/kbbr_repo/sys_rman.conf --stanza=kingbase --archive-copy --type=
```

```
full backup >> /u01/Kingbase/ES/V9/Server/log/sys_rman_backup_full.log 2>&1
0 4 */1 * * /u01/Kingbase/ES/V9/Server/bin/sys_rman --config=/home/kingbase/
kbbr_repo/sys_rman.conf --stanza=kingbase --archive-copy --type=incr backup >
> /u01/Kingbase/ES/V9/Server/log/sys_rman_backup_incr.log 2>&1
[kingbase@dbsvr ~]$
```

执行完脚本 sys_backup.sh start,将在 CentOS 7 操作系统中增加定时 crontab 任务,使用 sys_rman 进行定时备份。使用用户 kingbase,执行命令 crontab -e,可以看到这些新增的定时任务。

如果不再想按照参数文件 sys_backup.conf 中设置的备份计划定时自动进行 sys_rman 备份,可以使用用户 kingbase,执行下面的脚本命令删除定时备份任务:

```
[kingbase@dbsvr ~]$ /u00/Kingbase/ES/V9/kingbase/Server/bin/sys_backup.sh stop
# pre-condition: check the non-archived WAL files
Disable all sys_rman in crontab-daemon
[kingbase@dbsvr ~]$
```

执行完脚本 sys_backup.sh stop,将停止 CentOS 7 操作系统中设置的 sys_rman 定时 crontab 任务。使用用户 kingbase,执行命令 crontab -e,可以看到用户 kingbase 已经没有了定时自动执行 sys_rman 的备份任务了。

11.4.3 使用 sys_rman 备份数据库

使用 sys_rman 备份恢复工具,可以全量备份、差异备份和增量备份的方式,备份用户的生产环境中 KingbaseES 数据库。

1. 全量备份

使用用户 kingbase,执行下面的命令,全量备份数据库:

```
[kingbase@dbsvr ~]$ /u00/Kingbase/ES/V9/kingbase/Server/bin/sys_rman \
    --config=/dbbak/pbak/kbbr_repo/sys_rman.conf \
    --stanza=kingbase \
    --archive-copy \
    --type=full backup \
    >> /u00/Kingbase/ES/V9/kingbase/Server/log/sys_rman_backup_
full.log 2>&1
```

命令中的参数有如下含义。

(1) --config=/dbbak/pbak/kbbr_repo/sys_rman.conf:指定配置文件。
(2) --stanza=kingbase:指定标签在配置文件中找到对应的参数配置。
(3) --archive-copy:指定备份时同时备份恢复所需的 WAL 段文件。
(4) --type=full backup:指定备份类型为全量备份。

查看文件/u00/Kingbase/ES/V9/Server/log/sys_rman_backup_full.log 的内容可以了解命令的详细输出。

2. 差异备份

使用用户 kingbase,执行下面的命令,差异备份数据库:

```
[kingbase@dbsvr ~]$ /u00/Kingbase/ES/V9/kingbase/Server/bin/sys_rman \
                   --config=/dbbak/pbak/kbbr_repo/sys_rman.conf \
                   --stanza=kingbase \
                   --archive-copy \
                   --type=diff backup \
                   >> /u00/Kingbase/ES/V9/kingbase/Server/log/sys_rman_backup_
diff.log 2>&1
```

命令中的参数有如下含义，--type＝diff backup：指定备份类型为差异备份，只备份自上次全量备份以来改变过的数据。

查看文件/u00/Kingbase/ES/V9/Server/log/sys_rman_backup_diff.log 的内容可以了解命令的详细输出。

3. 增量备份

使用用户 kingbase，执行下面的命令，增量备份数据库（文件粒度）：

```
[kingbase@dbsvr ~]$ /u00/Kingbase/ES/V9/kingbase/Server/bin/sys_rman \
                   --config=/dbbak/pbak/kbbr_repo/sys_rman.conf \
                   --stanza=kingbase \
                   --archive-copy \
                   --type=incr backup \
                   >> /u00/Kingbase/ES/V9/kingbase/Server/log/sys_rman_backup_
incr.log 2>&1
```

命令中的参数有如下含义，--type＝incr backup：指定备份类型为增量备份，只备份上次全量备份、差异备份、增量备份以来改变过的数据。

查看文件/u00/Kingbase/ES/V9/Server/log/sys_rman_backup_incr.log 的内容可以了解命令的详细输出。

为了继续下面的 sys_rman 实战学习，使用用户 kingbase，再执行下面的命令两次（全量数据库备份）：

```
[kingbase@dbsvr ~]$ /u00/Kingbase/ES/V9/kingbase/Server/bin/sys_rman \
                   --config=/dbbak/pbak/kbbr_repo/sys_rman.conf \
                   --stanza=kingbase \
                   --archive-copy \
                   --type=full backup \
                   >> /u00/Kingbase/ES/V9/kingbase/Server/log/sys_rman_backup_
full.log 2>&1
```

11.4.4　管理 sys_rman 备份集

1. 查看备份集信息

使用用户 kingbase，执行下面的命令，可以查看 sys_rman 备份集信息：

```
[kingbase@dbsvr ~]$ sys_rman --config=/dbbak/pbak/kbbr_repo/sys_rman.conf --
stanza=kingbase info
stanza: kingbase
    status: ok
    cipher: none
```

```
        db (current)
            wal archive min/max (V009R001C001B0025): 000000010000000000000020/000000010000000000000002C

        full backup: 20240226-140048F
            timestamp start/stop: 2024-02-26 14:00:48 / 2024-02-26 14:01:04
            wal start/stop: 000000010000000000000022 / 000000010000000000000022
            database size: 1.9GB, database backup size: 1.9GB
            repo1: backup set size: 1.9GB, backup size: 1.9GB

        full backup: 20240226-140135F
            timestamp start/stop: 2024-02-26 14:01:35 / 2024-02-26 14:01:45
            wal start/stop: 000000010000000000000024 / 000000010000000000000024
            database size: 1.9GB, database backup size: 1.9GB
            repo1: backup set size: 1.9GB, backup size: 1.9GB

        diff backup: 20240226-140135F_20240226-140148D
            timestamp start/stop: 2024-02-26 14:01:48 / 2024-02-26 14:01:50
            wal start/stop: 000000010000000000000025 / 000000010000000000000026
            database size: 1.9GB, database backup size: 128MB
            repo1: backup set size: 1.9GB, backup size: 128MB
            backup reference list: 20240226-140135F

        incr backup: 20240226-140135F_20240226-140154I
            timestamp start/stop: 2024-02-26 14:01:54 / 2024-02-26 14:01:57
            wal start/stop: 000000010000000000000027 / 000000010000000000000028
            database size: 1.9GB, database backup size: 128MB
            repo1: backup set size: 1.9GB, backup size: 128MB
            backup reference list: 20240226-140135F

        full backup: 20240226-140203F
            timestamp start/stop: 2024-02-26 14:02:03 / 2024-02-26 14:02:08
            wal start/stop: 000000010000000000000029 / 000000010000000000000002A
            database size: 1.9GB, database backup size: 1.9GB
            repo1: backup set size: 1.9GB, backup size: 1.9GB

        full backup: 20240226-140210F
            timestamp start/stop: 2024-02-26 14:02:10 / 2024-02-26 14:02:15
            wal start/stop: 000000010000000000000002B / 000000010000000000000002C
            database size: 1.9GB, database backup size: 1.9GB
            repo1: backup set size: 1.9GB, backup size: 1.9GB
[kingbase@dbsvr /]$
```

2. 只保留最近的两次全量备份集

使用用户 kingbase，执行下面的命令，只保留最近两次的全量备份集：

```
[kingbase@dbsvr ~]$ sys_rman --config=/dbbak/pbak/kbbr_repo/sys_rman.conf \
                    --stanza=kingbase --repo1-retention-full=2 expire
#省略了一些输出
[kingbase@dbsvr ~]$
```

使用用户 kingbase，执行下面的命令，查看 sys_rman 备份集信息：

```
[kingbase@dbsvr ~]$ sys_rman --config=/dbbak/pbak/kbbr_repo/sys_rman.conf --stanza=kingbase info
stanza: kingbase
    status: ok
    cipher: none

    db (current)
        wal archive min/max (V009R001C001B0025): 000000010000000000000029/00000001000000000000002C

        full backup: 20240226-140203F
            timestamp start/stop: 2024-02-26 14:02:03 / 2024-02-26 14:02:08
            wal start/stop: 000000010000000000000029 / 00000001000000000000002A
            database size: 1.9GB, database backup size: 1.9GB
            repo1: backup set size: 1.9GB, backup size: 1.9GB

        full backup: 20240226-140210F
            timestamp start/stop: 2024-02-26 14:02:10 / 2024-02-26 14:02:15
            wal start/stop: 00000001000000000000002B / 00000001000000000000002C
            database size: 1.9GB, database backup size: 1.9GB
            repo1: backup set size: 1.9GB, backup size: 1.9GB
[kingbase@dbsvr ~]$
```

可以看到，执行完上面的命令后，只保留了最近的两个全量备份集。

3. 手动清除过期的备份集

使用用户 kingbase，执行下面的命令，清除过期的 sys_rman 备份集：

```
[kingbase@dbsvr ~]$ sys_rman --config=/dbbak/pbak/kbbr_repo/sys_rman.conf --stanza=kingbase expire
```

4. 备份集检查

sys_rman 备份集检查的作用是确保备份数据的完整性和可用性。

使用用户 kingbase 执行下面的命令对生成的备份集进行检查并报告问题：

```
[kingbase@dbsvr ~]$ sys_rman --config=/dbbak/pbak/kbbr_repo/sys_rman.conf --stanza=kingbase verify
2024-02-26 14:04:30.188 P00   INFO: verify command begin 2.27: --no-archive-statistics --band-width=0
 --cmd-ssh=/u00/Kingbase/ES/V9/Server/bin/sys_securecmd --config=/dbbak/pbak/kbbr_repo/sys_rman.conf --exec-id=5522-f895e88b
 --log-level-console=info --log-level-file=info --log-path=/u00/Kingbase/ES/V9/Server/log --log-subprocess
 --non-archived-space=1024 --process-max=4 --repo1-path=/dbbak/pbak/kbbr_repo --stanza=kingbase
2024-02-26 14:04:31.973 P00   INFO: Results:
             archive_id: 12-1, total WAL checked: 4, total valid WAL: 4
             missing: 0, checksum invalid: 0, size invalid: 0, other: 0
             backup: 20240226-140203F, status: valid, total files checked: 2349, total valid files: 2349
```

```
               missing: 0, checksum invalid: 0, size invalid: 0, other: 0
               backup: 20240226-140210F, status: valid, total files checked: 2349,
total valid files: 2349
               missing: 0, checksum invalid: 0, size invalid: 0, other: 0
WARN: if option 'archive-statistics' is not enabled, archived WAL file deletion
can't be detected
2024-02-26 14:04:31.973 P00   INFO: verify command end: completed successfully
(1789ms)
[kingbase@dbsvr ~]$
```

11.4.5　在生产数据库上执行完全恢复

在 KingbaseES 生产数据库上，可能会因为软硬件故障或人为原因，造成数据库集簇的某些文件（或目录）被错误删除或者损坏。如果不幸发生了这种情况，可以使用 sys_rman 备份集将 KingbaseES 数据库集簇恢复到故障点（无损恢复）。

1. 模拟生产环境的介质故障场景

为了测试 KingbaseES 数据库的完全恢复，首先使用用户 system，执行下面的命令和 SQL 语句，创建用于记录测试过程的信息表 dbinfo1 和 dbinfo2，用于帮助了解在生产数据库上的完全恢复：

```
[kingbase@dbsvr ~]$ ksql -d test -U system -c \
        "create table dbinfo1(info varchar(60),infodate date) tablespace user_ts;"
[kingbase@dbsvr ~]$ ksql -d test -U system -c \
        "create table dbinfo2(info varchar(60),infodate date) tablespace sys_
default;"
```

表 dbinfo1 在表空间 user_ts 中，用它记录发生误操作之前的时间，误操作是删除了表空间 user_ts 对应的目录，因此发生误操作后不能查到表 dbinfo1 的内容；表 dbinfo2 在系统默认的表空间 sys_default 中，即使发生误操作之后，用它也能记录 DBA 收到 KingbaseES 数据库发生故障的报告时间。发生误操作之后到报告给 DBA 的这段时间，生产数据库经历了大量的操作。

以用户 kingbase 的身份，执行下面的命令，先在 dbinfo1 中记录发生误操作的时间，然后删除表空间 user_ts 所在的目录，模拟出现运维事故：

```
[kingbase@dbsvr ~]$ ksql -d test -U system   \
        -c "insert into dbinfo1 values('before accidental operation',now());"
[kingbase@dbsvr ~]$ ksql -d test -U system -c "SELECT * FROM dbinfo1;"
           info            |      infodate
---------------------------+---------------------
 before accidental operation | 2024-02-26 14:11:34
(1 row)
[kingbase@dbsvr ~]$ rm -rf /u02/Kingbase/ES/V9/data/userts/user_ts/SYS_12_
202305151
```

此时，只要是涉及表空间 user_ts 中的 KingbaseES 数据库对象的操作都会报错：

```
[kingbase@dbsvr ~]$ ksql -d test -U kingbase -c "SELECT * FROM dbinfo1;"
```

```
ERROR:  could not open file "sys_tblspc/16385/SYS_12_202305151/14386/16504": No
such file or directory
#说明：表 dbinfo1 在表空间 user_ts 中，删除表空间 user_ts 对应的目录后，无法查看表
#dbinfo1 的内容
[kingbase@dbsvr ~]$
```

执行下面的命令模拟误操作后 KingbaseES 数据库还进行了大量的操作：

```
[kingbase@dbsvr ~]$ cat >workloadAfterMediaFail.sql<<EOF
create table dbwork(col1 varchar(100),col2 varchar(100),col3 varchar(100))
tablespace sys_default;
create index idx_dbwork_col1 on dbwork(col1) tablespace sys_default;
create index idx_dbwork_col2 on dbwork(col2) tablespace sys_default;
insert into dbwork select id::varchar,md5(id::varchar),md5(md5(id::varchar))
                from generate_series(1,1000000) as id;
EOF
[kingbase@dbsvr ~]$ ksql -d test -U system -f workloadAfterMediaFail.sql
[kingbase@dbsvr ~]$ ksql -d test -U system -c "SELECT count(*) FROM dbwork;"
  count
---------
 1000000
(1 row)
```

假设执行完上述操作后，DBA 收到了 KingbaseES 数据库发生了介质故障的报告，此时执行下面的 ksql 命令，在表 dbinfo2 中记录 DBA 收到数据库发生介质故障的报告时间：

```
[kingbase@dbsvr ~]$ ksql -d test -U system -c \
            "insert into dbinfo2 values('receive accidental report',now());"
[kingbase@dbsvr ~]$ ksql -d test -U kingbase -c "SELECT * FROM dbinfo2;"
         info             |      infodate
--------------------------+---------------------
 receive accidental report | 2024-02-26 14:20:13
(1 row)
(1 row)
```

2. 故障处理过程（使用最新的 sys_rman 全量备份集在生产数据库上进行还原和恢复）

为了保证能够恢复到故障点，使用用户 system，执行下面的命令，切换 WAL 日志：

```
[kingbase@dbsvr user_ts]$ ksql -d test -U system -c "SELECT sys_switch_wal();"
```

使用用户 kingbase，执行下面的命令，停止 KingbaseES 数据库：

```
[kingbase@dbsvr ~]$ sys_ctl stop
```

使用用户 kingbase，执行下面的命令，**使用最新的备份集**，复原被误删除的 KingbaseES 数据库集簇文件或目录（此处为表空间 user_ts 的目录）：

```
[kingbase@dbsvr ~]$ sys_rman --config=/dbbak/pbak/kbbr_repo/sys_rman.conf \
                    --stanza=kingbase --delta restore
```

其中，选项 --delta 表示让 KingbaseES 数据库自动确定可以保留集群目录中的哪些文件，以及需要从备份中恢复哪些文件。

以用户 kingbase 的身份，执行下面的命令重新启动 KingbaseES 数据库，启动 KingbaseES 数据库服务器时会自动执行 KingbaseES 数据库恢复操作（恢复到故障点）：

```
[kingbase@dbsvr ~]$ sys_ctl start
```

KingbaseES 数据库恢复到故障点之后，会自动执行 promote 操作（执行 SQL 语句"select sys_wal_replay_resume();"），数据库的时间线会增 1：

```
[kingbase@dbsvr ~]$ ksql -d test -U system -c "select timeline_id from sys_control_checkpoint();"
 timeline_id
-------------
           2
(1 row)
[kingbase@dbsvr ~]$
```

执行下面的命令验证已经完全恢复了数据库：

```
[kingbase@dbsvr ~]$ ls -ld /u02/Kingbase/ES/V9/data/userts/user_ts/SYS_12_202305151
drwx------. 3 kingbase dba 19 Feb 26 14:21 /u02/Kingbase/ES/V9/data/userts/user_ts/SYS_12_202305151
[kingbase@dbsvr ~]$ ksql -d test -U system -c "SELECT * FROM dbinfo1;"
           info             |      infodate
----------------------------+---------------------
 before accidental operation | 2024-02-26 14:11:34
(1 row)
[kingbase@dbsvr ~]$ ksql -d test -U system -c "SELECT * FROM dbinfo2;"
           info             |      infodate
----------------------------+---------------------
 receive accidental report  | 2024-02-26 14:20:13
(1 row)
[kingbase@dbsvr ~]$
```

恢复 KingbaseES 数据库集簇后，建议立即重做一次数据库全量物理备份：

```
[kingbase@dbsvr ~]$ ksql -d test -U system \
        -c "insert into dbinfo1 values('full backup after OK',now());"
[kingbase@dbsvr ~]$ /u00/Kingbase/ES/V9/kingbase/Server/bin/sys_rman \
        --config=/dbbak/pbak/kbbr_repo/sys_rman.conf \
        --stanza=kingbase \
        --archive-copy \
        --type=full backup \
        >> /u00/Kingbase/ES/V9/kingbase/Server/log/sys_rman_backup_full.log 2>&1
[kingbase@dbsvr ~]$
```

执行下面的命令再次查看 sys_rman 的备份集信息：

```
[kingbase@dbsvr ~]$ sys_rman --config=/dbbak/pbak/kbbr_repo/sys_rman.conf --stanza=kingbase info
stanza: kingbase
```

```
        status: ok
        cipher: none

    db (current)
        wal archive min/max (V009R001C001B0025): 000000010000000000000029/000000020000000000000038

        full backup: 20240226-140203F
            timestamp start/stop: 2024-02-26 14:02:03 / 2024-02-26 14:02:08
            wal start/stop: 000000010000000000000029 / 00000001000000000000002A
            database size: 1.9GB, database backup size: 1.9GB
            repo1: backup set size: 1.9GB, backup size: 1.9GB

        full backup: 20240226-140210F
            timestamp start/stop: 2024-02-26 14:02:10 / 2024-02-26 14:02:15
            wal start/stop: 00000001000000000000002B / 00000001000000000000002C
            database size: 1.9GB, database backup size: 1.9GB
            repo1: backup set size: 1.9GB, backup size: 1.9GB

        full backup: 20240226-142330F
            timestamp start/stop: 2024-02-26 14:23:30 / 2024-02-26 14:23:35
            wal start/stop: 000000020000000000000038 / 000000020000000000000038
            database size: 2GB, database backup size: 2GB
            repo1: backup set size: 2GB, backup size: 2GB
[kingbase@dbsvr ~]$
```

可以看到，在 sys_rman 备份集中的 WAL 信息，包含了时间线信息。

11.4.6 异机恢复 KingbaseES 数据库

把 Kingbase 生产数据库的 sys_rman 备份恢复到备份恢复测试机上，有以下作用：第一，可以在备份恢复测试机上，做一些无法在生产服务器上的测试；第二，由于软硬件方面的原因，在 Kingbase 生产数据库无法提供服务的情况下，可以使用备份恢复测试机作为应急 KinbaseES 生产数据库服务器。

在备份恢复测试机上，执行下面的命令将备份集同步到备份恢复测试机上：

```
[kingbase@dbtest ~]$ rsync -avz --delete kingbase@192.168.100.22:/dbbak/pbak/kbbr_repo/* /dbbak/pbak/kbbr_repo
#省略了许多输出
sent 173,514 bytes  received 40,930,595 bytes  6,323,709.08 bytes/sec
total size is 492,596,524  speedup is 11.98
[kingbase@dbtest ~]$
```

如果没有 KingbaseES 数据库在生产服务器与备份恢复测试机之间配置 rsync，也可以在备份恢复测试机上使用 scp 命令将备份集复制到备份恢复测试机上：

```
scp -r 192.168.100.22:/dbbak/pbak/kbbr_repo /dbbak/pbak
```

在备份恢复测试机上，执行下面的命令查看备份集的信息：

```
[kingbase@dbtest ~]$ sys_rman --config=/dbbak/pbak/kbbr_repo/sys_rman.conf --stanza=kingbase info
```

#省略了输出
[kingbase@dbtest bin]$
```

在备份恢复测试机上,执行下面的命令停止 KingbaseES 数据库系统:

```
[kingbase@dbtest ~]$ sys_ctl stop
```

在备份恢复测试机上,执行下面的命令删除备份恢复测试机上的 KingbaseES 数据库集簇:

```
[kingbase@dbtest ~]$ rm -rf /u0?/Kingbase/ES/V9/data
```

在备份恢复测试机上,执行下面的命令恢复 KingbaseES 数据库集簇:

```
[kingbase@dbtest ~]$ sys_rman --config=/dbbak/pbak/kbbr_repo/sys_rman.conf --stanza=kingbase restore
```

在备份恢复测试机上,执行下面的命令启动 KingbaseES 数据库系统:

```
[kingbase@dbtest ~]$ sys_ctl start
```

在备份恢复测试机上,执行下面的命令验证是否已经恢复了 KingbaseES 数据库集簇:

```
[kingbase@dbtest ~]$ ksql -d test -U system -c "SELECT * FROM dbinfo2;"
           info            |       infodate
---------------------------+---------------------
 receive accidental report | 2024-03-13 08:53:43
(1 row)
```

在备份恢复测试机上,执行下面的 ksql 命令和 SQL 语句查看恢复 KingbaseES 数据库后的时间线:

```
[kingbase@dbsvr ~]$ ksql -d test -U kingbase -c "select timeline_id from sys_control_checkpoint();"
 timeline_id
-------------
           3
(1 row)
[kingbase@dbsvr ~]$
```

由于在备份恢复测试机上使用的是时间线为 2 的 sys_rman 备份集进行的恢复,因此恢复后时间线的值变成了 3。

11.4.7 不完全恢复到指定时间点

1. 模拟生产环境中的常见误操作(更新表未使用条件,导致全表都更新成一样的值)

在生产数据库上,使用用户 kingbase,执行下面的命令和 SQL 语句,模拟出现了人为的误操作,不小心把数据库 test 中 tpch 模式下的表 orders 的 o_totalprice 字段值都设置为 10 000:

```
[kingbase@dbsvr ~]$ ksql -d test -U system
system@test=# SELECT now();
```
```

```
                now
-------------------------------
 2024-02-26 14:39:31.948962+08
(1 row)
system@test=# UPDATE tpch.orders SET o_totalprice=10000;
UPDATE 1500000
system@test=# SELECT o_orderkey, o_totalprice FROM tpch.orders LIMIT 3;
 o_orderkey  | o_totalprice
-------------+--------------
       4131  |     10000.00
       6951  |     10000.00
       8355  |     10000.00
(3 rows)
systeme@test=# \q
[kingbase@dbsvr ~]$
```

幸运的是,知道误操作之前的准确时间为 2024-02-26 14:39:31.948962+08。这样可以在备份恢复测试机上,将 KingbaseES 数据库恢复到这个指定的时间。

2. 故障处理和恢复过程

在生产数据库服务器 dbsvr 上,执行下面的命令切换 WAL 日志:

```
[kingbase@dbsvr ~]$ ksql -d test -U system -c "SELECT sys_switch_wal();"
```

在生产数据库服务器 dbsvr 上,执行下面的命令增量备份(文件粒度)数据库:

```
[kingbase@dbsvr ~]$ /u00/Kingbase/ES/V9/kingbase/Server/bin/sys_rman \
                --config=/dbbak/pbak/kbbr_repo/sys_rman.conf \
                --stanza=kingbase \
                --archive-copy \
                --type=incr backup \
                >> /u00/Kingbase/ES/V9/kingbase/Server/log/sys_rman_backup_inc.log 2>&1
```

在**备份恢复测试机 dbtest** 上,执行下面的命令将备份集同步到备份恢复测试机上:

```
[kingbase@dbtest ~]$ rsync -avz kingbase@192.168.100.22:/dbbak/pbak/kbbr_repo/* /dbbak/pbak/kbbr_repo
#省略了许多输出
[kingbase@dbtest ~]$
```

在备份恢复测试机 dbtest 上,执行下面的命令查看备份集的信息:

```
[kingbase@dbtest ~]$ sys_rman --config=/dbbak/pbak/kbbr_repo/sys_rman.conf --stanza=kingbase info
#省略了输出
[kingbase@dbtest bin]$
```

在备份恢复测试机上,执行下面的命令停止 KingbaseES 数据库:

```
[kingbase@dbtest ~]$ sys_ctl stop
```

在备份恢复测试机上,执行下面的命令删除备份恢复测试机上的 KingbaseES 数据库集簇：

```
[kingbase@dbtest ~]$ rm -rf /u0?/Kingbase/ES/V9/data
```

在备份恢复测试机上，执行下面的命令基于时间不完全恢复 KingbaseES 数据库集簇：

```
[kingbase@dbtest ~]$ sys_rman --config=/dbbak/pbak/kbbr_repo/sys_rman.conf --stanza=kingbase \
           --type=time  --target='2024-02-26 14:39:31.948962+08' --target-action=promote restore
```

参数 --target 设置了恢复截止时间为误操作之前的时间 2024-02-26 14：39：31.948962+08。
在备份恢复测试机上，执行下面的命令启动 KingbaseES 数据库：

```
[kingbase@dbtest ~]$ sys_ctl start
```

在备份恢复测试机上，执行下面的命令验证已经恢复了 KingbaseES 数据库：

```
[kingbase@dbtest ~]$ ksql -d test -U kingbase -c "SELECT o_orderkey, o_totalprice FROM tpch.orders LIMIT 3;"
 o_orderkey | o_totalprice
------------+--------------
          1 |    173665.47
          2 |     46929.18
          3 |    193846.25
(3 rows)
[kingbase@dbsvr ~]$
```

可以看到，表 tpch.orders 已经恢复到误操作之前的数据了。

被误操作的表 tpch.orders 使用 sys_dump 命令，从数据库备份恢复测试机上导出来，并复制到 KingbaseES 生产数据库服务器：

```
[kingbase@dbtest ~]$ cd
[kingbase@dbtest ~]$ sys_dump -U kingbase -d test -t tpch.orders -F p -f tpch.orders.sql \
                    --no-owner  --no-privileges
[kingbase@dbtest ~]$ scp tpch.orders.sql 192.168.100.22:/home/kingbase/
tpch.orders.sql                                        100% 2736     3.8MB/s   00:00
[kingbase@dbtest ~]$
```

关于 sys_dump 命令，将在第 11 章详细介绍它的使用方法。
在**生产数据库服务器**上，执行命令删除被误操作的表 tpch.orders：

```
[kingbase@dbsvr ~]$ ksql -d test -U system
system@test=# DROP TABLE tpch.orders;
ERROR:  cannot drop table tpch.orders because other objects depend on it
DETAIL:  constraint lineitem_l_orderkey_fkey on table tpch.lineitem depends on table tpch.orders
HINT:  Use DROP ... CASCADE to drop the dependent objects too.
system@test=#
```

可以看到，当前暂时不能删除表 tpch.orders，因为还存在一些外键约束，依赖于表 tpch.orders。可以先禁用这些约束，再将表 tpch.orders 清空：

```
system@test=# ALTER TABLE tpch.lineitem DISABLE CONSTRAINT lineitem_l_orderkey_fkey;
system@test=# TRUNCATE TABLE tpch.orders;
system@test=# \q
[kingbase@dbsvr ~]$
```

在生产数据库服务器上，执行下面的命令将表 tpch.orders 的数据恢复到生产数据库中：

```
[kingbase@dbsvr ~]$ ksql -d test -U system -f tpch.orders.sql -q
set_config
------------
 public
(1 row)
ksql:tpch.orders.sql:40: ERROR:  relation "orders" already exists
ksql:tpch.orders.sql:1500056: ERROR:  multiple primary keys for table "orders" are not allowed
ksql:tpch.orders.sql:1500063: ERROR:  relation "idx_orders_o_totalprince" already exists
ksql:tpch.orders.sql:1500071: ERROR:  constraint "orders_o_custkey_fkey" for relation "orders" already exists
[kingbase@dbsvr ~]$ ksql -d test -U kingbase -c "SELECT o_orderkey, o_totalprice FROM tpch.orders LIMIT 3;"
 o_orderkey | o_totalprice
---------+---------------
         1 |    173665.47
         2 |     46929.18
         3 |    193846.25
(3 rows)
[kingbase@dbsvr ~]$
```

在生产数据库服务器上，执行下面的命令重新启用约束 lineitem_l_orderkey_fkey：

```
[kingbase@dbsvr ~]$ ksql -d test -U system -c "ALTER TABLE tpch.lineitem ENABLE CONSTRAINT lineitem_l_orderkey_fkey;"
ALTER TABLE
[kingbase@dbsvr ~]$
```

至此，恢复了表 tpch.orders 的内容。

11.4.8 不完全恢复到指定事务号

1. 模拟生产环境中的常见误操作（更新表未使用条件，导致全表都更新成一样的值）

在生产数据库服务器上，执行下面的命令和 SQL 语句模拟出现了人为误操作，不小心把数据库 test 中 tpch 模式下的表 orders 的 o_totalprice 字段值都设置为 10 000 了：

```
[kingbase@dbsvr ~]$ ksql -d test -U system
system@test=# SELECT txid_current();
```

```
    txid_current
---------------
          1183
(1 row)
system@test=# UPDATE tpch.orders SET o_totalprice=10000;
UPDATE 1500000
system@test=# SELECT o_orderkey, o_totalprice FROM tpch.orders LIMIT 3;
 o_orderkey | o_totalprice
------------+--------------
       4131 |     10000.00
       6951 |     10000.00
       8355 |     10000.00
(3 rows)
system@test=# \q
[kingbase@dbsvr ~]$
```

幸运的是,知道误操作之前的准确事务号为1183。这样可以将数据库恢复到这个指定的事务号。

2. 故障处理和恢复过程

在**生产数据库服务器**上,执行下面的命令切换 WAL 日志:

```
[kingbase@dbsvr ~]$ ksql -d test -U system -c " select sys_switch_wal();"
```

在生产数据库服务器 dbsvr 上,执行下面的命令增量备份(文件粒度)数据库:

```
[kingbase@dbsvr ~]$ /u00/Kingbase/ES/V9/kingbase/Server/bin/sys_rman \
                   --config=/dbbak/pbak/kbbr_repo/sys_rman.conf \
                   --stanza=kingbase \
                   --archive-copy \
                   --type=incr backup \
                   >> /u00/Kingbase/ES/V9/kingbase/Server/log/sys_rman_backup_inc.log 2>&1
```

在**备份恢复测试机**上,执行下面的命令将备份集同步到备份恢复测试机上:

```
[kingbase@dbtest ~]$ rsync -avz --delete kingbase@192.168.100.22:/dbbak/pbak/kbbr_repo/*   /dbbak/pbak/kbbr_repo
# 省略了许多输出
[kingbase@test ~]$
```

在备份恢复测试机上,执行下面的命令查看备份集的信息:

```
[kingbase@dbtest ~]$ sys_rman --config=/dbbak/pbak/kbbr_repo/sys_rman.conf --stanza=kingbase info
stanza: kingbase
    status: ok
    cipher: none

    db (current)
        wal archive min/max (V009R001C001B0025): 000000010000000000000029/000000030000000000000039
```

```
full backup: 20240226-140203F
    timestamp start/stop: 2024-02-26 14:02:03 / 2024-02-26 14:02:08
    wal start/stop: 000000010000000000000029 / 00000001000000000000002A
    database size: 1.9GB, database backup size: 1.9GB
    repo1: backup set size: 1.9GB, backup size: 1.9GB

full backup: 20240226-140210F
    timestamp start/stop: 2024-02-26 14:02:10 / 2024-02-26 14:02:15
    wal start/stop: 00000001000000000000002B / 00000001000000000000002C
    database size: 1.9GB, database backup size: 1.9GB
    repo1: backup set size: 1.9GB, backup size: 1.9GB

full backup: 20240226-142330F
    timestamp start/stop: 2024-02-26 14:23:30 / 2024-02-26 14:23:35
    wal start/stop: 000000020000000000000038 / 000000020000000000000038
    database size: 2GB, database backup size: 2GB
    repo1: backup set size: 2GB, backup size: 2GB

incr backup: 20240226-142330F_20240226-144110I
    timestamp start/stop: 2024-02-26 14:41:10 / 2024-02-26 14:41:13
    wal start/stop: 000000020000000100000009 / 000000020000000100000009
    database size: 2.4GB, database backup size: 650.3MB
    repo1: backup set size: 2.4GB, backup size: 650.3MB
    backup reference list: 20240226-142330F

incr backup: 20240226-142330F_20240226-145324I
    timestamp start/stop: 2024-02-26 14:53:24 / 2024-02-26 14:53:27
    wal start/stop: 00000002000000010000001F / 000000020000000100000020
    database size: 2.4GB, database backup size: 757.8MB
    repo1: backup set size: 2.4GB, backup size: 757.8MB
    backup reference list: 20240226-142330F
[kingbase@dbtest ~]$
```

在备份恢复测试机上，执行下面的命令停止 KingbaseES 数据库管理系统：

```
[kingbase@dbtest ~]$ sys_ctl stop
```

在备份恢复测试机上，执行下面的命令删除备份恢复测试机上的 KingbaseES 数据库集簇：

```
[kingbase@dbtest ~]$ rm -rf /u0?/Kingbase/ES/V9/data
```

在备份恢复测试机上，执行下面的命令基于事务号不完全恢复 KingbaseES 数据库集簇：

```
[kingbase@dbtest ~]$ sys_rman --config=/dbbak/pbak/kbbr_repo/sys_rman.conf --stanza=kingbase \
    --type=xid --target='1183' --set='20240226-142330F' --target-action=promote restore
```

其中参数如下。

（1）选项--target 用于指定恢复到哪个事务号为止。

(2) 选项--set 用于指定使用哪个备份集进行恢复。

在备份恢复测试机上，执行下面的命令启动 KingbaseES 数据库管理系统：

[kingbase@test ~]$ sys_ctl start

在备份恢复测试机上，执行下面的命令验证已经恢复了 KingbaseES 数据库：

```
[kingbase@dbtest ~]$ ksql -d test -U kingbase -c "SELECT o_orderkey, o_totalprice
FROM tpch.orders LIMIT 3;"
 o_orderkey | o_totalprice
---------+--------------
        1 |    173665.47
        2 |     46929.18
        3 |    193846.25
(3 rows)
[kingbase@dbtest ~]$
```

可以看到，表 tpch.orders 已经恢复到误操作之前的数据了。

后面的处理与 11.4.7 节中，将表 tpch.orders 从备份恢复测试机上导出，然后导入到生产数据库服务器的过程一样，不再赘述。

第 12 章

逻辑备份与恢复

KingbaseES 数据库逻辑备份与恢复是 KingbaseES 数据库管理的关键组成部分,目的是确保数据的完整性和可用性,保护数据免受损坏、误删除或其他意外事件的影响。

KingbaseES 数据库的**逻辑备份**是指将 KingbaseES 数据库中的数据和结构导出为 SQL 脚本或备份文件形式存储的逻辑文件;**逻辑恢复**则是将这些备份文件重新加载到 KingbaseES 数据库中,从而恢复 KingbaseES 数据库的数据和结构。

KingbaseES 数据库逻辑备份与恢复的优点是:①灵活性,用户可以选择备份整个 KingbaseES 数据库、某些指定的表或者仅备份 KingbaseES 数据库中的数据而不包括表结构;②跨平台、跨版本兼容性,由于逻辑备份通常与特定 KingbaseES 数据库管理系统的版本无关,因此很容易在不同的硬件平台、KingbaseES 数据库版本之间进行备份和恢复。

KingbaseES 数据库逻辑备份与恢复的主要缺点是:①对于大型 KingbaseES 数据库,在逻辑备份期间可能会严重影响 KingbaseES 生产数据库的性能;②进行 KingbaseES 数据库逻辑恢复的时间比较长,由此造成的用户业务暂停可能无法接受。

KingbaseES 数据库提供了两套逻辑备份与恢复工具。

(1) sys_dump 和 restore 导入导出工具。

(2) Oracle 兼容的 exp 和 imp 导入导出工具。

本章的实验环境可以使用 1.4.5 节制作的虚拟机备份文件 dbserver-3DB-BestPractice.rar。

12.1　sys_dump 和 sys_restore

使用 sys_dump 和 sys_dumpall 命令,可以把 KingbaseES 数据库里的数据,以 SQL 语句的方式导出到备份文件中。使用 sys_restore 命令执行备份文件中的 SQL 语句,可以将 KingbaseES 数据库的逻辑备份恢复到目标 KingbaseES 数据库中。关于 sys_dump、sys_dumpall 和 sys_restore 命令的语法,请参考官方文档。

12.1.1　数据库的逻辑备份与恢复

本节以 KingbaseES 数据库为单位,进行逻辑备份,下面是备份和恢复单个 KingbaseES 数据库的例子。

例 12.1　以 plain text 格式备份数据库 test;删除数据库 test;使用数据库 test 的备份

文件恢复数据库 test。

（1）执行下面的命令备份数据库 test：

[kingbase@dbsvr ~]$ sys_dump -U system -d test -F p -f test_bk.sql

其中,选项-F p 表示生成 plain text 格式的数据库备份文件。

执行完上面的命令,将数据库 test 备份到文件 test_bk.sql 中。

（2）执行下面的命令删除并重建数据库 test：

[kingbase@dbsvr ~]$ ksql -d kingbase -U system -c "drop database test;"
[kingbase@dbsvr ~]$ ksql -d kingbase -U system -c "create database test tablespace user_ts;"

（3）执行下面的命令使用包含表空间信息的数据库备份 test_bk.sql 进行恢复：

[kingbase@dbsvr ~]$ ksql -d test -U system -f test_bk.sql -q

（4）执行下面的命令查看恢复后数据库 test 的信息：

```
[kingbase@dbsvr ~]$ ksql -d test -U system -c "\l+ test"
                                    List of databases
 Name |  Owner   | Encoding |  Collate   |   Ctype    | Access privileges |  Size   | Tablespace | Description
------+----------+----------+------------+------------+-------------------+---------+------------+-------------
 test | kingbase | UTF8     | zh_CN.UTF-8| zh_CN.UTF-8|                   | 1818 MB | user_ts    |
(1 row)
[kingbase@dbsvr ~]$
```

例 12.2 以 tar 格式备份数据库 test；删除数据库 test；使用数据库 test 的备份文件恢复数据库 test。

（1）执行下面的命令备份数据库 test,在备份中包含创建数据库的命令,也包含表空间信息,并且在创建前删除数据库中的对象：

[kingbase@dbsvr ~]$ sys_dump -U system -d test --no-owner -C -c -F t -f test_bk21.tar

其中的参数如下。

① 选项-F t 表示生成 tar 格式的数据库备份。tar 格式的数据库备份,需要使用 sys_restore 命令进行恢复。

② 选项-C 表示在备份中包括 create database 命令。

③ 选项-c 表示重新创建前,先删除数据库对象。

④ 选项--no-owner 表示不备份数据库对象属主的信息。

（2）不想在备份中包含表空间信息,可以执行下面的命令：

[kingbase@dbsvr ~]$ sys_dump -U system -d test --no-owner -C -c -F t --no-tablespaces \
                   -f test_bk22.tar

其中,选项--no-tablespaces 表示不备份数据库对象表空间的信息。

（3）执行下面的命令,先删除数据库 test,然后再使用包含表空间信息的数据库 test 备份文件 test_bk21.tar 进行恢复：

```
[kingbase@dbsvr ~]$ ksql -d kingbase -U system -c "drop database test;"
[kingbase@dbsvr ~]$ sys_restore -C  -d kingbase -U system  test_bk21.tar
[kingbase@dbsvr ~]$ ksql -d test -U system -c "\l+ test"
                                    List of databases
 Name | Owner | Encoding |  Collate   |   Ctype    | Access privileges | Size  | Tablespace | Description
------+-------+----------+------------+------------+-------------------+-------+------------+------------
 test | system| UTF8     | zh_CN.UTF-8| zh_CN.UTF-8|                   |1797 MB| user_ts    |
(1 row)
[kingbase@dbsvr ~]$
```

可以看到,数据库 test 恢复后存储在 user_ts 表空间。

(4) 执行下面的命令,先删除数据库 test,然后再使用不包含表空间信息的数据库 test 备份文件 test_bk22.tar 进行恢复:

```
[kingbase@dbsvr ~]$ ksql -d kingbase -U system -c "drop database test;"
[kingbase@dbsvr ~]$ sys_restore -C  -d kingbase -U system  test_bk22.tar
[kingbase@dbsvr ~]$ ksql -d test -U system -c "\l+ test"
                                    List of databases
 Name | Owner | Encoding |  Collate   |   Ctype    | Access privileges | Size  | Tablespace  | Description
------+-------+----------+------------+------------+-------------------+-------+-------------+------------
 test | system| UTF8     | zh_CN.UTF-8| zh_CN.UTF-8|                   |1797 MB| sys_default |
(1 row)
[kingbase@dbsvr ~]$
```

可以看到,数据库 test 恢复后存储在 sys_default 表空间。

例 12.3 以用户定义的 dmp 格式备份数据库 test,生成的数据库 test 的备份文件为 test.bk3.dmp;创建数据库 newdb1;使用数据库 test 的备份文件,将数据恢复到数据库 newdb1 中。

(1) 执行下面的命令备份数据库 test,生成的备份文件为 test_bk3.dmp,在备份中包含创建数据库的命令,并且在创建前删除数据库中的对象:

```
[kingbase@dbsvr ~]$ sys_dump -U kingbase -d test --no-privileges -C -c -F c -f test_bk3.dmp
```

其中的参数如下。

① 选项 -F c 表示生成 custom 格式的数据库备份(后缀名使用 .dmp)。用户定义的 dmp 格式的数据库备份,需要使用 sys_restore 命令进行恢复。

② 选项 --no-privileges 表示不备份权限(grant/revoke)。

(2) 执行下面的命令创建数据库 newdb1:

```
[kingbase@dbsvr ~]$ ksql -d test -U system -c "create database newdb1 tablespace user_ts;"
```

(3) 执行下面的命令生成 toc:

```
[kingbase@dbsvr ~]$ sys_restore --no-owner  -l -f  ./toc ./test_bk3.dmp
```

其中,选项 -l 表示以单个事务的方式恢复数据。

(4) 执行下面的命令,根据生成的 toc 进行恢复:

```
[kingbase@dbsvr ~]$ sys_restore --no-owner -d newdb1 -U system -F c -L ./toc ./test_bk3.dmp
```

其中,选项-L 表示使用列表文件来指定恢复过程中要包含或排除的对象。

(5) 执行下面的命令查看恢复后的数据库 newdb1 信息:

```
[kingbase@dbsvr ~]$ ksql -d test -U system -c "\l+ newdb1"
                                List of databases
   Name  | Owner  | Encoding |  Collate   |   Ctype    | Access privileges | Size   | Tablespace | Description
---------+--------+----------+------------+------------+-------------------+--------+------------+-------------
 newdb1  | system | UTF8     | zh_CN.UTF-8| zh_CN.UTF-8|                   | 1797 MB| user_ts    |
(1 row)
[kingbase@dbsvr ~]$
```

可以看到,执行完上述操作后,已经使用用户 system 将数据库 test 恢复到了数据库 newdb 中。

例 12.4 使用数据库 test 的逻辑备份文件 test.bk3.dmp,将数据库 test 的一部分恢复到新的数据库 newdb2 中。

(1) 执行下面的命令创建数据库 newdb2:

```
[kingbase@dbsvr ~]$ ksql -d test -U system -c "create database newdb2 tablespace user_ts;"
```

(2) 执行下面的命令生成文件 toc:

```
[kingbase@dbsvr ~]$ sys_restore --no-owner -l -f ./toc ./test_bk3.dmp
```

(3) 使用 vi 编辑器编辑文件 toc,用分号注释掉不需要的部分(tpch 模式下的表 lineitem 的数据):

```
[kingbase@dbsvr ~]$ vi toc
```

找到下面的行:

```
6052; 0 16688 TABLE DATA tpch lineitem system
```

在这一行前加上表示注释的分号";",意思是不恢复表 lineitem 的数据,只恢复空表。

```
;6052; 0 16688 TABLE DATA tpch lineitem system
```

(4) 执行下面的命令进行部分恢复:

```
[kingbase@dbsvr ~]$ sys_restore --no-owner -d newdb2 -U system -F c -L ./toc ./test_bk3.dmp
```

(5) 执行下面的命令查看恢复后的数据库 newdb2 信息:

```
[kingbase@dbsvr ~]$ ksql -d test -U system -c "\l+ newdb2"
                                List of databases
   Name  | Owner  | Encoding |  Collate   |   Ctype    | Access privileges | Size   | Tablespace | Description
---------+--------+----------+------------+------------+-------------------+--------+------------+-------------
 newdb2  | system | UTF8     | zh_CN.UTF-8| zh_CN.UTF-8|                   | 563 MB | user_ts    |
(1 row)
```

可以看到,执行完上述操作后,已经使用用户 system 将数据库 test 恢复到了数据库 newdb2 中。

(6) 执行下面的命令验证只进行了部分恢复:

```
[kingbase@dbsvr ~]$ ksql -d newdb2 -U system -c \
                    "SELECT l_orderkey, l_linenumber FROM newdb2.tpch.lineitem"
 l_orderkey | l_linenumber
------------+--------------
(0 rows)
```

可以看到，表 newdb2.tpch.lineitem 目前是空表，也就是说，只从数据库 test 的逻辑备份中恢复了部分数据。

例 12.5 以目录格式备份数据库 test，备份到操作系统目录/dbbak/lbak/test_bk，然后删除数据库 test；再使用目录格式的备份恢复数据库 test。

（1）执行下面的命令以目录的方式备份数据库 test：

```
[kingbase@dbsvr ~]$ sys_dump -U system -d test -F d -f /dbbak/lbak/test_bk
```

其中的参数如下。

① 选项-F d 表示生成目录格式的数据库备份。目录格式的数据库逻辑备份，需要使用 sys_restore 命令进行恢复。

② 选项-f 指定目标目录名。

（2）执行下面的命令删除并重建数据库 test：

```
[kingbase@dbsvr ~]$ ksql -d kingbase -U system -c "drop database test;"
[kingbase@dbsvr ~]$ ksql -d kingbase -U system -c "create database test tablespace user_ts;"
```

（3）执行下面的命令使用刚生成的目录格式的数据库 test 备份恢复数据库 test：

```
[kingbase@dbsvr ~]$ sys_restore -U system -d test  /dbbak/lbak/test_bk
```

（4）执行下面的命令查看恢复后的数据库 test：

```
[kingbase@dbsvr ~]$ ksql -d test -U system
system@test=# \l+ test
                                       List of databases
 Name | Owner | Encoding |  Collate   |   Ctype    | Access privileges |  Size   | Tablespace | Description
------+-------+----------+------------+------------+-------------------+---------+------------+-------------
 test | system| UTF8     | zh_CN.UTF-8| zh_CN.UTF-8|                   | 1818 MB | user_ts    |
(1 row)
system@test=# \dt tpch.*
         List of relations
 Schema |  Name   | Type  | Owner
--------+---------+-------+--------
 tpch   | customer| table | system
--省略了一些输出
(10 rows)
system@test=# \q
[kingbase@dbsvr ~]$
```

12.1.2 模式的逻辑备份与恢复

本节以模式为单位，进行逻辑备份。下面是备份和恢复单个模式的例子。

例 12.6 以 SQL 脚本的格式，备份数据库 test 的模式 tpch，并将模式 tpch 的备份恢复到数据库 newdb3 中。

(1) 使用用户 kingbase，执行下面的命令备份数据库 test 下的模式 tpch：

```
[kingbase@dbsvr ~]$ sys_dump -U system -d test -n tpch -F p -f test_tpch.sql \
                    --no-owner  --no-tablespaces  --no-privileges
```

其中，选项 -n 用来指定要备份的数据库模式名。

(2) 执行下面的命令创建数据库 newdb3：

```
[kingbase@dbsvr ~]$ ksql -d test -U system -c "create database newdb3 tablespace user_ts;"
```

(3) 执行下面的命令在数据库 newdb3 中恢复模式 tpch：

```
[kingbase@dbsvr ~]$ ksql -d newdb3 -U system  -f test_tpch.sql -q
```

(4) 执行下面的命令验证已经在数据库 newdb3 中恢复了模式 tpch：

```
[kingbase@dbsvr ~]$ ksql -d newdb3 -U system
system@newdb3=# \dn tpch
List of schemas
 Name | Owner
------+--------
 tpch | system
(1 row)
system@newdb3=# \dt tpch.*
         List of relations
 Schema |   Name   | Type  | Owner
--------+----------+-------+--------
 tpch   | customer | table | system
--省略了一些输出
(8 rows)
system@newdb3=# \q
[kingbase@dbsvr ~]$
```

可以看到，已经在数据库 newdb3 中恢复了模式 tpch 及其中的模式对象。

例 12.7 以 customer(dmp 文件)格式，备份数据库 test 的模式 tpch，并将它恢复到数据库 newdb4 中。

(1) 执行下面的命令备份数据库 test 下的模式 tpch：

```
[kingbase@dbsvr ~]$ sys_dump -U system -d test -n tpch -F c -f test_tpch.dmp \
                    --no-owner  --no-tablespaces  --no-privileges
```

(2) 执行下面的命令创建数据库 newdb4：

```
[kingbase@dbsvr ~]$ ksql -d test -U system -c "create database newdb4 tablespace user_ts;"
```

(3) 执行下面的命令在数据库 newdb4 中恢复模式 tpch：

```
[kingbase@dbsvr ~]$ sys_restore --no-owner -d newdb4 -U system  test_tpch.dmp
```

(4) 执行下面的命令验证已经在数据库 newdb4 中恢复了模式 tpch：

```
[kingbase@dbsvr ~]$ ksql -d newdb4 -U system
system@newdb4=# \dn tpch
List of schemas
 Name  | Owner
-------+--------
 tpch  | system
(1 row)
system@newdb4=# \dt tpch.*
          List of relations
 Schema |   Name   | Type  | Owner
--------+----------+-------+--------
 tpch   | customer | table | system
--省略了一些输出
(8 rows)
system@newdb4=# \q
[kingbase@dbsvr ~]$
```

可以看到,已经在数据库 newdb4 中恢复了模式 tpch 及其中的模式对象。

12.1.3 表的逻辑备份与恢复

本节以表为单位,进行逻辑备份。下面是备份和恢复表的例子。

例 12.8 以 SQL 脚本的格式,备份数据库 test 中,模式 tpch 下的表 orders,并将它恢复到数据库 newdb5 的 tpch 模式中;恢复到数据库 newdb6 的 public 模式中。

(1) 执行下面的命令备份数据库 test 模式 tpch 下的表 orders,生成的 test_tpch_orders.1.sql 备份文件将使用 COPY 来还原表数据:

```
[kingbase@dbsvr ~]$ sys_dump -d test -U system -t tpch.orders -F p -f test_tpch_orders.1.sql \
                 --no-owner --no-tablespaces  --no-privileges
```

其中,选项 -t 用来指定要备份的表的名字。

(2) 执行下面的命令备份数据库 test 模式 tpch 下的表 orders,生成的 test_tpch_orders.2.sql 备份文件将使用 INSERT 语句来还原表数据:

```
[kingbase@dbsvr ~]$ sys_dump -d test -U system -t tpch.orders -F p -f test_tpch_orders.2.sql \
                 --no-owner --no-tablespaces  --no-privileges --column-inserts
```

其中,选项 --column-inserts 用来指定使用 INSERT 语句还原表数据,而不是 COPY。

(3) 执行下面的命令创建数据库 newdb5,并在数据库 newdb5 中创建模式 tpch:

```
[kingbase@dbsvr ~]$ ksql -d kingbase -U kingbase -c "create database newdb5 tablespace user_ts;"
[kingbase@dbsvr ~]$ ksql -d newdb5 -U kingbase -c "create schema tpch;"
```

(4) 执行下面的命令在数据库 newdb5 的 tpch 模式下恢复表 orders:

```
[kingbase@dbsvr ~]$ ksql -d  newdb5 -U system -f  test_tpch_orders.1.sql -q
set_config
------------
```

```
          public
(1 row)
ksql:test_tpch_orders.1.sql:1500069: ERROR:  relation "tpch.customer" does not exist
[kingbase@dbsvr ~]$
```

(5) 执行下面的命令验证已经在数据库 newdb5 的模式 tpch 中恢复了表 orders：

```
[kingbase@dbsvr ~]$ ksql -d newdb5 -U system -c "SELECT count(*) FROM tpch.orders;"
  count
---------
 1500000
(1 row)
[kingbase@dbsvr ~]$
```

如果想把表 orders 恢复到数据库 newdb6 的 public 模式中，可以执行下面的步骤。
(1) 执行下面的命令创建数据库 newdb6：

```
[kingbase@dbsvr ~]$ ksql -d kingbase -U system -c "create database newdb6 tablespace user_ts;"
```

(2) 使用 vi 编辑器编辑文件 test_tpch_orders.1.sql，删除文件中所有的模式前缀 tpch.（在命令模式下，按 Esc 键，输入字符串：%s/tpch\.//g，然后再按 Enter 键）。

(3) 执行下面的命令在数据库 newdb6 的 pulic 模式下恢复表 orders：

```
[kingbase@dbsvr ~]$ ksql -d  newdb6 -U system -f  test_tpch_orders.1.sql -q
set_config
------------
 public
(1 row)
ksql:test_tpch_orders.1.sql:1500069: ERROR:  relation "customer" does not exist
[kingbase@dbsvr ~]$
```

(4) 执行下面的命令验证已经在数据库 newdb6 的 public 模式中恢复了表 orders：

```
[kingbase@dbsvr ~]$ ksql -d newdb6 -U system -c "SELECT count(*) FROM orders;"
  count
---------
 1500000
(1 row)
[kingbase@dbsvr ~]$
```

例 12.9 以 customer 格式（dmp 文件），备份数据库 test 中，模式 tpch 下的表 orders，并将它恢复到数据库 newdb7 的 tpch 模式中。

(1) 执行下面的命令备份数据库 test 的模式 tpch 下的表 orders：

```
[kingbase@dbsvr ~]$ sys_dump -U system -d test -t tpch.orders -F c -f test_tpch_orders.dmp \
                    --no-owner  --no-tablespaces  --no-privileges
[kingbase@dbsvr ~]$
```

（2）执行下面的命令创建数据库 newdb7，并在数据库 newdb7 创建模式 tpch：

```
[kingbase@dbsvr ~]$ ksql -d kingbase -U system -c "create database newdb7 tablespace user_ts;"
[kingbase@dbsvr ~]$ ksql -d newdb7 -U system -c "create schema tpch;"
```

（3）执行下面的命令在数据库 newdb7 的 tpch 模式下恢复表 orders：

```
[kingbase@dbsvr ~]$ sys_restore --no-owner -d newdb7 -U system test_tpch_orders.dmp
sys_restore: while PROCESSING TOC:
sys_restore: from TOC entry 5515; 2606 17034 FK CONSTRAINT orders orders_o_custkey_fkey system
sys_restore: error: could not execute query: ERROR:  relation "tpch.customer" does not exist
Command was: ALTER TABLE ONLY tpch.orders
    ADD CONSTRAINT orders_o_custkey_fkey FOREIGN KEY (o_custkey) REFERENCES tpch.customer(c_custkey);
sys_restore: warning: errors ignored on restore: 1
[kingbase@dbsvr ~]$
```

（4）执行下面的命令验证已经在数据库 newdb7 的 tpch 模式中恢复了表 orders：

```
[kingbase@dbsvr ~]$ ksql -d newdb7 -U system -c "SELECT count(*) FROM tpch.orders;"
  count
---------
 1500000
(1 row)
[kingbase@dbsvr ~]$
```

12.1.4　逻辑备份用户和表空间定义

执行下面的命令备份角色（用户）和表空间定义：

```
[kingbase@dbtest ~]$ sys_dumpall -U system -f db_roleANDtablespace.sql --globals-only
```

执行下面的命令只备份角色（用户）定义：

```
[kingbase@dbtest ~]$ sys_dumpall -U system -f db_role.sql --roles-only
```

执行下面的命令只备份表空间定义：

```
[kingbase@dbtest ~]$ sys_dumpall -U system --no-owner -f db_tablespaces.sql --tablespaces-only
```

DBA 可以直接使用 Linux 的 cat 命令或 vi 编辑器，查看这些备份生成的 SQL 脚本文件。在必要时，可以在目标 KingbaseES 数据库上执行这些 SQL 脚本，恢复 KingbaseES 数据库的用户和表空间。

12.2　Oracle 兼容的 exp 和 imp

KingbaseES 数据库还提供了 Oracle 兼容的 exp 和 imp 逻辑备份与恢复工具，方便熟悉 Oracle 数据库的用户使用。

Oracle 兼容的 exp 和 imp 逻辑备份与恢复工具都需要使用服务名。附录 B.3.7 中已经配置了本节需要使用的服务名 sntest、snkingbase 和 sntestdb。

关于 KingbaseES 数据库的 exp 和 imp 命令的语法，请参考官方手册。

12.2.1　导出/导入数据库

例 12.10　使用 KingbaseES 数据库的 exp 和 imp 命令导出数据库 test，生成的备份文件为 test_full.dmp；删除数据库 test；重建数据库 test；使用数据库 test 的逻辑备份文件恢复数据库 test。

（1）执行下面的命令导出数据库 test，生成的备份文件为 test_full.dump：

```
[kingbase@dbsvr ~]$ mkdir -p /dbbak/lbak/oracle
[kingbase@dbsvr ~]$ cd /dbbak/lbak/oracle
[kingbase@dbsvr oracle]$ exp system/Passw0rd@sntest file=test_full.dmp full=y
exp: NOTICE:  undefine audit event report.
exp: NOTICE:  undefine audit event report.
Export terminated successfully without warnings.
[kingbase@dbsvr oracle]$
```

exp 命令默认导出服务名中的数据库（示例中服务名 sntest，访问的是数据库 test）。

（2）执行下面的命令删除并重建数据库 test：

```
[kingbase@dbsvr oracle]$ ksql -d kingbase -U system -c "drop database test;"
[kingbase@dbsvr oracle]$ ksql -d kingbase -U system -c "create database test;"
```

（3）执行下面的命令将逻辑备份文件 test_full.dump 导入数据库 test：

```
[kingbase@dbsvr oracle]$ imp system/Passw0rd@sntest  file=test_full.dmp  full=y
imp: NOTICE:  undefine audit event report.
Import terminated successfully without warnings.
[kingbase@dbsvr oracle]$
```

（4）执行下面的命令可以验证已经成功导入数据：

```
[kingbase@dbsvr oracle]$ ksql -d test -U system -c "\dt tpch.*;"
         List of relations
 Schema |    Name    | Type  | Owner
--------+------------+-------+--------
 tpch   | customer   | table | system
--省略了一些输出
(10 rows)
[kingbase@dbsvr oracle]$
```

12.2.2 导出/导入用户模式

例 12.11 使用 KingbaseES 数据库的 exp 和 imp 命令导出数据库 test 中的模式 tpch，生成的备份文件为 tpch.dmp；在 KingbaseES 数据库中新建一个数据库模式 tpchnew；将 tpch 模式的逻辑备份恢复到新创建的模式 tpchnew 中。

(1) 执行下面的命令导出数据库 test 的 tpch 模式：

```
[kingbase@dbsvr ~]$ cd /dbbak/lbak/oracle
[kingbase@dbsvr oracle]$ exp system/Passw0rd@sntest file=tpch.dmp owner=tpch
exp: NOTICE:   undefine audit event report.
exp: NOTICE:   undefine audit event report.
Export terminated successfully without warnings.
[kingbase@dbsvr ~]$
```

其中，exp 命令的选项，选项 owner 指定了要导出的数据库模式。

(2) 执行下面的命令为 KingbaseES 数据库创建一个新模式 tpchnew：

```
[kingbase@dbsvr oracle]$ ksql -d kingbase -U system -c "create schema tpchnew;"
```

(3) 执行下面的命令将之前备份的 tpch 模式，导入 KingbaseES 数据库的模式 tpchnew 中：

```
[kingbase@dbsvr oracle]$ imp system/Passw0rd@snkingbase file=tpch.dmp \
                        fromuser=tpch touser=tpchnew
imp: NOTICE:   undefine audit event report.
Import terminated successfully without warnings.
[kingbase@dbsvr oracle]$
```

其中，imp 命令的选项如下。

① fromuser 选项指定导入的源模式名。
② touser 选项指定导入的目标模式名。

(4) 执行下面的命令可以验证已经成功导入数据：

```
[kingbase@dbsvr oracle]$ ksql -d kingbase -U system -c "\dt tpchnew.*;"
        List of relations
 Schema  |  Name    | Type  | Owner
---------+----------+-------+--------
 tpchnew | customer | table | system
--省略了一些输出
(8 rows)
[kingbase@dbsvr oracle]$
```

12.2.3 导出/导入表

例 12.12 使用 KingbaseES 数据库的 exp 和 imp 命令导出数据库 test 模式 tpch 下的表 orders、customer、nation、region，生成备份文件为 tablesbk.dmp；把其中的表 orders 导入到数据库 testdb 的 tpch 模式下；把所有的表导入到数据库 testdb 的 public 模式下。

(1) exp 命令的 tables 选项指定了要导出的表名。执行下面的命令导出数据库 test 的模式 tpch 下的表 orders、customer、nation、region：

```
[kingbase@dbsvr ~]$ cd /dbbak/lbak/oracle
[kingbase@dbsvr oracle]$ exp system/Passw0rd@sntest file=tablesbk.dmp \
        tables=test.tpch.orders,test.tpch.customer,test.tpch.nation,test.
tpch.region
exp: NOTICE:  undefine audit event report.
exp: NOTICE:  undefine audit event report.
Export terminated successfully without warnings.
[kingbase@dbsvr oracle]$
```

其中，exp 命令的选项中，选项 tables 指定了要导出的表名（可以是多个表）。

（2）执行下面的命令创建数据库 testdb，在数据库 testdb 中创建模式 tpch：

```
[kingbase@dbsvr oracle]$ ksql -d test   -U system -c "CREATE DATABASE testdb;"
[kingbase@dbsvr oracle]$ ksql -d testdb -U system -c "CREATE SCHEMA    tpch;"
```

（3）执行下面的命令将表 orders 导入数据库 testdb 的 tpch 模式下：

```
[kingbase@dbsvr oracle]$ imp system/Passw0rd@sntestdb file=tablesbk.dmp tables
=orders
imp: NOTICE:  undefine audit event report.
Import terminated successfully without warnings.
[kingbase@dbsvr oracle]$
```

其中，imp 命令的选项中，选项 tables 指定了要导入的表名（可以是多个表）。

（4）执行下面的命令可以验证已经成功将表 orders 导入数据库 testdb 的 tpch 模式：

```
[kingbase@dbsvr oracle]$ ksql -d testdb -U system -c "\dt tpch.*;"
       List of relations
 Schema |  Name  | Type  | Owner
--------+--------+-------+--------
 tpch   | orders | table | system
(1 row)
[kingbase@dbsvr oracle]$
```

（5）执行下面的命令将备份的 4 张表 orders、customer、nation、region 导入数据库 testdb 的 public 模式下：

```
[kingbase@dbsvr oracle]$ imp kingbase/Passw0rd@sntestdb file=tablesbk.dmp \
                   fromuser=tpch touser=public
imp: NOTICE:  undefine audit event report.
Import terminated successfully without warnings.
[kingbase@dbsvr oracle]$
```

其中，imp 命令的选项如下。

① 选项 fromuser 指定了要导入的表的源模式名。

② 选项 touser 指定了要导入的表的目标模式名。

（6）执行下面的命令可以验证已经成功导入数据：

```
[kingbase@dbsvr oracle]$ ksql -d testdb -U system -c "\dt public.*;"
       List of relations
```

```
 Schema |   Name   | Type | Owner
--------+----------+------+--------
 public | customer | table | system
 public | nation   | table | system
 public | orders   | table | system
 public | region   | table | system
(4 rows)
[kingbase@dbsvr oracle]$
```

第 13 章 闪回查询与闪回表

KingbaseES 数据库基于插件 kdb_flashback，向用户提供了与 Oracle 数据库类似的闪回功能，DBA 或用户可以快速地从误操作中恢复数据。目前 KingbaseES 数据库提供的闪回技术如下。

(1) 闪回回收站。

(2) 闪回查询(时间戳和 csn)。

(3) 闪回表。

本章的实验环境可以使用 1.4.5 节制作的虚拟机备份文件 dbserver-3DB-BestPractice.rar。

13.1 配置 KingbaseES 数据库的闪回功能

要配置 KingbaseES 数据库的闪回功能，需要在 KingbaseES 数据库服务器上修改 KingbaseES 数据库的系统启动参数文件 /u00/Kingbase/ES/V9/data/kingbase.conf。

(1) 确保参数 shared_preload_libraries 中包含 kdb_flashback 插件：

```
shared_preload_libraries = 'liboracle_parser, synonym, plsql, force_view, kdb_flashback,plugin_debugger, plsql_plugin_debugger, plsql_plprofiler, kdb_ora_expr, sepapower, dblink, sys_kwr, sys_spacequota, sys_stat_statements, backtrace, kdb_utils_function, auto_bmr, sys_squeeze, src_restrict'
```

(2) 设置参数 track_commit_timestamp 为 on：

```
track_commit_timestamp = on
```

(3) 增加以下两个参数：

```
kdb_flashback.enable_flashback_query=on
kdb_flashback.db_recyclebin=on
```

修改完以上参数后，请重新启动 KingbaseES 数据库。

13.2 闪回回收站

闪回回收站为用户提供一种误删表后还原表的一种手段。回收站需用户定期维护，避免回收站膨胀，用户可以通过 PURGE 操作对回收站进行维护。

对闪回回收站的操作如下。

（1）清空闪回回收站。清空回收站时，回收站视图 recyclebin 和系统表 sys_recyclebin 中相关对象将被清除。

（2）删除闪回回收站中的一个表。

（3）从闪回回收站中恢复被误删除的表。

13.2.1 清空闪回回收站

清空闪回回收站，将清除闪回回收站中所有的对象。下面是清空闪回回收站的测试：

```
[kingbase@dbtest ~]$ ksql -d test -U system
system@test=# SET search_path TO tpch;
system@test=# --创建测试表 orders1、customer1 和 nation1
system@test=# CREATE TABLE orders1 AS SELECT * from orders LIMIT 10;
system@test=# CREATE TABLE customer1 AS SELECT * from customer  LIMIT 10;
system@test=# CREATE TABLE nation1 AS SELECT * from nation LIMIT 10;
system@test=# --查看闪回回收站中的内容
system@test=# SELECT * FROM recyclebin;
 oid | original_name | droptime | type
----+-----------+-------+------
(0 rows)
system@test=# --删除测试表 orders1、customer1 和 nation1
system@test=# DROP TABLE orders1;
system@test=# DROP TABLE customer1;
system@test=# DROP TABLE nation1;
system@test=# --查看闪回回收站中的内容
system@test=# SELECT * FROM recyclebin;
  oid  | original_name |          droptime           | type
-----+-----------+-----------------------+-------
 16760 | orders1       | 2024-03-13 10:47:39.516321+08 | TABLE
 16763 | customer1     | 2024-03-13 10:47:39.517176+08 | TABLE
 16766 | nation1       | 2024-03-13 10:47:39.517699+08 | TABLE
(3 rows)
system@test=# --清空闪回回收站中所有的内容
system@test=# purge recyclebin;
PURGE
system@test=# --查看闪回回收站中的内容
system@test=# SELECT * FROM recyclebin;
 oid | original_name | droptime | type
----+-----------+-------+------
(0 rows)
```

13.2.2 删除闪回回收站中的一个表

下面是删除闪回回收站中的一个表的例子：

```
system@test=# --创建测试表 orders1 和 customer1
system@test=# CREATE TABLE orders1 AS SELECT * from orders LIMIT 10;
system@test=# CREATE TABLE customer1 AS SELECT * from customer  LIMIT 10;
system@test=# --查看闪回回收站中的内容
```

```
system@test=# SELECT * FROM recyclebin;
 oid | original_name | droptime | type
-----+---------------+----------+------
(0 rows)
system@test=# --删除测试表 orders1 和 customer1
system@test=# DROP TABLE orders1;
system@test=# DROP TABLE customer1;
system@test=# --查看闪回回收站中的内容
system@test=# SELECT * FROM recyclebin;
  oid  | original_name |           droptime            | type
-------+---------------+-------------------------------+-------
 16769 | orders1       | 2024-03-13 10:49:51.092186+08 | TABLE
 16772 | customer1     | 2024-03-13 10:49:51.092902+08 | TABLE
(2 rows)
system@test=# --执行下面的 SQL 语句系列,将闪回回收站中的表 orders1 删除
system@test=# PURGE TABLE orders1;
system@test=# --查看闪回回收站中的内容
system@test=# SELECT * FROM recyclebin;
  oid  | original_name |           droptime            | type
-------+---------------+-------------------------------+-------
 16772 | customer1     | 2024-03-13 10:49:51.092902+08 | TABLE
(1 rows)
```

可以看到,回收站中的表 orders1 被删除了。

在闪回回收站中,可以保存多个同名的表:

```
system@test=# --创建测试表 customer1 然后再将其删除,重复两次
system@test=# CREATE TABLE customer1 AS SELECT * from customer LIMIT 10;
system@test=# DROP TABLE customer1;
system@test=# CREATE TABLE customer1 AS SELECT * from customer LIMIT 10;
system@test=# DROP TABLE customer1;
system@test=# --查看闪回回收站中的内容
system@test=# SELECT * FROM recyclebin;
  oid  | original_name |           droptime            | type
-------+---------------+-------------------------------+-------
 16772 | customer1     | 2024-03-13 10:49:51.092902+08 | TABLE
 16775 | customer1     | 2024-03-13 10:51:29.555765+08 | TABLE
 16778 | customer1     | 2024-03-13 10:51:29.557940+08 | TABLE
(3 rows)
```

可以看到,当前在闪回回收站中,有 3 个名为 customer1 的表,不过它们的 oid 是不同的。

此时可以执行下面的命令清除回收站中的 customer1:

```
system@test=# PURGE TABLE customer1;
PURGE
system@test=# SELECT * FROM recyclebin;
  oid  | original_name |           droptime            | type
-------+---------------+-------------------------------+-------
 16775 | customer1     | 2024-03-13 10:51:29.555765+08 | TABLE
 16778 | customer1     | 2024-03-13 10:51:29.557940+08 | TABLE
```

```
 (2 rows)
system@test=# PURGE TABLE customer1;
PURGE
system@test=# SELECT * FROM recyclebin;
  oid  | original_name |          droptime             | type
-------+---------------+-------------------------------+-------
 16778 | customer1     | 2024-03-13 10:51:29.557940+08 | TABLE
(1 row)
system@test=# PURGE TABLE customer1;
PURGE
system@test=# SELECT * FROM recyclebin;
 oid | original_name | droptime | type
-----+---------------+----------+------
(0 rows)
```

可以看到,在闪回回收站清除多个同名的表时,会按照先进先出的策略,来删除同名的表。

13.2.3 从闪回回收站恢复被误删除的表

下面是一个从闪回回收站恢复被误删除的表的例子:

```
system@test=# --创建测试表 customer1
system@test=# CREATE TABLE customer1 AS SELECT * from customer  LIMIT 10;
system@test=# --模拟误删除了表 customer1
system@test=# DROP TABLE customer1;
system@test=# --查看闪回回收站中的内容
system@test=# SELECT * FROM recyclebin;
  oid  | original_name |          droptime             | type
-------+---------------+-------------------------------+-------
 16781 | customer1     | 2024-03-13 10:54:41.490845+08 | TABLE
(1 row)
system@test=# --从闪回回收站那儿恢复表 customer1
system@test=# flashback table customer1 to before drop ;
system@test=# --查看表 customer1 的内容
system@test=# SELECT c_name,c_address from customer1;
       c_name        |            c_address
---------------------+------------------------------------
 Customer#000000001  | IVhzIApeRb ot,c,E
--省略了一些输出
(10 rows)
```

可以看到,表 customer1 的内容已经被恢复了。

13.3 闪回查询

闪回查询的作用主要有以下 3 方面。

(1) 进行历史数据的分析。

(2) 查看数据的变更历史。

(3) 确定闪回表的精确时间点。

闪回查询能返回用户指定历史时刻的快照数据，历史快照时刻指定方式有以下两种。

(1) 基于时间戳(timestamp)的闪回查询。

(2) 基于 csn 的闪回查询。

13.3.1　基于时间戳的闪回查询

下面是基于时间戳的闪回查询的例子。

(1) 执行下面的 SQL 语句构建测试环境：

```
system@test=# drop table if exists fb_example;
NOTICE:  table "fb_example" does not exist, skipping
DROP TABLE
system@test=# purge table fb_example;
ERROR:  table "fb_example" doesn't in recyclebin.
system@test=# create table fb_example(id int, name varchar(100));
system@test=# insert into fb_example values(1, 'name1');
system@test=# insert into fb_example values(2, 'name2');
system@test=# insert into fb_example values(3, 'name3');
system@test=# --时间点 tp0
system@test=# \set 'tp0' `date "+%Y-%m-%d %H:%M:%S.%N"`
system@test=# select * from fb_example;
 id | name
----+-------
  1 | name1
  2 | name2
  3 | name3
(3 rows)

system@test=# --时间点 tp1
system@test=# \set 'tp1' `date "+%Y-%m-%d %H:%M:%S.%N"`
system@test=# update fb_example set name = 'name1_update' where id=1;
system@test=# --时间点 tp2
system@test=# \set 'tp2' `date "+%Y-%m-%d %H:%M:%S.%N"`
system@test=# update fb_example set name = 'name2_update' where id=2;
system@test=# --时间点 tp3
system@test=# \set 'tp3' `date "+%Y-%m-%d %H:%M:%S.%N"`
system@test=# update fb_example set name = 'name3_update' where id=3;
```

(2) 执行下面 SQL 语句查看在时间戳 tp0 时刻表 fb_example 的内容：

```
system@test=# select * from fb_example as of timestamp :'tp0';
 id | name
----+-------
  1 | name1
  2 | name2
  3 | name3
(3 rows)
```

(3) 执行下面 SQL 语句查看在时间点 tp1 时刻表 fb_example 的内容：

```
system@test=# select * from fb_example as of timestamp :'tp1';
```

```
 id | name
---+-------
  1 | name1
  2 | name2
  3 | name3
(3 rows)
```

（4）执行下面 SQL 语句查看在时间点 tp2 时刻表 fb_example 的内容：

```
system@test=# select * from fb_example as of timestamp :'tp2';
 id |     name
---+--------------
  2 | name2
  3 | name3
  1 | name1_update
(3 rows)
```

（5）执行下面 SQL 语句查看在时间点 tp3 时刻表 fb_example 的内容：

```
system@test=# select * from fb_example as of timestamp :'tp3';
 id |     name
---+--------------
  3 | name3
  1 | name1_update
  2 | name2_update
(3 rows)
```

（6）执行下面 SQL 语句查看当前表 fb_example 的内容：

```
system@test=# select * from fb_example;
 id |     name
---+--------------
  1 | name1_update
  2 | name2_update
  3 | name3_update
(3 rows)
```

13.3.2 基于 CSN 的闪回查询

执行下面的 SQL 语句查看闪回版本的 csn 起始信息：

```
system@test=# select versions_startscn, versions_endcsn, *
test-#          from fb_example versions between csn minvalue and maxvalue;
 versions_startscn | versions_endcsn | id |     name
-------------------+-----------------+----+--------------
       65536000023 |     65536000026 |  1 | name1
       65536000024 |     65536000027 |  2 | name2
       65536000025 |     65536000028 |  3 | name3
       65536000026 |                 |  1 | name1_update
       65536000027 |                 |  2 | name2_update
       65536000028 |                 |  3 | name3_update
(6 rows)
```

执行下面的 SQL 命令查看指定 csn 号时刻的表 fa_example 的信息：

```
system@test=# select * from fb_example as of csn 65536000023;
 id | name
----+-------
  1 | name1
(1 row)
system@test=# select * from fb_example as of csn 65536000024;
 id | name
----+-------
  1 | name1
  2 | name2
(2 rows)
system@test=# select * from fb_example as of csn 65536000025;
 id | name
----+-------
  1 | name1
  2 | name2
  3 | name3
(3 rows)
system@test=# select * from fb_example as of csn 65536000026;
 id |    name
----+--------------
  2 | name2
  3 | name3
  1 | name1_update
(3 rows)
system@test=# select * from fb_example as of csn 65536000027;
 id |    name
----+--------------
  3 | name3
  1 | name1_update
  2 | name2_update
(3 rows)
system@test=# select * from fb_example as of csn 65536000028;
 id |    name
----+--------------
  1 | name1_update
  2 | name2_update
  3 | name3_update
(3 rows)
```

13.4 闪回表

闪回表到指定的时间点，实际上为用户提供了还原数据的一种手段。如果说闪回查询能够帮助用户查询历史数据，那么闪回表实际上是帮助用户彻底地找回数据。

完成表闪回的 SQL 语句，语法如下：

```
FLASHBACK TABLE TO [ Timestamp | CSN ]
```

下面是一个基于 csn 号闪回表的例子。执行下面的 SQL 语句查看闪回版本的 csn 起始信息：

```
system@test=# select versions_startscn, versions_endcsn, *
test-#         from fb_example versions between csn minvalue and maxvalue;
 versions_startscn | versions_endcsn | id |    name
----------------+-------------+---+--------------
----------------+-------------+---+--------------
      65536000023 |     65536000026 | 1 | name1
      65536000024 |     65536000027 | 2 | name2
      65536000025 |     65536000028 | 3 | name3
      65536000026 |                 | 1 | name1_update
      65536000027 |                 | 2 | name2_update
      65536000028 |                 | 3 | name3_update
(6 rows)
system@test=# select * from fb_example as of csn 65536000026;
 id |    name
---+--------------
  2 | name2
  3 | name3
  1 | name1_update
(3 rows)
system@test=# --基于 csn 号闪回表 fb_example
system@test=# FLASHBACK TABLE  fb_example TO csn 65536000026;
system@test=# -- 查看闪回后的表 fb_example 的数据
system@test=# select * from fb_example;
 id |    name
---+--------------
  1 | name1_update
  2 | name2
  3 | name3
(3 rows)
```

基于时间戳的闪回表，DBA 需要知道准确的时间戳信息，请读者自己构造一个测试用例。

13.5　闪回技术的限制

闪回查询和闪回表技术依赖于历史数据，如果历史数据因为 vacuum、truncate、rewrite 等操作被回收掉，那么会导致无法闪回到这些操作之前的时刻。因此推荐用户在期望使用闪回查询的时候对 vacuum 相关参数做一定的调整（关闭表级的 autovacuum，推荐调大 vacuum_defer_cleanup_age 的值以降低历史数据被回收的机会）。

目前闪回查询和闪回表技术在 vacuum、truncate 和部分 ddl 操作之后，不允许闪回到这些操作之前。

闪回查询应用于视图或物化视图，应该尽量避免对于常量时间戳和 csn 的使用，因为可能会引发 dump 和 restore 的失败。

第 14 章

KingbaseES 主备集群

单机 KingbaseES 数据库服务器会由于各种原因(如软硬件故障、人为误操作)导致数据库服务中断;此外单机 KingbaseES 数据库服务器的性能有可能不能满足客户的需求。KingbaseES 主备集群为用户解决这两个问题提供了一个稳定的高可用解决方案。对于 KingbaseES 数据库用户来说,KingbaeES 主备集群相当于一台单机 KingbaseES 数据库服务器,对外服务的 IP 地址是集群的虚拟 IP(virtual ip,vip)。

本章主要介绍金仓数据库 KingbaseES 主备集群在 Linux 操作系统中的配置和使用方法。

14.1　KingbaseES 主备集群简介

KingbaseES 主备集群有两种使用方式:金仓数据守护集群和金仓读写分离集群。

金仓数据守护集群(kingbase data watch,KDW)是由主库、备库和守护进程组成的集群,主库提供数据库读写服务,备库和主库通过流复制同步数据作为备份,守护进程检查各个数据库状态及环境状态,当主库故障后可以进行故障转移将备库提升为主库继续对外提供服务,确保主备集群持续提供服务。

金仓读写分离集群(KingbaseRWC)在金仓数据守护集群的基础上增加了对应用透明的读写负载均衡能力。相比数据守护集群,该类集群中所有备库均可对外提供查询能力,从而减轻了主库的读负载压力,可实现更高的事务吞吐率。对于应用系统中查询等只读操作远多于写入操作的情况,KingbaseRWC 集群可以实现在高并发、高压力下性能不下降。读写分离集群通过配置 JDBC 来实现。

本节主要介绍 KingbaseES 主备集群服务器端的配置和使用,对金仓数据守护集群和金仓读写分离集群都适用。

14.1.1　KingbaseES 主备集群的拓扑结构

一主两备的 KingbaseES 主备集群的拓扑结构,如图 14-1 所示,KingbaseES 主备集群通常由一个主节点、多个备节点及见证节点组成。

(1) **主节点**(primary):KingbaseES 主备集群中的主节点,处理客户端的读写请求,并负责数据的写操作和同步。

（2）**备节点**（**standby**）：在 KingbaseES 主备集群中充当主节点的备份节点。备节点通过复制主节点的数据来保持与主节点的数据同步，并在主节点故障时接管主节点的角色。

（3）**见证节点**（**witness**）：在 KingbaseES 主备集群中，见证节点是一种可选的、特殊类型的节点，其作用是帮助实现故障转移、参与选举过程、协助确定新的主节点，但不参与实际的数据库复制。引入见证节点可以增强 KingbaseES 主备集群的可靠性和决策能力，确保在主节点故障或网络问题时能够准确地进行故障转移，并避免数据冲突和不一致的问题。

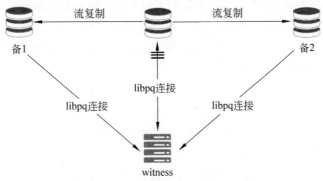

图 14-1　KingbaseES 主备集群的拓扑图

KingbaseES 主备集群中主节点和备节点都是 KingbaseES 数据库节点，每个 KingbaseES 数据库节点都有一个完整的 KingbaseES 数据库副本，它们通过数据复制来实现数据的同步。

KingbaseES 主备集群支持**故障转移**（**fail over**），当主库（计划外）发生故障时，自动切换到备机，实现了库级和中心级的自动容灾功能。同时也支持便捷高效的主备集群主动切换（**switch over**），切换过程仅需秒级完成，为计划内维护等切换需求提供了高效灵活的实施方法。

KingbaseES 主备集群通过使用**虚拟 IP**，将始终运行于主库服务器，当主库或主库所在服务器发生故障，虚拟 IP 将自动随 KingbaseES 数据库切换而漂移至新主库服务器，可以实现对外部客户端透明的高可用性和故障转移。虚拟 IP 简化了用户端的配置和管理，提供了更好的用户体验和系统可靠性。

14.1.2　KingbaseES 主备集群的组件

与单机 KingbaseES 数据库相比，KingbaseES 主备集群中的数据库节点会增加以下组件：用于数据库复制的进程、repmgrd 守护进程、kbha 守护进程及远程通信工具等。

1. 用于数据库复制的进程

在主备集群的**主服务器**上有 walsender 进程，负责将 WAL 日志数据发送给备服务器，实现 WAL 日志的复制。在主备集群的**从服务器**上有 walreceiver 进程，负责接收来自主服务器的 WAL 数据，并将其写入备服务器的磁盘。startup 进程负责将接收到的 WAL 数据应用到备库中，实现数据的同步。主备库之间的流复制连接由主库端的 walsender 进程和备库端的 walreceiver 进程建立，每个备库在主库上都有一个 walsender 进程为它传送 WAL 日志。

KingbaseES 主备集群的数据复制流程说明如下。

(1) 数据库 walwriter 进程把产生的 WAL 日志写入磁盘。

(2) WAL 日志写入磁盘后,由 walsender 进程将新增的日志读取出来并通过流的形式发送给备库。

(3) 备库 walreceiver 进程将接收到的 WAL 日志写入本地 WAL 段文件。

(4) 备库 startup 进程读取 WAL 日志进行重放,完成数据同步。

在 KingbaseES 数据库中,有两种不同的复制机制:同步复制和异步复制。

同步复制是指主备数据最终一致,发生任何故障都不会丢失数据。从上述流程看,要达到实时同步复制,需要主库等待 WAL 日志完成流程(3)或(4),主库才能继续。同步复制保证了数据最终一致,但牺牲了部分性能。

异步复制不能保证主备数据最终一致,一旦主库发生故障,备库可能没有同步完成主库已经写入的 WAL 日志,导致丢失部分数据。从上述流程看,异步主备情况下,主库只需要 WAL 日志完成流程(1)或(2)就可以继续,显然异步复制比同步复制拥有更好的性能。

选择同步复制还是异步复制取决于应用的需求和性能要求。如果应用需要高度数据一致性和可用性,且可以容忍较大的性能开销和延迟,可以选择同步复制;如果应用对数据一致性要求较低,但更关注性能和低延迟,可以选择异步复制。

在主库上的系统视图 sys_stat_replication 中可以查询到数据复制的信息,视图中的每条记录对应一个主库的 walsender 进程。

2. repmgrd 守护进程

每个 KingbaseES 数据库节点都有一个 **repmgrd 守护进程**,主要负责 KingbaseES 数据库的状态检查和故障处理,如故障自动切换、故障自动恢复等。主节点的 repmgrd 只监视本地 KingbaseES 数据库,备节点的 repmgrd 会同时监控本地 KingbaseES 数据库和主 KingbaseES 数据库。所有节点的状态信息保存在表 repmgr.nodes 里,并通过流复制同步到所有备节点。

KingbaseES 集群主节点上的 repmgrd 守护进程,每隔 monitor_interval_secs(默认 2s)会检查一次本地 KingbaseES 数据库是否可连接;主库的 repmgrd 进程检查备库状态,是通过从主库查询视图 sys_stat_replication 直接调取流复制状态来判断备库是否正常。当备库发生故障或由于主备节点间的网络发生中断,导致流复制连接中断后,主库端的 walsender 进程退出,视图 sys_stat_replication 的记录被清除,主库的 repmgrd 进程会判断备库下线。

备节点的 repmgrd 守护进程,每隔 monitor_interval_secs(默认 2s)会检查本地 KingbaseES 数据库和主 KingbaseES 数据库是否可连接。检查失败后,备库会尝试重连主库,重连尝试失败则会触发备库的切换动作。控制重连次数和重连尝试间隔的参数为 reconnect_attempts 和 reconnect_interval,表示第一次检查失败后,最多执行 reconnect_attempts 次的重连尝试,每两次尝试间隔 reconnect_interval 秒。

repmgr 命令行工具用于执行管理任务的命令行工具,如设置备库,将备库提升为主库、主备切换、显示集群状态等。

3. kbha 守护进程

每个 KingbaseES 数据库节点都有一个 **kbha 守护进程**,负责确保本节点上的 repmgrd

守护进程一直处于运行状态及一些环境的检查,如信任网关检查、存储检测等。如果 repmgrd 守护进程因为某种原因退出,则 kbha 守护进程将重新启动 repmgrd 守护进程。kbha 守护进程还能对环境进行监控,如监控集群节点的磁盘状态、信任网关的状态。

4. 远程通信工具

sys_securecmdd 是 KingbaseES 主备集群中自带的工具,用于集群管理、在监控时安全地执行命令。**服务器端**运行 sys_securecmdd,其配置文件为 sys_HAscmdd.conf,默认监听 8890 端口,接受 sys_securecmdd 客户端连接。服务器端还可以运行 sys_secureftp,用于接收文件。**客户端**运行 sys_securecmd,用于连接服务端。

14.2 安装 KingbaseES 主备集群

14.2.1 规划一个 KingbaseES 主备集群

计划安装一个有 3 个节点的 KingbaseES 主备集群,具体内容如下。
(1) primary 节点:IP 地址为 192.168.100.21/24,服务器名为 node01。
(2) standby 节点:IP 地址为 192.168.100.22/24,服务器名为 node02。
(3) witness 节点:IP 地址为 192.168.100.23/24,服务器名为 node03。
(4) KingbaseES 主备集群将为用户提供一个虚拟 IP,IP 地址为 192.168.100.20/24。
(5) 集群的信任网关 IP 地址为 192.168.100.1/24。
(6) KingbaseES 数据库的二进制安装目录位于/u00/Kingbase/ES/V9/kingbase 下。
(7) KingbaseES 数据库的数据目录位于/u00/Kingbase/ES/V9/data 下。
(8) KingbaseES 数据库的归档日志位于/u04/Kingbase/ES/V9/archivelog 下。
(9) KingbaseES 数据库的用户表空间未来可以创建在如下目录。
① /u01/Kingbase/ES/V9/data/userts。
② /u02/Kingbase/ES/V9/data/userts。
③ /u03/Kingbase/ES/V9/data/userts。

14.2.2 准备安装主备集群的服务器

准备 3 台 x86-64 服务器,依次命名为 node01、node02 和 node03,每台服务器上都有 4 块硬盘。
(1) 硬盘/dev/sda,容量为 900GB,用于安装 CentOS 7.9 操作系统。
(2) 硬盘/dev/sdb,容量为 1200GB,用于创建卷组 dbvg,存储数据库数据。
(3) 硬盘/dev/sdc,容量为 1200GB,用于创建卷组 archvg,存储数据库归档日志。
(4) 硬盘/dev/sdd,容量为 1200GB,用于创建卷组 bakvg,存储数据库备份。
如果读者无法获得物理服务器,可以使用 VMware 虚拟机来模拟。
首先,按照附录 A 在 node01(网卡 ens33 的 IP 地址为 192.168.100.21/24)、node02(网卡 ens33 的 IP 地址为 192.168.100.22/24)和 node03(网卡 ens33 的 IP 地址为 192.168.100.23/24)这 3 台服务器上,安装好 CentOS 7.9 操作系统。
然后,在 node01、node02 和 node03 这 3 台服务器上,以用户 root 的身份,执行下面的

命令创建卷组、逻辑卷和 xfs 文件系统:

```
# 初始化物理卷
pvcreate /dev/sdb
pvcreate /dev/sdc
pvcreate /dev/sdd
# 创建卷组 dbvg, 并在卷组 dbvg 上创建逻辑卷 u01lv
vgcreate dbvg /dev/sdb
lvcreate -L 200G -n u00lv dbvg
lvcreate -L 200G -n u01lv dbvg
lvcreate -L 200G -n u02lv dbvg
lvcreate -L 200G -n u03lv dbvg
# 创建卷组 dbarchvg, 并在卷组 archvg 上创建逻辑卷 archlv
vgcreate archvg /dev/sdc
lvcreate -L 800G -n u04lv archvg
# 创建卷组 dbbakvg, 并在卷组 bakvg 上创建逻辑卷 baklv
vgcreate bakvg /dev/sdd
lvcreate -L 800G -n baklv bakvg
# 在逻辑卷上创建 xfs 文件系统
mkfs.xfs /dev/dbvg/u00lv
mkfs.xfs /dev/dbvg/u01lv
mkfs.xfs /dev/dbvg/u02lv
mkfs.xfs /dev/dbvg/u03lv
mkfs.xfs /dev/archvg/u04lv
mkfs.xfs /dev/bakvg/baklv
```

接着，创建文件系统挂接点目录:

```
mkdir /u00
mkdir /u01
mkdir /u02
mkdir /u03
mkdir /u04
mkdir /dbbak
```

接下来，使用 vi 编辑器编辑文件 /etc/fstab, 在文件末尾添加以下的行:

```
/dev/mapper/dbvg-u00lv      /u00        xfs     defaults    0 0
/dev/mapper/dbvg-u01lv      /u01        xfs     defaults    0 0
/dev/mapper/dbvg-u02lv      /u02        xfs     defaults    0 0
/dev/mapper/dbvg-u03lv      /u03        xfs     defaults    0 0
/dev/mapper/archvg-u04lv    /u04        xfs     defaults    0 0
/dev/mapper/bakvg-baklv     /dbbak      xfs     defaults    0 0
```

紧接着，执行下面的命令将这些逻辑卷挂接到指定的目录上:

```
mount -a
```

最后，执行命令 df -h 验证所有的逻辑卷都已经挂接到文件系统上了:

```
Filesystem                  Size    Used Avail Use% Mounted on
#省略了一些输出
/dev/mapper/dbvg-u00lv      200G    33M  200G   1% /u00
```

```
/dev/mapper/dbvg-u01lv      200G   33M   200G   1%   /u01
/dev/mapper/dbvg-u02lv      200G   33M   200G   1%   /u02
/dev/mapper/dbvg-u03lv      200G   33M   200G   1%   /u03
/dev/mapper/archvg-u04lv    800G   33M   800G   1%   /u04
/dev/mapper/bakvg-baklv     800G   33M   800G   1%   /dbbak
```

14.2.3 安装主备集群的准备工作

1. 配置集群节点的操作系统

为每个集群节点的操作系统,完成如下配置。

(1) 创建用于安装 KingbaseES 主备集群的操作系统用户和用户组。

在集群节点 node01、node02 和 node03 上,以用户 root 的身份,执行下面的命令创建用户组 dba 和用户 kingbase,并将用户 kingbase 的密码设置为 kingbase123(在生产环境中需要设置为强度更高的用户密码):

```
groupadd dba -g 3000
groupadd kingbase  -g 3001
useradd   kingbase  -g 3000 -G 3001 -u 3001
echo "kingbase123"|passwd --stdin kingbase
id kingbase
userdel -r test
```

上面的最后一条命令删除了 CentOS 安装过程中创建的用户 test。

(2) 配置 /etc/hosts 文件。

在集群节点 node01、node02 和 node03 上,用户 root 执行下面的命令在文件 /etc/hosts 中添加 IP 地址——主机名映射记录:

```
cat >>/etc/hosts<<EOF
192.168.100.21    node01
192.168.100.22    node02
192.168.100.23    node03
192.168.100.24    node04
192.168.100.19    dbtest
192.168.100.18    dbclient
EOF
```

(3) 创建安装 KingbaseES 主备集群所需的目录。

在集群节点 node01、node02 和 node03 上,用户 root 执行下面的命令创建目录 /opt/media,用于存放 KingbaseES 数据库软件介质:

```
mkdir -p /opt/media
chown -R kingbase:dba /opt/media
```

在集群节点 node01、node02 和 node03 上,用户 root 执行下面的命令创建下面的目录并修改目录的所有者和所属组,KingbaseES 数据库将使用这些目录:

```
mkdir -p /u00/Kingbase/ES/V9                    # KingbaseES 客户端二进制
mkdir -p /u00/Kingbase/ES/V9/kingbase           # KingbaseES 数据库二进制
mkdir -p /u00/Kingbase/ES/V9/data               # KingbaseES 数据库
```

```
mkdir -p /u01/Kingbase/ES/V9/data              # KingbaseES 数据库
mkdir -p /u02/Kingbase/ES/V9/data              # KingbaseES 数据库
mkdir -p /u03/Kingbase/ES/V9/data              # KingbaseES 数据库
mkdir -p /u04/Kingbase/ES/V9/archivelog        # KingbaseES 归档
mkdir -p /dbbak/pbak
mkdir -p /dbbak/lbak
chown -R kingbase:dba /u00/Kingbase
chown -R kingbase:dba /u01/Kingbase
chown -R kingbase:dba /u02/Kingbase
chown -R kingbase:dba /u03/Kingbase
chown -R kingbase:dba /u04/Kingbase
chown -R kingbase:dba /dbbak
```

（4）修改操作系统内核参数。

在集群节点 node01、node02 和 node03 上，使用用户 root，修改内核参数（文件系统、共享内存、信号量和网络等）：

```
cat>>/etc/sysctl.conf<<EOF
fs.aio-max-nr= 1048576
fs.file-max= 6815744
kernel.shmall= 2097152
kernel.shmmax= 17179869184
kernel.shmmni= 4096
kernel.sem= 5010 641280 5010 256
net.ipv4.ip_local_port_range= 9000 65500
net.core.rmem_default= 262144
net.core.rmem_max= 4194304
net.core.wmem_default= 262144
net.core.wmem_max= 1048576
EOF
```

（5）修改用户资源限制参数。

在集群节点 node01、node02 和 node03 上，使用用户 root，配置用户 kingbase 的堆栈限制、打开文件数限制、用户最大进程数限制：

```
cat>>/etc/security/limits.conf<<EOF
kingbase soft nofile 1048576
kingbase hard nofile 1048576
kingbase soft nproc  131072
kingbase hard nproc  131072
kingbase soft stack  10240
kingbase hard stack  32768
kingbase soft core   unlimited
kingbase hard core   unlimited
EOF
```

（6）修改 RemoveIPC 参数。

systemd-logind 服务中当一个用户退出系统后会删除所有有关的 IPC 对象，该特性由 /etc/systemd/logind.conf 文件中的 RemoveIPC 参数控制。某些操作系统默认打开该参数，会造成程序信号丢失等问题。

在集群节点 node01、node02 和 node03 上，用户 root 使用 vi 编辑器编辑文件/etc/systemd/logind.conf，将下面的行：

```
#RemoveIPC=no
```

修改为：

```
RemoveIPC=no
```

（7）配置磁盘 I/O 调度策略。

为了让 KingbaseES 数据库获得更好的磁盘 I/O 性能，对于 HDD，推荐使用 deadline 调度算法（好处是后期算法保证请求不饿死）；对于固态硬盘、SAN 网络或 RAID、虚拟化环境，推荐使用 noop 调度算法（基于存储本身的磁盘 I/O 优化调度，无须操作系统干预）。

执行下面的命令查看/dev/sdb 的磁盘 I/O 调度策略：

```
cat /sys/block/sdb/queue/scheduler
noop [deadline] cfq
```

如果只想临时将/dev/sdb 的磁盘 I/O 调度策略设置为 noop，可以执行下面的命令：

```
echo noop > /sys/block/sdb/queue/scheduler
```

如果需要永久地将/dev/sdb 的磁盘 I/O 调度策略设置为 noop，可以以用户 root 的身份，使用 vi 编辑器编辑文件/etc/rc.d/rc.local，在文件的末尾增加下面的一行：

```
echo noop > /sys/block/sdb/queue/scheduler
```

重新启动系统后，可以执行命令 cat /sys/block/sdb/queue/scheduler 查看/dev/sdb 的磁盘 I/O 调度策略。

（8）关闭 SELinux。

在集群节点 node01、node02 和 node03 上，用户 root 执行下面的命令关闭 CentOS 的 SELinux：

```
getenforce
sed -i 's/^SELINUX=.*/SELINUX=disabled/' /etc/selinux/config
setenforce 0
getenforce
```

2. 为主备集群配置时间同步服务

KingbaseES 主备集群各节点的时间必须保证高度一致，才能避免发生一些与绝对时间相关的问题。因此需要执行下面的步骤，为 KingbaseES 主备集群配置 chronyd 时间同步服务。

（1）使用用户 root，在 node01、node02 和 node03 上执行下面的命令打开 chronyd 服务的端口：

```
firewall-cmd --permanent --add-service=ntp
firewall-cmd --reload
firewall-cmd --list-all
```

(2) 使用用户 root，在 node01、node02 和 node03 上执行下面的命令停止 ntpd 服务：

```
systemctl stop ntpd
systemctl disable ntpd
```

(3) 将集群节点 node01 配置为 chrony 服务器。
在 node01 节点上，用户 root 使用 vi 编辑器修改文件 /etc/chrony.conf 中的两处信息。
① 指定时间基准服务器。
找到文件中的如下位置：

```
# Use public servers from the pool.ntp.org project.
# Please consider joining the pool (http://www.pool.ntp.org/join.html).
server 0.centos.pool.ntp.org iburst
server 1.centos.pool.ntp.org iburst
server 2.centos.pool.ntp.org iburst
server 3.centos.pool.ntp.org iburst
```

删除上面的行，添加下面的行：

```
# Use public servers from the pool.ntp.org project.
# Please consider joining the pool (http://www.pool.ntp.org/join.html).
server 0.cn.pool.ntp.org iburst
server 1.cn.pool.ntp.org iburst
server 2.cn.pool.ntp.org iburst
server 3.cn.pool.ntp.org iburst
```

② 指定允许的客户端网络。
找到文件中的如下位置：

```
# Allow NTP client access from local network.
#allow 192.168.0.0/16
```

直接添加下面的行：

```
# Allow NTP client access from local network.
#allow 192.168.0.0/16
allow 192.168.100.0/24
```

在 node01 上，使用用户 root，执行下面的命令设置系统开机后自动启动 chronyd 服务，并在此刻启动 chronyd 服务：

```
[root@node01 ~]# systemctl enable chronyd
[root@node01 ~]# systemctl start chronyd
```

重新启动集群节点 node01：

```
[root@node01 ~]# reboot
```

重新启动 node01 后，使用用户 root，执行下面的命令查看 node01 服务器的时钟同步源：

```
[root@node01 ~]# chronyc sources -v
210 Number of sources = 4
```

```
  .-- Source mode  '^' = server, '=' = peer, '#' = local clock.
 / .- Source state '*' = current synced, '+' = combined , '-' = not combined,
| /   '?' = unreachable, 'x' = time may be in error, '~' = time too variable.
||                                          .- xxxx [ yyyy ] +/- zzzz
||      Reachability register (octal) -.   |  xxxx = adjusted offset,
||      Log2(Polling interval) --.     |   |  yyyy = measured offset,
||                                \    |   |  zzzz = estimated error.
||                                 |   |   \
MS Name/IP address            Stratum Poll Reach LastRx Last sample
===============================================================================
^+ ntp6.flashdance.cx              2   7   337    112   -16ms[  -16ms] +/-  145ms
^* makaki2.miuku.net               3   7   377    117   +28ms[  +29ms] +/-   74ms
^+ a.chl.la                        2   6   377     48  -2702us[-2702us] +/-   92ms
^+ electrode.felixc.at             2   7   377    115  +1428us[+1428us] +/-   91ms
[root@node01 ~]#
```

(4) 配置主备集群其他的节点作为 chronyd 服务的客户机。

在 node02 上，以用户 root 的身份，使用 vi 编辑器修改文件/etc/chrony.conf，只需修改 1 处。

找到文件中的如下位置：

```
# Use public servers from the pool.ntp.org project.
# Please consider joining the pool (http://www.pool.ntp.org/join.html).
server 0.centos.pool.ntp.org iburst
server 1.centos.pool.ntp.org iburst
server 2.centos.pool.ntp.org iburst
server 3.centos.pool.ntp.org iburst
```

删除上面的行，添加下面的行：

```
# Use public servers from the pool.ntp.org project.
# Please consider joining the pool (http://www.pool.ntp.org/join.html).
server 192.168.100.21 iburst
```

在 node02 上，以用户 root 的身份，执行下面的命令设置系统开机后自动启动 chronyd 服务，并在此刻启动 chronyd 服务：

```
[root@node02 ~]# systemctl enable chronyd
[root@node02 ~]# systemctl stop chronyd
[root@node02 ~]# systemctl start chronyd
```

在 node02 上，以用户 root 的身份，执行下面的命令查看时钟同步源：

```
[root@node02 ~]# chronyc sources -v
210 Number of sources = 1

  .-- Source mode  '^' = server, '=' = peer, '#' = local clock.
 / .- Source state '*' = current synced, '+' = combined , '-' = not combined,
| /   '?' = unreachable, 'x' = time may be in error, '~' = time too variable.
||                                          .- xxxx [ yyyy ] +/- zzzz
```

```
||       Reachability register (octal) -.     |   xxxx = adjusted offset,
||       Log2(Polling interval) --.      |    |   yyyy = measured offset,
||                                \      |    |   zzzz = estimated error.
||                                 |     |    \
MS Name/IP address         Stratum Poll Reach LastRx Last sample
===============================================================================
^* node01                     4    6    17    30   +1197ns[  +40us] +/-    68ms
[root@node02 ~]#
```

重新启动集群节点 node02：

```
[root@node02 ~]# reboot
```

在 node01 上，用户 root 执行下面的命令查看 chrony 服务器有哪些客户机：

```
[root@node01 ~]# chronyc clients
Hostname                    NTP   Drop Int IntL Last   Cmd  Drop Int  Last
===========================================================================
node02                       8     0   6   -    28     0    0    -    -
[root@node01 ~]#
```

重复上面的步骤，配置 node03 作为 chronyd 服务的客户机。

3. 准备 KingbaseES 数据库安装介质

安装之前，使用用户 root，将 KingbaseES 数据库的安装介质和许可证上传到 node01、node02 和 node03 的 /root 目录下。

在 node01、node02 和 node03 上，以用户 root 的身份，执行下面的命令将 KingbaseES 数据库的 ISO 文件挂接到 CentOS 7 的文件系统的 /mnt/iso 目录上：

```
mkdir -p /mnt/iso #创建挂接点目录/mnt/iso
cd /opt/media
#挂接 KingbaseES 安装介质 ISO 文件
mount -t iso9660 -o loop KingbaseES_V009R001C001B0025_Lin64_install.iso /mnt/iso
```

在 node01、node02 和 node03 上，用户 root 执行下面的命令解压 KingbaseES 数据库软件许可证：

```
unzip license_企业版.zip
cd license_34148/
cp license_34148_0.dat /home/kingbase
```

4. 安装 KingbaseES 数据库客户端

在安装 KingbaseES 主备集群的时候，需要首先在 KingbaseES 主备集群的所有的节点上，按照 2.1.1 节安装一个 KingbaseES 数据库的客户端，因为 KingbaseES 主备集群的软件安装介质位于 KingbaseES 数据库客户端二进制安装目录的子目录 ClientTools/guitools/DeployTools/zip 中。

```
[kingbase @ node01 ~] $ cd /u00/Kingbase/ES/V9/client/ClientTools/guitools/DeployTools/zip
[kingbase@node01 zip]$ ls -l
```

```
total 317496
-rwxrwxr-x 1 kingbase dba    179967 Oct 30 21:07 cluster_install.sh
-rw-rw-r-- 1 kingbase dba 322323560 Oct 30 21:07 db.zip
-rw-rw-r-- 1 kingbase dba     14409 Oct 30 21:07 install.conf
-rw-rw-r-- 1 kingbase dba   2584328 Oct 30 21:07 securecmdd.zip
-rwxrwxr-x 1 kingbase dba      7374 Oct 30 21:07 trust_cluster.sh
[kingbase@node01 zip]$
```

5. 分发主备集群的软件安装介质

在 node01、node02、node03 上，以用户 kingbase 的身份，执行下面的命令分发安装 KingbaseES 集群的软件介质：

```
mkdir -p /u00/Kingbase/ES/V9/kingbase
cd /u00/Kingbase/ES/V9/client/ClientTools/guitools/DeployTools/zip
cp * /u00/Kingbase/ES/V9/kingbase
cd /u00/Kingbase/ES/V9/kingbase
unzip db.zip
```

6. 分发主备集群的软件许可证

在 node01、node02、node03 上，以用户 kingbase 的身份，执行下面的命令将 KingbaseES 数据库软件许可证分发到各个安装节点上：

```
cd
cp license_34148_0.dat /u00/Kingbase/ES/V9/kingbase/bin/license.dat
```

7. 运行 sys_securecmd 服务

部署 KingbaseES 主备集群，可以使用 ssh 免密服务。在有些用户环境下，不允许使用 ssh 免密服务，此时可以使用电科金仓提供的 sys_securecmd 服务。

在 node01、node02、node03 上，使用用户 root，执行下面的命令部署 sys_securecmd 服务：

```
firewall-cmd --permanent --add-port=8890/tcp
firewall-cmd --reload
cd /u00/Kingbase/ES/V9/kingbase/bin
./sys_HAscmdd.sh init
./sys_HAscmdd.sh start
```

8. 允许数据库访问端口通过防火墙

在集群的所有节点上，使用用户 root，执行下面的命令允许数据库访问端口通过防火墙：

```
firewall-cmd --zone=public --add-port=54321/tcp --permanent
firewall-cmd --reload
firewall-cmd --list-all
```

9. 重新启动所有的集群节点

在 node01、node02 和 node03 上，使用用户 root，执行下面的命令重新启动服务器：

```
reboot
```

14.2.4　安装主备集群

首先，在 KingbaseES 主备集群的第 1 个节点 node01 上，用户 kingbase 使用 vi 编辑器编辑主备集群的安装配置文件 /u00/Kingbase/ES/V9/kingbase/install.conf，修改 [install] 部分中的项：

```
on_bmj=0
all_ip=(192.168.100.21 192.168.100.22)
witness_ip="192.168.100.23"
production_ip=()
local_disaster_recovery_ip=()
remote_disaster_recovery_ip=()
install_dir="/u00/Kingbase/ES/V9"   #会自动在该目录后面加上 kingbase 子目录
zip_package="/u00/Kingbase/ES/V9/kingbase/db.zip"
license_file=(license.dat)

db_password="Passw0rd"

trusted_servers="192.168.100.1"
data_directory="/u00/Kingbase/ES/V9/data"
virtual_ip="192.168.100.20/24"
net_device=(ens33)
net_device_ip=(192.168.100.21 192.168.100.22)
ipaddr_path="/usr/sbin"
arping_path="/usr/sbin"
ping_path="/usr/bin"
super_user="root"
execute_user="kingbase"
deploy_by_sshd=0
use_scmd=1
reconnect_attempts="10"
reconnect_interval="6"
recovery="standby"
ssh_port="22"
scmd_port="8890"
auto_cluster_recovery_level='1'
use_check_disk='off'
synchronous=''
```

其中的参数如下。

(1) 参数 deploy_by_sshd=0 表示使用电科金仓的 sys_securecmd 服务部署集群，如果要使用 ssh 免密服务进行部署，则需要设置参数 deploy_by_sshd=1。

(2) 参数 synchronous 可以设置为以下值。

① **quorum**：第 1 个执行 WAL 日志重放的备节点将作为同步节点。

② **sync**：将同步备节点名字列表中的第 1 个连接到主节点的备节点作为同步节点。

③ **all**：当所有的备节点，能连接到主节点时，等价于 sync；如果没有任何备节点能连接到主节点时，等价于 async。

④ **async**：所有的备节点都运行在异步流复制模式。

然后，在 KingbaseES 主备集群的第 1 个节点 node01 上，用户 kingbase 执行下面的命令，开始部署 KingbaseES 主备集群：

```
[kingbase@node01 ~]$ cd /u00/Kingbase/ES/V9/kingbase
[kingbase@node01 kingbase]$ sh cluster_install.sh
```

14.2.5 部署完主备集群后的操作

1. 修改操作系统用户 kingbase 的初始化环境文件

在 node01、node02、node03 上，以用户 kingbase 的身份，使用 vi 编辑器编辑初始化环境文件 /home/kingbase/.bashrc，在文件末尾加上如下的行：

```
export KINGBASE_HOME=/u00/Kingbase/ES/V9/kingbase
export PATH=$KINGBASE_HOME/bin:$PATH
export LD_LIBRARY_PATH=$KINGBASE_HOME/lib:$LD_LIBRARY_PATH
export KINGBASE_DATA=/u00/Kingbase/ES/V9/data
export KINGBASE_PORT=54321
```

为了使刚添加的环境变量起作用，需要让用户 kingbase 退出后重新登录到 CentOS。

2. 创建 KingbaseES 数据库用户密码文件 /home/kingbase/.kbpass

在 node01、node02、node03 上，使用用户 kingbase，执行下面的命令创建数据库用户密码文件 /home/kingbase/.kbpass：

```
[kingbase@dbsvr ~]$ cat >/home/kingbase/.kbpass<<EOF
192.168.100.21:54321:*:system:Passw0rd
192.168.100.22:54321:*:system:Passw0rd
192.168.100.23:54321:*:system:Passw0rd
localhost:54321:*:system:Passw0rd
EOF
[kingbase@dbsvr ~]$ chmod 600 /home/kingbase/.kbpass
```

3. 创建工具脚本 kbps

为了方便地查看 KingbaseES 数据库中的所有进程，可以在 node01、node02、node03 上，以用户 root 的身份，使用 vi 编辑器创建一个工具脚本文件 /usr/bin/kbps，将下面的两行复制到文件中：

```
#/bin/bash
ps -ef |grep $(head -1 /u00/Kingbase/ES/V9/data/kingbase.pid) |grep -v grep
```

然后使用用户 root，执行下面的命令让脚本 /usr/bin/kbps 拥有执行权限：

```
[root@dbsvr ~]# chmod 755 /usr/bin/kbps
```

14.3 管理 KingbaseES 主备集群

14.3.1 获取集群信息的指令

（1）命令 FC1：查看集群的状态。

在集群的任意节点上，使用用户 kingbase，执行下面的命令可以获取集群的信息：

```
[kingbase@node01 ~]$ repmgr cluster show
 ID | Name  | Role    | Status      | Upstream | Location | Priority | Timeline | LSN_Lag | Connection string
----+-------+---------+-------------+----------+----------+----------+----------+---------+--------------------
  1 | node1 | primary | * running   |          | default  | 100      | 1        |         | host=192.168.100.21 user=esrep dbname=esrep port=54321 connect_timeout=10 keepalives=1 keepalives_idle=2 keepalives_interval=2 keepalives_count=3 tcp_user_timeout=9000
  2 | node2 | standby |   running   | node1    | default  | 100      | 1        | 0 bytes | host=192.168.100.22 user=esrep dbname=esrep port=54321 connect_timeout=10 keepalives=1 keepalives_idle=2 keepalives_interval=2 keepalives_count=3 tcp_user_timeout=9000
  3 | node3 | witness | * running   | node1    | default  | 0        | n/a      |         | host=192.168.100.23 user=esrep dbname=esrep port=54321 connect_timeout=10 keepalives=1 keepalives_idle=2 keepalives_interval=2 keepalives_count=3 tcp_user_timeout=9000
[kingbase@node01 ~]$ repmgr cluster show --compact
 ID | Name  | Role    | Status      | Upstream | Location | Prio. | TLI | LSN_Lag
----+-------+---------+-------------+----------+----------+-------+-----+---------
  1 | node1 | primary | * running   |          | default  | 100   | 1   |
  2 | node2 | standby |   running   | node1    | default  | 100   | 1   | 0 bytes
  3 | node3 | witness | * running   | node1    | default  | 0     | n/a |
[kingbase@node01 ~]$
```

（2）命令 FC2：查看集群各节点上 repmgrd 守护进程的状态。

在集群的任意节点上，使用用户 kingbase，执行下面的命令可以获取集群各节点上 repmgrd 守护进程的状态信息：

```
[kingbase@node01 ~]$ repmgr service status
 ID | Name  | Role    | Status      | Upstream | repmgrd | PID   | Paused? | Upstream last seen
----+-------+---------+-------------+----------+---------+-------+---------+--------------------
  1 | node1 | primary | * running   |          | running | 12418 | no      | n/a
  2 | node2 | standby |   running   | node1    | running | 7054  | no      | 0 second(s) ago
  3 | node3 | witness | * running   | node1    | running | 8502  | no      | 1 second(s) ago
[kingbase@node01 ~]$ repmgr service status --detail
 ID | Name  | Role    | Status      | Upstream | Location | Priority | repmgrd | PID   | Paused? | Upstream last seen
----+-------+---------+-------------+----------+----------+----------+---------+-------+---------+--------------------
  1 | node1 | primary | * running   |          | default  | 100      | running | 12418 | no      | n/a
  2 | node2 | standby |   running   | node1    | default  | 100      | running | 7054  | no      | 1 second(s) ago
  3 | node3 | witness | * running   | node1    | default  | 0        | running | 8502  | no      | 0 second(s) ago
[kingbase@node01 ~]$
```

（3）命令 FC3：查看集群各节点上，KingbaseES 数据库的状态。

在集群节点上，使用用户 kingbase，执行下面的命令查看 KingbaseES 数据库是否正在运行：

```
[kingbase@node01 ~]$ ps -ef|grep "kingbase:"|grep -v grep
kingbase  11192  11105  0 11:18 ?        00:00:00 kingbase: logger
kingbase  11108  11105  0 11:18 ?        00:00:00 kingbase: checkpointer
kingbase  11109  11105  0 11:18 ?        00:00:00 kingbase: background writer
kingbase  11110  11105  0 11:18 ?        00:00:00 kingbase: walwriter
kingbase  11111  11105  0 11:18 ?        00:00:00 kingbase: autovacuum launcher
kingbase  11112  11105  0 11:18 ?        00:00:00 kingbase: archiver    last was 000000010000000000000003.00000028.backup
kingbase  11113  11105  0 11:18 ?        00:00:01 kingbase: stats collector
kingbase  11114  11105  0 11:18 ?        00:00:00 kingbase: kwr collector
kingbase  11115  11105  0 11:18 ?        00:00:00 kingbase: ksh writer
kingbase  11116  11105  0 11:18 ?        00:00:00 kingbase: ksh collector
kingbase  11117  11105  0 11:18 ?        00:00:00 kingbase: logical replication launcher
```

```
kingbase    11413    11105   0 11:18 ?        00:00:00 kingbase: walsender esrep 192.168.100.22
(32323) streaming 0/40052E8
kingbase    12416    11105   0 11:19 ?        00:00:02 kingbase: esrep esrep 192.168.100.21
(29417) idle
kingbase    12525    11105   0 11:19 ?        00:00:01 kingbase: esrep esrep 192.168.100.22
(32339) idle
kingbase    12540    11105   0 11:19 ?        00:00:01 kingbase: esrep esrep 192.168.100.23
(11910) idle
kingbase    12877    11105   0 11:19 ?        00:00:00 kingbase: esrep esrep 192.168.100.21
(29501) idle
[kingbase@node01 ~]$
```

如果看到上面的输出，则说明 KingbaseES 数据库已经运行在这个节点上了。

也可以执行安装时创建的 kbps 脚本：

```
[kingbase@node01 ~]$ kbps
kingbase    11105       1   0 11:18 ?        00:00:02 /u00/Kingbase/cluster/kingbase/bin/
kingbase -D /u00/Kingbase/cluster/kingbase/data
kingbase    11192    11105   0 11:18 ?        00:00:00 kingbase: logger
kingbase    11108    11105   0 11:18 ?        00:00:00 kingbase: checkpointer
kingbase    11109    11105   0 11:18 ?        00:00:00 kingbase: background writer
kingbase    11110    11105   0 11:18 ?        00:00:00 kingbase: walwriter
kingbase    11111    11105   0 11:18 ?        00:00:00 kingbase: autovacuum launcher
kingbase    11112    11105   0 11:18 ?        00:00:00 kingbase: archiver     last was
000000010000000000000003.00000028.backup
kingbase    11113    11105   0 11:18 ?        00:00:01 kingbase: stats collector
kingbase    11114    11105   0 11:18 ?        00:00:00 kingbase: kwr collector
kingbase    11115    11105   0 11:18 ?        00:00:00 kingbase: ksh writer
kingbase    11116    11105   0 11:18 ?        00:00:00 kingbase: ksh collector
kingbase    11117    11105   0 11:18 ?        00:00:00 kingbase: logical replication launcher
kingbase    11413    11105   0 11:18 ?        00:00:00 kingbase: walsender esrep 192.168.100.22
(32323) streaming 0/40052E8
kingbase    12416    11105   0 11:19 ?        00:00:02 kingbase: esrep esrep 192.168.100.21
(29417) idle
kingbase    12525    11105   0 11:19 ?        00:00:01 kingbase: esrep esrep 192.168.100.22
(32339) idle
kingbase    12540    11105   0 11:19 ?        00:00:01 kingbase: esrep esrep 192.168.100.23
(11910) idle
kingbase    12877    11105   0 11:19 ?        00:00:00 kingbase: esrep esrep 192.168.100.21
(29501) idle
[kingbase@node01 ~]$
```

（4）命令 FC4：查看集群各节点上，守护进程 repmgrd 是否正在运行。

在集群节点上，使用用户 kingbase，执行下面的命令查看守护进程 repmgrd 是否正在运行：

```
[kingbase@node01 ~]$ ps -ef|grep repmgrd|grep -v grep
kingbase    12418       1   0 11:19 ?        00:00:09 /u00/Kingbase/cluster/kingbase/
bin/repmgrd -d -v -f /u00/Kingbase/cluster/kingbase/bin/../etc/repmgr.conf
[kingbase@node01 ~]$
```

（5）命令 FC5：查看集群各节点上，守护进程 kbha 是否正在运行。

在集群节点上，使用用户 kingbase，执行下面的命令查看守护进程 kbha 是否正在运行：

```
[kingbase@node01 ~]$ ps -ef|grep "kbha"|grep -v grep
kingbase  22113      1  0 16:39 ?        00:00:02 /u01/Kingbase/cluster/kingbase/
bin/kbha -A daemon -f /u01/Kingbase/cluster/kingbase/bin/../etc/repmgr.conf
[kingbase@node01 ~]$
```

(6) 命令 FC6：查看集群各节点上，守护进程 sys_securecmdd 是否正在运行。

在集群节点上，使用用户 kingbase，执行命令查看守护进程 sys_securecmdd 是否正在运行：

```
[kingbase@node01 ~]$ ps -ef|grep "sys_securecmdd"|grep -v grep
root       1700      1  0 11:14 ?        00:00:00 sys_securecmdd: /u00/Kingbase/
cluster/kingbase/bin/sys_securecmdd -f /etc/.kes/securecmdd_config [listener] 0
of 128-256 startups
[kingbase@node01 ~]$
```

(7) 命令 FC7：查看集群的备用节点上，检查 data 目录下是否存在 standby.signal 文件。

在集群的备用节点（当前是 node02）上，使用用户 kingbase，执行命令查看 data 目录下是否存在 standby.signal 文件：

```
[kingbase@node02 ~]$ ls -l $KINGBASE_DATA/standby.signal
-rw------- 1 kingbase dba 20 Feb 15 11:18 /u00/Kingbase/cluster/kingbase/data/
standby.signal
[kingbase@node02 ~]$ cat $KINGBASE_DATA/standby.signal
# created by repmgr
[kingbase@node02 ~]$
```

(8) 命令 FC8：在集群的主节点上，查看流复制状态。

在集群的主节点（当前是 node01）上，使用用户 kingbase，执行命令查询流复制状态：

```
[kingbase@node01 ~]$ ksql -h 192.168.100.21 -U system -d kingbase
kingbase=# \x
Expanded display is on.
kingbase=# select * from sys_stat_replication;
-[ RECORD 1 ]----+-----------------------------
pid              | 11413
usesysid         | 16385
usename          | esrep
application_name | node2
client_addr      | 192.168.100.22
client_hostname  |
client_port      | 32323
backend_start    | 2024-02-15 11:18:45.762245+08
backend_xmin     |
state            | streaming
sent_lsn         |
write_lsn        | 0/40052E8
flush_lsn        | 0/40052E8
replay_lsn       | 0/40052E8
write_lag        |
```

```
flush_lag        |
replay_lag       |
sync_priority    | 1
sync_state       | quorum
reply_time       | 2024-02-15 12:03:11.427817+08
kingbase=# \q
[kingbase@node01 ~]$
```

如果集群的流复制处于正常状态，将显示所有集群备机的信息。

（9）命令 FC9：在集群的主节点上，查看虚拟 IP 的情况。

在集群的主节点（当前是 node01）上，使用用户 kingbase，执行命令查看集群使用的虚拟 IP 信息：

```
[kingbase@node01 ~]$ ip a
#省略了一些输出
2: ens33: <BROADCAST,MULTICAST,UP,LOWER_UP> mtu 1500 qdisc fq_codel state UP group default qlen 1000
    link/ether 00:50:56:39:aa:4a brd ff:ff:ff:ff:ff:ff
    inet 192.168.100.21/24 brd 192.168.100.255 scope global noprefixroute ens33
       valid_lft forever preferred_lft forever
    inet 192.168.100.20/24 scope global secondary ens33:3
       valid_lft forever preferred_lft forever
    inet6 fe80::8d4e:d00e:ee75:26/64 scope link noprefixroute
       valid_lft forever preferred_lft forever
#省略了一些输出
[kingbase@node01 ~]$
```

可以看到 ens33 网卡上，有一个 IP 地址的值为 192.168.100.20/24，它是 KingbaseES 主备集群的虚拟 IP 地址。

14.3.2 停止主备集群

在 KingbaseES 主备集群的任一节点（可以是主节点、备节点，或者是见证节点）上，以用户 kingbase 的身份，执行下面的命令停止主备集群的所有节点（示例是在见证节点 node03 上）：

```
[kingbase@node03 ~]$ sys_monitor.sh stop
#省略了许多输出
2024-02-15 12:07:33 Done.
[kingbase@node03 ~]$
```

14.3.3 启动主备集群

在 KingbaseES 主备集群的任一节点（可以是主节点、备节点，或者是见证节点）上，以用户 kingbase 的身份，执行下面的命令启动主备集群的所有节点（示例是在备节点 node02 上）：

```
[kingbase@node02 ~]$ sys_monitor.sh start
#省略了许多输出
2024-02-15 12:11:03 Done.
[kingbase@node02 ~]$
```

KingbaseES 主备集群启动后,处于正常运行的状态:

```
[kingbase@node01 ~]$ repmgr service status
 ID | Name  | Role    | Status    | Upstream | repmgrd | PID   | Paused? | Upstream last seen
----+-------+---------+-----------+----------+---------+-------+---------+--------------------
 1  | node1 | primary | * running |          | running | 26260 | no      | n/a
 2  | node2 | standby |   running | node1    | running | 16365 | no      | 0 second(s) ago
 3  | node3 | witness | * running | node1    | running | 84173 | no      | 1 second(s) ago
[kingbase@node01 ~]$ ip addr show ens33 | grep inet |grep -v inet6
    inet 192.168.100.21/24 brd 192.168.100.255 scope global noprefixroute ens33
    inet 192.168.100.20/24 scope global secondary ens33:3
[kingbase@node01 ~]$
```

可以看到,当前集群的主节点在 node01 上,并且集群的浮动 IP 也在节点 node01 的网卡 ens33 上。

14.3.4 让节点重新加入集群

假设此时,KingbaseES 主备集群的主节点 node1,因为某种原因突然停机了。在集群节点 node01 上,以用户 root 的身份,执行 poweroff 关机命令,来仿真这种情况:

```
[root@node01 ~]# poweroff
Connection to 192.168.100.21 closed by remote host.
Connection to 192.168.100.21 closed.
```

稍等一会儿(3min),集群将备节点 node02 切换为主节点,并且继续运行,为用户提供 KingbaseES 数据库服务。此时在节点 node02 上,执行下面的命令查看集群的状态:

```
[kingbase@node02 ~]$ repmgr service status
 ID | Name  | Role    | Status     | Upstream | repmgrd | PID   | Paused? | Upstream last seen
----+-------+---------+------------+----------+---------+-------+---------+--------------------
 1  | node1 | primary | - failed   | ?        | n/a     | n/a   | n/a     | n/a
 2  | node2 | primary | * running  |          | running | 16365 | no      | n/a
 3  | node3 | witness | * running  | node2    | running | 84173 | no      | 0 second(s) ago

[WARNING] following issues were detected
  - unable to  connect to node "node1" (ID: 1)

[HINT] execute with --verbose option to see connection error messages
[kingbase@node02 ~]$ ip addr show ens33 | grep inet |grep -v inet6
    inet 192.168.100.22/24 brd 192.168.100.255 scope global noprefixroute ens33
    inet 192.168.100.20/24 scope global secondary ens33:3
[kingbase@node02 ~]$
```

可以看到,node1 节点处于 failed 状态,repmgrd 服务没有运行,node2 被切换为主节点,并且主备集群的浮动 IP 也被切换到了节点 node02 的网卡 ens33 上了。

此时重新启动 node01 服务器,用户 kingbase 执行下面的命令查看节点 node01 上网卡 ens33 上的 IP 地址:

```
[kingbase@node01 ~]$ ip addr show ens33 | grep inet |grep -v inet6
    inet 192.168.100.21/24 brd 192.168.100.255 scope global noprefixroute ens33
[kingbase@node01 ~]$
```

可以看到,此时 node01 的 ens33 网卡上没有配置集群的浮动 IP。

在服务器 node02 上,用户 kingbase 执行下面的命令查看集群的状态:

```
[kingbase@node02 ~]$ repmgr service status
 ID | Name  | Role    | Status    | Upstream | repmgrd | PID   | Paused? | Upstream last seen
----+-------+---------+-----------+----------+---------+-------+---------+--------------------
 1  | node1 | primary | - failed  | ?        | n/a     | n/a   | n/a     | n/a
 2  | node2 | primary | * running |          | running | 16365 | no      | n/a
 3  | node3 | witness | * running | node2    | running | 84173 | no      | 0 second(s) ago

[WARNING] following issues were detected
  - unable to  connect to node "node1" (ID: 1)

[HINT] execute with --verbose option to see connection error messages
[kingbase@node02 ~]$ ip addr show ens33 | grep inet |grep -v inet6
    inet 192.168.100.22/24 brd 192.168.100.255 scope global noprefixroute ens33
    inet 192.168.100.20/24 scope global secondary ens33:3
[kingbase@node02 ~]$
```

可以看到,KingbaseES 主备集群的状态没有发生任何变化。

在服务器 node01 上,用户 kingbase 执行下面的命令让节点 node1(IP 地址是 192.168.100.21)重新加入集群(此时集群的主节点为 node2,IP 地址是 192.168.100.22):

```
[kingbase@node01 ~]$ repmgr -h 192.168.100.22 -U esrep -d esrep -p 54321 node rejoin --force-rewind
#省略了输出
[kingbase@node01 ~]$
```

在服务器 node02 上,用户 kingbase 执行下面的命令查看集群当前的状态:

```
[kingbase@node02 ~]$ repmgr service status
 ID | Name  | Role    | Status    | Upstream | repmgrd | PID   | Paused? | Upstream last seen
----+-------+---------+-----------+----------+---------+-------+---------+--------------------
 1  | node1 | standby |   running | node2    | running | 4065  | no      | 1 second(s) ago
 2  | node2 | primary | * running |          | running | 16365 | no      | n/a
 3  | node3 | witness | * running | node2    | running | 84173 | no      | 0 second(s) ago
[kingbase@node02 ~]$ ip addr show ens33 | grep inet |grep -v inet6
    inet 192.168.100.22/24 brd 192.168.100.255 scope global noprefixroute ens33
    inet 192.168.100.20/24 scope global secondary ens33:3
[kingbase@node02 ~]$
```

可以看到,node1 已经重新加入了集群,node1 重新加入集群后,作为集群的备节点。

14.3.5 集群主备切换

如果想将主备集群的主节点切换回 node01,可以在当前的备节点 node1 上,用户 kingbase 执行下面的命令进行集群主备节点切换:

```
[kingbase@node01 ~]$ repmgr standby switchover
(省略了一些输出)
[INFO] unpausing repmgrd on node "node1" (ID 1)
[INFO] unpause node "node1" (ID 1) successfully
[INFO] unpausing repmgrd on node "node2" (ID 2)
[INFO] unpause node "node2" (ID 2) successfully
```

```
[INFO] unpausing repmgrd on node "node3" (ID 3)
[INFO] unpause node "node3" (ID 3) successfully
[NOTICE] STANDBY SWITCHOVER has completed successfully
[kingbase@node01 ~]$ repmgr service status
 ID | Name  | Role    | Status      | Upstream | repmgrd | PID   | Paused? | Upstream last seen
----+-------+---------+-------------+----------+---------+-------+---------+--------------------
 1  | node1 | primary | * running   |          | running | 4065  | no      | n/a
 2  | node2 | standby |   running   | node1    | running | 16365 | no      | 0 second(s) ago
 3  | node3 | witness | * running   | node1    | running | 84173 | no      | 1 second(s) ago
[kingbase@node01 ~]$ ip addr show ens33 | grep inet |grep -v inet6
    inet 192.168.100.21/24 brd 192.168.100.255 scope global noprefixroute ens33
    inet 192.168.100.20/24 scope global secondary ens33:3
[kingbase@node01 ~]$
```

可以看到，节点 node1 已经成为集群的主节点，并且 KingbaseES 主备集群的浮动 IP 地址 192.168.100.20/24，也切换到节点 node01 的网卡 ens33 上了。

14.3.6 重做备用节点

有时候，需要重做集群的备用节点的数据库。

首先，在要重做的备节点（此时是节点 node2），使用用户 kingbase 执行下面的命令停止数据库：

```
[kingbase@node02 ~]$ sys_ctl stop
waiting for server to shut down.... done
server stopped
[kingbase@node02 ~]$
```

在 node02 上，使用 kingbase 用户执行命令将主库（IP 地址为 192.168.100.21）的数据克隆到备库（节点 node2，IP 地址是 192.168.100.22）：

```
[kingbase@node02 ~]$ repmgr -h 192.168.100.21 -U esrep -d esrep -p 54321  standby clone -F
[NOTICE] destination directory "/u00/Kingbase/cluster/kingbase/data" provided
[INFO] connecting to source node
[DETAIL] connection string is: host=192.168.100.21 user=esrep port=54321 dbname=esrep
[DETAIL] current installation size is 82 MB
[NOTICE] checking for available walsenders on the source node (2 required)
[NOTICE] checking replication connections can be made to the source server (2 required)
[WARNING] directory "/u00/Kingbase/cluster/kingbase/data " exists but is not empty
[NOTICE] - F/- - isForce provided - deleting existing data directory "/u00/Kingbase/cluster/kingbase/data"
[INFO] creating replication slot as user "esrep"
[NOTICE] starting backup (using sys_basebackup)...
[HINT] this may take some time; consider using the -c/--fast-checkpoint option
[INFO] executing:
  /u00/Kingbase/cluster/kingbase/bin/sys_basebackup -l "repmgr base backup"  -D /u00/Kingbase/cluster/kingbase/data -h 192.168.100.21 -p 54321 -U esrep -X stream -S repmgr_slot_2
```

```
[NOTICE] standby clone (using sys_basebackup) complete
[NOTICE] you can now start your Kingbase server
[HINT] for example: sys_ctl -D /u00/Kingbase/cluster/kingbase/data start
[HINT] after starting the server, you need to re-register this standby with "
repmgr standby register --force" to update the existing node record
[kingbase@node02 ~]$
```

上面命令中的参数-F,表示覆盖原有的数据目录。

在 node02 上,使用用户 kingbase,执行下面的命令启动数据库:

```
[kingbase@node02 ~]$ sys_ctl start
```

在 node02 上,使用用户 kingbase,执行下面的命令将重做的备机实例注册到集群:

```
[kingbase@node02 ~]$ repmgr standby register -F
[INFO] connecting to local node "node2" (ID: 2)
[INFO] connecting to primary database
[INFO] standby registration complete
[NOTICE] standby node "node2" (ID: 2) successfully registered
[kingbase@node02 ~]$
```

上面命令中的参数-F,表示强制将节点注册到集群。

此时,在 node02 上,使用用户 kingbase,执行下面的命令查看集群的状态:

```
[kingbase@node02 ~]$ repmgr service status
 ID | Name  | Role    | Status    | Upstream | repmgrd | PID   | Paused? | Upstream last seen
----+-------+---------+-----------+----------+---------+-------+---------+--------------------
 1  | node1 | primary | * running |          | running | 4065  | no      | n/a
 2  | node2 | standby |   running | node1    | running | 16365 | no      | 0 second(s) ago
 3  | node3 | witness | * running | node1    | running | 84173 | no      | 1 second(s) ago
[kingbase@node02 ~]$
```

14.3.7 为集群添加新的备节点

在添加集群节点前,需要确认当前集群的主节点信息。在集群的任一节点(示例是在节点 node01)上,使用用户 kingbase,执行下面的命令查看集群的状态:

```
[kingbase@node01 ~]$ repmgr service status
 ID | Name  | Role    | Status    | Upstream | repmgrd | PID   | Paused? | Upstream last seen
----+-------+---------+-----------+----------+---------+-------+---------+--------------------
 1  | node1 | primary | * running |          | running | 4065  | no      | n/a
 2  | node2 | standby |   running | node1    | running | 16365 | no      | 1 second(s) ago
 3  | node3 | witness | * running | node1    | running | 84173 | no      | 1 second(s) ago
[kingbase@node01 ~]$
```

可以看到,当前 KingbaseES 主备集群的主节点是 node01,IP 地址为 192.168.100.21;当前集群有 3 个节点,ID 值分别为 1、2 和 3。

计划把新节点 node04(IP 地址为 192.168.100.24/24)添加到 KingbaseES 主备集群中,并且设置新节点的 ID 值为 4。为了完成这个任务,请按照 14.2.2 节、14.2.3 节,准备集群的新节点 node04(将 node04 配置成 chronyd 服务的客户机)。

然后,在要添加的新节点 node04 上,以用户 kingbase 的身份,使用 vi 编辑器编辑主备集群的安装配置文件/u00/Kingbase/ES/V9/kingbase/install.conf，在文件的数据库初始化用户配置部分，修改如下的项：

```
# database initializes user configuration
db_user="system"                    # the user name of database
db_password="Passw0rd"              # the password of database.
```

修改[expand]部分中的项：

```
[expand]
expand_type="0"
primary_ip="192.168.100.21"
expand_ip="192.168.100.24"
node_id="4"
install_dir="/u00/Kingbase/ES/V9"      #自动在该目录后面加上 kingbase 子目录
zip_package="/u00/Kingbase/ES/V9/kingbase/db.zip"
net_device=(ens33)                     # if virtual_ip set,it must be set
net_device_ip=(192.168.100.24)         # if virtual_ip set,it must be set
license_file=(license.dat)
deploy_by_sshd="0"
ssh_port="22"
scmd_port="8890"
```

其中的参数如下。

（1）参数 expand_type 的值为 0 表示添加的新节点是备节点，值为 1 表示添加的新节点是见证节点。

（2）参数 deploy_by_sshd=0 表示使用金仓的 sys_securecmd 服务部署集群，如果要使用 ssh 免密服务进行部署，则需要设置参数 deploy_by_sshd=1；

接着，要确保新增加的节点 node04 上，KingbaseES 数据库管理系统的数据目录不存在：

```
[kingbase@node04 ~]$ rm -rf /u00/Kingbase/ES/V9/data
```

接下来，使用用户 kingbase，在新添加的节点 node04 上，执行下面的命令为集群添加新节点：

```
[kingbase@node04 ~]$ cd /u00/Kingbase/ES/V9/kingbase
[kingbase@node04 kingbase]$ ./cluster_install.sh expand
```

执行完成后，在 node01 上，使用用户 kingbase，执行下面的命令查看集群的状态：

```
[kingbase@node01 ~]$ repmgr service status
 ID | Name  | Role    | Status      | Upstream | repmgrd | PID   | Paused? | Upstream last seen
----+-------+---------+-------------+----------+---------+-------+---------+--------------------
 1  | node1 | primary | * running   |          | running | 4065  | no      | n/a
 2  | node2 | standby |   running   | node1    | running | 16365 | no      | 0 second(s) ago
 3  | node3 | witness | * running   | node1    | running | 84173 | no      | 1 second(s) ago
 4  | node4 | standby |   running   | node1    | running | 79833 | no      | 0 second(s) ago
[kingbase@node01 ~]$
```

最后，在 node04 上，按照 14.2.5 节添加节点后的配置。

14.3.8 从集群中删除备节点

在 node01 上,使用用户 kingbase,执行命令查看集群的状态:

```
[kingbase@node01 ~]$ repmgr service status
 ID | Name  | Role    | Status      | Upstream | repmgrd | PID   | Paused? | Upstream last seen
----+-------+---------+-------------+----------+---------+-------+---------+--------------------
 1  | node1 | primary | * running   |          | running | 4065  | no      | n/a
 2  | node2 | standby |   running   | node1    | running | 16365 | no      | 0 second(s) ago
 3  | node3 | witness | * running   | node1    | running | 84173 | no      | 1 second(s) ago
 4  | node4 | standby |   running   | node1    | running | 79833 | no      | 0 second(s) ago
[kingbase@node01 ~]$
```

可以看到 ID 为 4 的节点,名为 node4,是集群的备节点。

假设想从集群中删除这个 ID 为 4 的备节点。需要在 node04 上,以用户 kingbase 的身份,使用 vi 编辑器编辑集群配置文件/u00/Kingbase/ES/V9/kingbase/install.conf,在文件的数据库初始化用户配置部分,修改如下的项:

```
# database initializes user configuration
db_user="system"                    # the user name of database
db_password="Passw0rd"              # the password of database.
```

在文件的[shrink]部分,配置如下的项:

```
## config of drop a standby/witness node
[shrink]
shrink_type="0"
primary_ip="192.168.100.21"
shrink_ip="192.168.100.24"
node_id="4"
install_dir="/u00/Kingbase/ES/V9"   #会自动在该目录后面加上 kingbase 子目录
deploy_by_sshd="0"
ssh_port="22"
scmd_port="8890"
```

然后,在节点 node04 上,使用 root 用户,执行下面的命令从集群中删除节点 node04:

```
[kingbase@node04 ~]$ cd /u00/Kingbase/ES/V9/kingbase/
[kingbase@node04 kingbase]$ ./cluster_install.sh shrink
```

最后,在 node01 上,使用用户 kingbase,执行下面的命令查看集群的状态:

```
[kingbase@node01 ~]$ repmgr service status
 ID | Name  | Role    | Status      | Upstream | repmgrd | PID   | Paused? | Upstream last seen
----+-------+---------+-------------+----------+---------+-------+---------+--------------------
 1  | node1 | primary | * running   |          | running | 21711 | no      | n/a
 2  | node2 | standby |   running   | node1    | running | 13889 | no      | 0 second(s) ago
 3  | node3 | witness | * running   | node1    | running | 15840 | no      | 1 second(s) ago
[kingbase@node01 ~]$
```

可以看到,ID 为 4 的节点 node4 已经被删除了。

14.3.9 从集群中删除见证节点

在 node01 上，使用用户 kingbase，执行下面的命令查看集群的状态：

```
[kingbase@node01 ~]$ repmgr service status
 ID | Name  | Role    | Status     | Upstream | repmgrd | PID   | Paused? | Upstream last seen
----+-------+---------+------------+----------+---------+-------+---------+-------------------
 1  | node1 | primary | * running  |          | running | 10742 | no      | n/a
 2  | node2 | standby |   running  | node1    | running | 9709  | no      | 1 second(s) ago
 3  | node3 | witness | * running  | node1    | running | 9155  | no      | 1 second(s) ago
[kingbase@node01 ~]$
```

可以看到 ID 为 3 的节点，名为 node3，是集群的见证节点。

假设想从集群中删除这个 ID 为 3 的见证节点。需要在 node03 上，以用户 kingbase 的身份，使用 vi 编辑器编辑集群配置文件 /u00/Kingbase/ES/V9/kingbase/install.conf，在文件的数据库初始化用户配置部分，写入数据库用户 system 的密码 Passw0rd：

```
# database initializes user configuration
db_user="system"                     # the user name of database
db_password="Passw0rd"               # the password of database.
```

在文件的[shrink]部分，配置如下的项：

```
## config of drop a standby/witness node
[shrink]
shrink_type="1"                #参数 shrink_type 的值为 1 表示要删除的节点是见证节点。
primary_ip="192.168.100.21"
shrink_ip="192.168.100.23"
node_id="3"
install_dir="/u00/Kingbase/ES/V9"  #会自动在该目录后面加上 kingbase 子目录
deploy_by_sshd="0"
```

其中，参数 deploy_by_sshd＝0 表示使用电科金仓的 sys_securecmd 服务部署集群，如果要使用 ssh 免密服务进行部署，则需要设置参数 deploy_by_sshd＝1；

然后，在节点 node03 上，使用用户 kingbase，执行下面的命令从集群中删除见证节点 node03：

```
[kingbase@node03 ~]$ cd /u00/Kingbase/ES/V9/kingbase
[kingbase@node03 kingbase]$ ./cluster_install.sh shrink
```

最后，在 node01 上，使用用户 kingbase，执行下面的命令查看集群的状态：

```
[kingbase@node01 ~]$ repmgr service status
 ID | Name  | Role    | Status     | Upstream | repmgrd | PID   | Paused? | Upstream last seen
----+-------+---------+------------+----------+---------+-------+---------+-------------------
 1  | node1 | primary | * running  |          | running | 4065  | no      | n/a
 2  | node2 | standby |   running  | node1    | running | 16365 | no      | 0 second(s) ago
[kingbase@node01 ~]$
```

可以看到，ID 为 3 的见证节点 node3 已经被删除了。

14.3.10 为集群添加见证节点

假设想为集群添加 1 个 ID 为 3 的见证节点。需要在 node03 上,以用户 kingbase 的身份,使用 vi 编辑器编辑集群配置文件 /u00/Kingbase/ES/V9/kingbase/install.conf,在文件的数据库初始化用户配置部分,修改下面的项:

```
# database initializes user configuration
db_user="system"                    # the user name of database
db_password="Passw0rd"              # the password of database.
```

修改[expand]部分中的项:

```
[expand]
expand_type="1"
primary_ip="192.168.100.21"
expand_ip="192.168.100.23"
node_id="3"
install_dir="/u00/Kingbase/ES/V9"    #会自动在该目录后面加上 kingbase 子目录
zip_package="/u00/Kingbase/ES/V9/kingbase/db.zip"
net_device=(ens33)                   # if virtual_ip set,it must be set
net_device_ip=(192.168.100.23)       # if virtual_ip set,it must be set
license_file=(license.dat)
deploy_by_sshd="0"
ssh_port="22"
scmd_port="8890"
```

然后,要确保新增加的节点 node03 上,KingbaseES 数据库管理系统的数据目录不存在:

```
[kingbase@node03 ~]$ rm -rf /u00/Kingbase/ES/V9/data
```

接下来,在节点 node03 上,使用用户 kingbase,执行下面的命令为集群添加见证节点 node03:

```
[kingbase@node03 ~]$ cd /u00/Kingbase/ES/V9/kingbase
[kingbase@node03 kingbase]$ ./cluster_install.sh expand
```

最后,在 node01 上,使用用户 kingbase,执行下面的命令查看集群的状态:

```
[kingbase@node01 ~]$ repmgr service status
 ID | Name  | Role    | Status    | Upstream | repmgrd | PID   | Paused? | Upstream last seen
----+-------+---------+-----------+----------+---------+-------+---------+--------------------
 1  | node1 | primary | * running |          | running | 10742 | no      | n/a
 2  | node2 | standby |   running | node1    | running | 9709  | no      | 0 second(s) ago
 3  | node3 | witness | * running | node1    | running | 25443 | no      | 0 second(s) ago
[kingbase@node01 ~]$
```

可以看到,已经为集群添加了 ID 为 3 的见证节点。

为了继续下面的实战,请在 node03 上,执行下面的命令删除集群的见证节点:

```
[kingbase@node03 ~]$ cd /u00/Kingbase/ES/V9/kingbase
[kingbase@node03 kingbase]$ ./cluster_install.sh shrink
```

14.4　主备集群 sys_rman 备份实战

14.4.1　REPO 外部部署

sys_rman 的 REPO 外部部署模式，如图 14-2 所示。外部部署的服务器 KingbaseES 数据库的安装目录结构，必须与主备集群节点的完全一致。简单起见，在之前进行过主备集群节点扩展的服务器 node04 上，部署 sys_rman 的 REPO 存储库，因为 node04 上已经具有了与主备集群节点相同的 KingbaseES 数据库的目录结构。外部存储库位于 node04 节点的目录/dbbak/pbak/kbbr_repo 下。

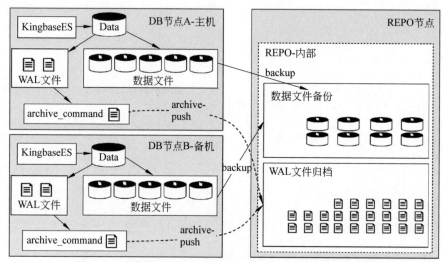

图 14-2　主备集群 sys_rman 存储库 REPO 外部部署拓扑图

在节点 node01、node02 和 node04 上，使用用户 root，执行下面的命令创建目录/dbbak/pbak：

```
mkdir -p /dbbak/pbak
chown kingbase.dba /dbbak/pbak/
```

1. 配置 REPO 外部存储的 sys_rman 备份

在驻留外部存储库 REPO 的节点 node04 上，使用用户 kingbase，执行下面的命令为 sys_rman 准备配置文件/u00/Kingbase/cluster/kingbase/bin/sys_backup.conf：

```
[kingbase@node04 ~]$ cp /u00/Kingbase/ES/V9/kingbase/share/sys_backup.conf /u00/Kingbase/ES/V9/kingbase/bin
[kingbase@node04 ~]$ vi /u00/Kingbase/ES/V9/kingbase/bin/sys_backup.conf
```

修改如下的行：

```
_target_db_style="cluster"
_one_db_ip="192.168.100.21"          #主节点的 IP
_repo_ip="192.168.100.24"            #外部存储库 REPO 的 IP
```

```
_stanza_name="kingbase"
_os_user_name="kingbase"
_repo_path="/dbbak/pbak/kbbr_repo"
_single_data_dir="/u00/Kingbase/ES/V9/data"
_single_bin_dir="/u00/Kingbase/ES/V9/kingbase/bin"
_single_db_user="system"
_single_db_port="54321"
# on means sys_securecmd, off means normal ssh
_use_scmd=on
```

使用用户 kingbase,在驻留外部存储库 REPO 的节点 node04 上,执行脚本 sys_backup.sh,生成 sys_rman 的配置文件 sys_rman.conf:

```
[kingbase@node04 ~]$ /u00/Kingbase/ES/V9/kingbase/bin/sys_backup.sh init
#省略了输出
[kingbase@node04 ~]$
```

执行完脚本 sys_backup.sh,会在驻留外部存储库 REPO 的节点 node04 上生成 sys_rman 的配置文件 sys_rman.conf,并进行一次全量备份。此外,还会在主备集群的 primary 节点和 standby 节点的目录/dbbak/pbak/kbbr_repo 下生成 sys_rman 的配置文件 sys_rman.conf,在主节点 node01 上,执行下面的命令:

```
[kingbase@node01 ~]$ ls -l /dbbak/pbak/kbbr_repo/
total 4
-rw-r--r-- 1 kingbase dba 554 Feb 15 16:56 sys_rman.conf
[kingbase@node01 ~]$
```

在备节点 node02 上,执行下面的命令:

```
[kingbase@node02 ~]$ ls -l /dbbak/pbak/kbbr_repo/
total 4
-rw-r--r-- 1 kingbase dba 554 Feb 15 16:56 sys_rman.conf
[kingbase@node02 ~]$
```

在 KingbaseES 主备集群的主节点 node01 上,使用用户 kingbase,执行下面的命令将 sys_rman 的配置文件 sys_rman.conf,复制到驻留 REPO 外部存储的服务器 node04 的目录/home/kingbase 下:

```
[kingbase@node01 ~]$ scp /dbbak/pbak/kbbr_repo/sys_rman.conf node04:/home/kingbase/
kingbase@node04's password: (输入密码 kingbase123)
sys_rman.conf                                    100%   475   639.2KB/s   00:00
[kingbase@node01 ~]$
```

2. 在外部存储库 REPO 驻留的服务器上备份主备集群数据库

无论 KingbaseES 主备集群的主节点在 node01 上,还是在 node02 上,要使用 sys_rman 手工备份 KingbaseES 数据库,都必须在驻留外部存储库 REPO 的节点 node04 上,执行下面的命令进行数据库备份:

```
[kingbase@node04 kingbase]$ /u00/Kingbase/ES/V9/kingbase/bin/sys_rman \
```

```
                    --config=/dbbak/pbak/kbbr_repo/sys_rman.conf \
                    --stanza=kingbase \
                    --archive-copy \
                    --type=full backup \
                    >> /u00/Kingbase/ES/V9/kingbase/log/sys_rman_backup_
full.log 2>&1
```

3. 查看数据库备份信息

可以在主备集群的节点上,或者是 REPO 外部存储的服务器上,使用用户 kingbase,执行下面的命令查看数据库的备份信息:

```
[kingbase@node04 ~]$ sys_rman --config=/dbbak/pbak/kbbr_repo/sys_rman.conf --stanza=kingbase info
stanza: kingbase
    status: ok
    cipher: none

    db (current)
        wal archive min/max (V009R001C001B0025): 000000030000000000000010/000000030000000000000014

        full backup: 20240215-165607F
            timestamp start/stop: 2024-02-15 16:56:07 / 2024-02-15 16:56:12
            wal start/stop: 000000030000000000000012 / 000000030000000000000012
            database size: 99.6MB, database backup size: 99.6MB
            repo1: backup set size: 99.6MB, backup size: 99.6MB

        full backup: 20240215-170340F
            timestamp start/stop: 2024-02-15 17:03:40 / 2024-02-15 17:03:44
            wal start/stop: 000000030000000000000014 / 000000030000000000000014
            database size: 99.6MB, database backup size: 99.6MB
            repo1: backup set size: 99.6MB, backup size: 99.6MB
[kingbase@node04 ~]$
```

4. 外部 REPO 服务器作为备份恢复测试机

可以把驻留外部存储库 REPO 的节点 node04 当作 1 台备份恢复测试机,将主备集群的 sys_rman 备份恢复在 node04 上,恢复过程如下。

(1) 如果 node04 之前启动了 kingbaseES 数据库,可以执行下面的命令停止 KingbaseES 数据库:

```
[kingbase@node04 ~]$ sys_ctl stop
```

(2) 在 node04 上,执行下面的命令删除 KingbaseES 数据库集簇:

```
[kingbase@node04 ~]$ rm -rf /u0?/Kingbase/ES/V9/data
```

(3) 在 REPO 外部部署的服务器上,执行下面的命令恢复 KingbaseES 数据库集簇:

```
[kingbase@node04 ~]$ /u00/Kingbase/ES/V9/kingbase/bin/sys_rman \
                    --config=/home/kingbase/sys_rman.conf \
                    --stanza=kingbase restore
```

（4）在 REPO 外部部署的服务器上，执行下面的命令启动 KingbaseES 数据库管理系统：

```
[kingbase@node04 ~]$ sys_ctl start
```

5. 在主备集群的节点上恢复数据库

也可以将主备集群的 sys_rman 备份恢复到主备集群的节点上。在主备集群的节点 node01 上恢复 sys_rman 的过程如下。

（1）停止 KingbaseES 集群：

```
[kingbase@node01 ~]$ sys_monitor.sh stop
```

（2）删除原有的数据库目录：

```
[kingbase@node01 ~]$ rm -rf /u0?/Kingbase/ES/V9/data
```

（3）恢复数据库：

```
[kingbase@node01 ~]$ /u00/Kingbase/ES/V9/kingbase/bin/sys_rman \
            --config=/dbbak/pbak/kbbr_repo/sys_rman.conf \
            --stanza=kingbase restore
```

（4）启动集群：

```
[kingbase@node01 ~]$ sys_monitor.sh start
```

（5）查看集群的状态：

```
[kingbase@node01 ~]$ repmgr service status
[WARNING] node "node2" not found in "pg_stat_replication"
 ID | Name  | Role    | Status    | Upstream | repmgrd     | PID | Paused? | Upstream last seen
----+-------+---------+-----------+----------+-------------+-----+---------+--------------------
 1  | node1 | primary | * running |          | not running | n/a | n/a     | n/a
 2  | node2 | standby |   running | ! node1  | not running | n/a | n/a     | n/a

[WARNING] following issues were detected
  - node "node2" (ID: 2) is not attached to its upstream node "node1" (ID: 1)

[kingbase@node01 ~]$
```

可以看到，当前集群的 repmgrd 没有运行。

（6）分别在 node01 和 node02 上，使用用户 kingbase，执行下面的命令启动 repmgrd：

```
repmgrd -f /u00/Kingbase/ES/V9/kingbase/etc/repmgr.conf
```

（7）查看集群的状态：

```
[kingbase@node01 ~]$ repmgr service status
[WARNING] node "node2" not found in "pg_stat_replication"
 ID | Name  | Role    | Status    | Upstream | repmgrd | PID   | Paused? | Upstream last seen
----+-------+---------+-----------+----------+---------+-------+---------+--------------------
 1  | node1 | primary | * running |          | running | 11233 | no      | n/a
 2  | node2 | standby |   running | ! node1  | running | 11351 | no      | 1 second(s) ago
[WARNING] following issues were detected
```

```
- node "node2" (ID: 2) is not attached to its upstream node "node1" (ID: 1)
[kingbase@node01 ~]$
```

可以看到,当前主备集群的 repmgrd 已经处于 running 状态了,但是 node2 还不是一个正常运行的备节点。为此需要重做备节点。

在 node02 上,执行下面的命令重做备节点 node2:

```
[kingbase@node02 ~]$ # 停止数据库
[kingbase@node02 ~]$ sys_ctl stop
#省略了输出
[kingbase@node02 ~]$ #克隆主库到备节点 node2,-F 选项表示强制覆盖原有的数据目录
[kingbase@node02 ~]$ repmgr -h 192.168.100.21 -U esrep -d esrep -p 54321  standby clone -F
#省略了输出
[kingbase@node02 ~]$ # 启动数据库
[kingbase@node02 ~]$ sys_ctl start
#省略了输出
[kingbase@node02 ~]$ # 重新将节点 node2 实例注册到集群,-F 选项表示强制将节点注册到集群
[kingbase@node02 ~]$ repmgr standby register -F
[INFO] connecting to local node "node2" (ID: 2)
[INFO] connecting to primary database
[INFO] standby registration complete
[NOTICE] standby node "node2" (ID: 2) successfully registered
[kingbase@node02 ~]$
```

在 node01 上,执行下面的命令查看集群的状态:

```
[kingbase@node01 ~]$ repmgr service status
 ID | Name  | Role    | Status    | Upstream | repmgrd | PID    | Paused? | Upstream last seen
----+-------+---------+-----------+----------+---------+--------+---------+--------------------
 1  | node1 | primary | * running |          | running | 39112  | no      | n/a
 2  | node2 | standby |   running | node1    | running | 116285 | no      | 1 second(s) ago
[kingbase@node01 ~]$ ip addr show ens33 | grep inet |grep -v inet6
    inet 192.168.100.21/24 brd 192.168.100.255 scope global noprefixroute ens33
    inet 192.168.100.20/24 scope global secondary ens33:3
[kingbase@node01 ~]$
```

可以看到,当前集群处于正常状态了,主节点 node01 上的网卡 ens33 上配置有浮动 IP。

14.4.2 REPO 内部部署

sys_rman 的 REPO 内部部署模式,如图 14-3 所示。主备集群的 sys_rman 备份的 REPO 存储库驻留在集群节点 node01 上。

在两个节点的 KingbaseES 主备集群上,计划以如下方式配置 sys_rman 备份。

(1) node01 是主节点,sys_rman 的 REPO 存储库驻留在 node01 上,位于目录/dbbak/pbak/kbbr_repo 下。

(2) node02 是备节点。

1. 配置 REPO 内部存储的 sys_rman 备份

在主备集群的所有节点上(本例为 node01 和 node02),使用用户 root,执行下面的命令

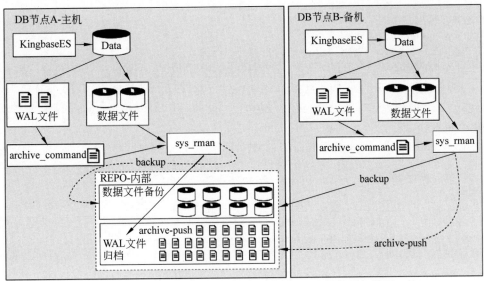

图 14-3　主备集群 sys_rman 存储库 REPO 内部部署拓扑图

创建目录/dbbak/pbak：

```
rm -rf /dbbak/pbak/
mkdir -p /dbbak/pbak
chown kingbase.dba /dbbak/pbak/
```

使用用户 kingbase，在主备集群的节点 node01 上，执行下面的命令为 sys_rman 准备配置文件/u00/Kingbase/ES/V9/kingbase/bin/sys_backup.conf：

```
[kingbase@node01 ~]$ cp /u00/Kingbase/ES/V9/kingbase/share/sys_backup.conf /u00/Kingbase/ES/V9/kingbase/bin
[kingbase@node01 ~]$ vi /u00/Kingbase/ES/V9/kingbase/bin/sys_backup.conf
```

修改如下的行：

```
_target_db_style="cluster"
_one_db_ip="192.168.100.21"          #主节点的 IP
_repo_ip="192.168.100.21"            #内部存储库 REPO 的 IP

_stanza_name="kingbase"
_os_user_name="kingbase"
_repo_path="/dbbak/pbak/kbbr_repo"
_single_data_dir="/u00/Kingbase/ES/V9/data"
_single_bin_dir="/u00/Kingbase/ES/V9/kingbase/bin"
_single_db_user="system"
_single_db_port="54321"
# on means sys_securecmd, off means normal ssh
_use_scmd=on
```

现在，使用用户 kingbase，在主备集群的节点 node01 上，执行脚本 sys_backup.sh，生成 sys_rman 的配置文件 sys_rman.conf：

```
[kingbase@node01 ~]$ /u00/Kingbase/ES/V9/kingbase/bin/sys_backup.sh init
#省略了输出
[kingbase@node01 ~]$
```

执行完脚本 sys_backup.sh,将在主备集群的 primary 节点的目录/dbbak/pbak/kbbr_repo 下,生成 sys_rman 的配置文件 sys_rman.conf,并进行一次全量备份;同时会在主备集群的 standby 节点的目录/dbbak/pbak/kbbr_repo 下,生成 sys_rman 的配置文件 sys_rman.conf:

在节点 node01 上,执行下面的命令:

```
[kingbase@node01 ~]$ ls -l /dbbak/pbak/kbbr_repo/
total 4
drwxr-x--- 3 kingbase dba   22 Feb 15 17:50 archive
drwxr-x--- 3 kingbase dba   22 Feb 15 17:50 backup
-rw-r--r-- 1 kingbase dba  869 Feb 15 17:50 sys_rman.conf
[kingbase@node01 ~]$
```

在备节点 node02 上,执行命令:

```
[kingbase@node02 ~]$ ls -l /dbbak/pbak/kbbr_repo/
total 4
-rw-r--r-- 1 kingbase dba 554 Feb 15 17:50 sys_rman.conf
[kingbase@node02 ~]$
```

2. 备份主备集群

无论 KingbaseES 主备集群的主节点在 node01 上,还是在 node02 上,要使用 sys_rman 手工备份 KingbaseES 主备集群,必须在 REPO 驻留的节点(本例为节点 node01)执行下面的命令进行数据库备份:

```
[kingbase@node01 ~]$ /u00/Kingbase/ES/V9/kingbase/bin/sys_rman \
        --config=/dbbak/pbak/kbbr_repo/sys_rman.conf \
        --stanza=kingbase \
        --archive-copy \
        --type=full backup \
        >> /u00/Kingbase/ES/V9/kingbase/log/sys_rman_backup_full.log 2>&1
```

3. 查看数据库备份信息

可以在 node01 或者 node02 上,执行下面的命令查看主备集群的 sys_rman 备份集信息:

```
[kingbase@node01 ~]$ sys_rman --config=/dbbak/pbak/kbbr_repo/sys_rman.conf --stanza=kingbase info
stanza: kingbase
    status: ok
    cipher: none

    db (current)
        wal archive min/max (V009R001C001B0025): 000000040000000000000018/000000040000000000000001C
```

```
        full backup: 20240215-175029F
            timestamp start/stop: 2024-02-15 17:50:29 / 2024-02-15 17:50:31
            wal start/stop: 000000040000000000000001A / 000000040000000000000001A
            database size: 99.7MB, database backup size: 99.7MB
            repo1: backup set size: 99.7MB, backup size: 99.7MB

        full backup: 20240215-175422F
            timestamp start/stop: 2024-02-15 17:54:22 / 2024-02-15 17:54:25
            wal start/stop: 000000040000000000000001C / 000000040000000000000001C
            database size: 99.7MB, database backup size: 99.7MB
            repo1: backup set size: 99.7MB, backup size: 99.7MB
[kingbase@node01 ~]$
```

4. 在主备集群的节点上恢复数据库

在主备集群的节点上恢复 KingbaseES 数据库的过程,和 14.4.1 节中的"5.在主备集群的节点上恢复数据库"的步骤一样。

5. 在备份恢复测试机上恢复 sys_rman 备份

还是把 node04 服务器作为备份恢复测试机。

在 node01 上,执行下面的步骤,将 sys_rman 备份恢复到备份恢复测试机 node04 上。

(1) 将 node01 上的 sys_rman 备份复制到备份恢复测试机 node04。在 node01 上执行下面的命令:

```
[kingbase@node01 pbak]$ cd /dbbak/pbak
[kingbase@node01 pbak]$ tar cf kbbr_repo.tar kbbr_repo/
[kingbase@node01 pbak]$ scp  kbbr_repo.tar 192.168.100.24:/dbbak/pbak/
kingbase@192.168.100.24's password: #输入用户 kingbase 的密码 kingbase123
kbbr_repo.tar                                    100%   332MB 146.2MB/s   00:02
[kingbase@node01 pbak]$
```

在 node04 上执行下面的命令:

```
[kingbase@node04 ~]$   cd /dbbak/pbak/
[kingbase@node04 pbak]$ rm -rf kbbr_repo
[kingbase@node04 pbak]$ tar xf kbbr_repo.tar
[kingbase@node04 pbak]$
```

(2) 如果 node04 之前启动了 KingbaseES 数据库,可以执行下面的命令停止 KingbaseES:

```
[kingbase@node04 ~]$ sys_ctl stop
```

(3) 在 node04 上,执行下面的命令删除原有的数据库目录:

```
[kingbase@node04 ~]$ rm -rf /u0?/Kingbase/ES/V9/data
```

(4) 在 node04 上,执行下面的命令恢复 KingbaseES 数据库集簇:

```
[kingbase@node04 ~]$ /u00/Kingbase/ES/V9/kingbase/bin/sys_rman \
            --config=/dbbak/pbak/kbbr_repo/sys_rman.conf \
```

```
          --stanza=kingbase restore
```

(5) 在 node04 上,执行下面的命令启动 KingbaseES 数据库管理系统:

```
[kingbase@node04 ~]$ sys_ctl start
```

至此,把 KingbaseES 主备集群的 sys_rman 数据库备份恢复到了备份恢复测试机 node04 上。

附录 A

安装 CentOS 7 操作系统

A.1 准备服务器硬件

可用把 KingbaseES 数据库部署在物理服务器或云服务器上。如果决定使用物理服务器来直接部署 KingbaseES 数据库,建议如下。

(1) **CPU**:KingbaseES 数据库服务器的 CPU,可以选择 Intel 公司或 AMD 公司生产的 x86-64 架构处理器,也可以选择国产的 Kunpeng64、Aarch64、Mips64、Loongarch64 等架构的处理器。

(2) **物理内存**:KingbaseES 数据库服务器的物理内存,一般需要根据用户的数据总量(考虑到未来 5~8 年的情况)、用户客户端连接的数量等因素,进行容量规划。由于当前服务器的内存并不昂贵,因此一定要为 KingbaseES 服务器配置足够容量的物理内存,避免因内存成为 KingbaseES 服务器的性能瓶颈。

(3) **存储**:KingbaseES 数据库服务器,一般会有内置的 RAID 控制器。特别提醒要为内置 RAID 控制器配置尽量大的高速缓存 cache,这样可以获得更好的性能。

① 如果有外置的磁盘阵列用于存储 KingbaseES 数据库的数据,可以为服务器配置 4 块内置硬盘,例如,2 块磁盘做 raid1,用于安装操作系统;1 块磁盘用作 raid 的热冗余(host spare);1 块磁盘用于存储 KingbaseES 数据库备份。

② 如果没有外置的磁盘阵列用于存储 KingbaseES 数据库的数据,那么服务器应该配置至少 12 块内置硬盘,例如,2 块磁盘做 raid1,用于安装操作系统;8 块磁盘做 raid10,用于存储 KingbaseES 数据库的数据;1 块磁盘用于 raid 的热冗余;1 块磁盘用于数据库备份。

(4) **网络**:建议为 KingbaseES 数据库服务器配置多个 10Gb/s 以太网卡,或者性能更好的 25Gb/s、40Gb/s、56Gb/s、100Gb/s 以太网卡,这样可以满足大量客户访问 KingbaseES 数据库服务器的需要。

如果 KingbaseES DBA 决定在私有云或公有云上部署 KingbaseES 数据库服务器,相对于直接使用硬件服务器,DBA 只需要设置 CPU 的类型和数量、磁盘存储的规格和容量、物理内存容量、网络带宽的要求。DBA 可以根据 2~3 年的数据和访问情况,来规划这些需求。

KingbaseES 数据库 DBA 常会使用属于个人的 KingbaseES 数据库测试环境,通常可以在自己的笔记本或 PC 上,使用 VMware Workstation 虚拟化软件,运行虚拟机,安装

KingbaseES 数据库测试环境。

可以从 VMware 官网上下载 VMware Workstation Pro 虚拟化软件的测试版，在本书写作时，可以从官网上下载的最新版本是 VMware Workstation 17 Pro。

如图 A-1 所示，安装完 VMware Workstation 17 Pro 虚拟化软件之后，会在 Windows 10 宿主机上新增两块虚拟网卡。

（1）**网卡 VMnet1**（**host-only** 类型）：建议在 Windows 10 宿主机上，将这个网卡的 IP 地址设置为 192.168.100.1，子网掩码设置为 255.255.255.0。

（2）**网卡 VMnet8**（**NAT** 类型）：不需要进行任何配置，默认采用 DHCP 获取 IP 配置。

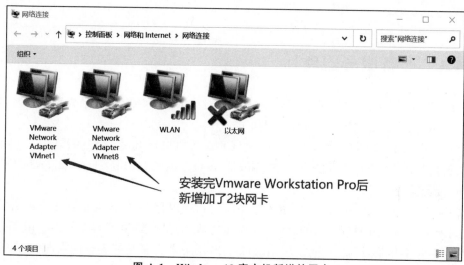

图 A-1　Windows 10 宿主机新增的网卡

如图 A-2 所示，安装完 VMware Workstation 17 Pro 虚拟化软件之后，在 Windows 10 宿主机还会新增 3 台虚拟交换机。

（1）**host-only** 类型的虚拟交换机：Windows 10 宿主机上的 host-only 类型的虚拟网卡 VMnet1 会自动连接到 host-only 类型的交换机上。

（2）**NAT** 类型的虚拟交换机：Windows 10 宿主机上 NAT 类型的虚拟网卡 VMnet8 会自动连接到 NAT 类型的交换机上。

（3）**bridge** 类型的虚拟交换机：Windows 10 宿主机上的物理网卡（无线网卡或以太网卡），会自动连接到 bridge 类型的交换机上。

如图 A-2 所示，虚拟机（无论是 Linux，还是 Windows）的网卡，可以是以下 3 种类型。

（1）**虚拟机上的 host-only 类型的虚拟网卡**：会自动连接到 Windows 10 宿主机上的 host-only 类型的虚拟交换机上。

（2）**虚拟机上的 NAT 类型的虚拟网卡**：会自动连接到 Windows 10 宿主机上的 NAT 类型的虚拟交换机上。

（3）**虚拟机上的 bridge 类型的虚拟网卡**：会自动连接到 Windows 10 宿主机上的 bridge 类型的虚拟交换机上。

如图 A-2 所示，安装完 VMware Workstation 17 Pro 虚拟化软件之后，还会启动一个 VMware DHCP 服务，它只能用来为虚拟网卡（Windows 10 宿主机或虚拟机）自动配置 IP。

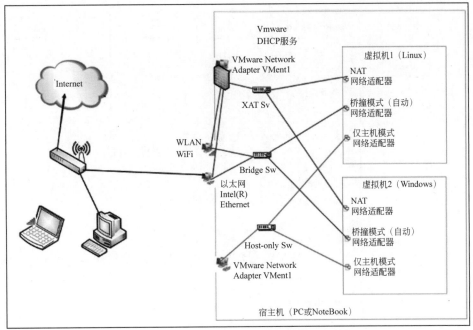

图 A-2　Windows 10 宿主机新增的虚拟交换机

在 Windows 10 宿主机上运行 VMware Workstation 虚拟机软件，选择"文件"→"新建虚拟机"命令来创建一个 CentOS 7 虚拟机，具体创建步骤如图 A-3～图 A-8 所示。

图 A-3　创建虚拟机画面 1

完成上述操作后，将创建 1 台配置如图 A-9 所示的虚拟机。这个配置太低，还需要修改虚拟机的内存大小、处理器个数，并为虚拟机光盘驱动器装载 CentOS 7 的安装介质 ISO 文件。

如图 A-9 所示，单击"编辑虚拟机设置"命令，将虚拟机的配置修改为 16GB 内存、4 个处理器、1 个 900GB 的硬盘、2 个网卡(1 个 NAT，1 个 host-only)，如图 A-10 所示。

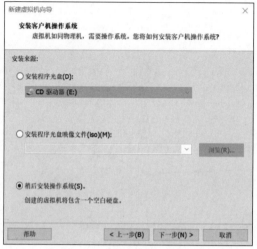

图 A-4　创建虚拟机画面 2

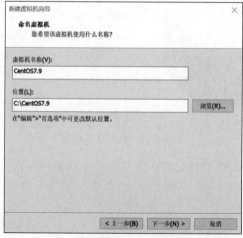

图 A-5　创建虚拟机画面 3

图 A-6　创建虚拟机画面 4

附录 A 安装 CentOS 7 操作系统

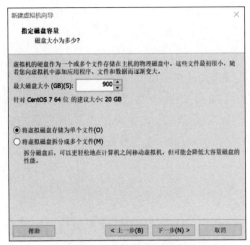

图 A-7 创建虚拟机画面 5　　　图 A-8 创建虚拟机画面 6

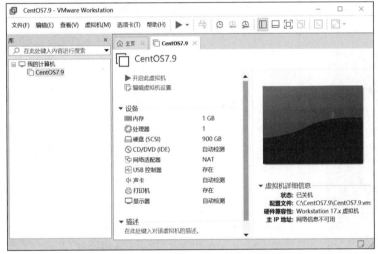

图 A-9　CentOS 7 虚拟机的初始配置

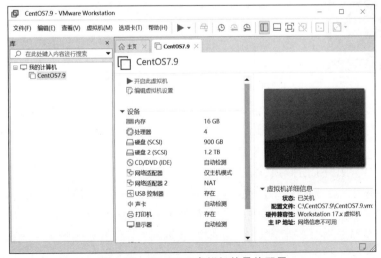

图 A-10　CentOS 7 虚拟机的最终配置

A.2　下载 CentOS 7

读者可以从 URL 地址 https://www.centos.org/centos-linux 下载 CentOS 7 操作系统介质。不过直接从这个 URL 地址下载，速度可能会很慢。建议从 URL 地址 http://mirrors.neusoft.edu.cn/centos/7.9.2009/isos/x86_64 下载，速度会快很多。下载后将介质复制到 Windows 10 操作系统的 C 盘根目录下。

A.3　安装 CentOS 7

如图 A-10 所示，选择"编辑虚拟机设置"选项，为 CentOS 虚拟机的虚拟机光驱装载刚下载的 CentOS-7-x86_64-Everything-2009.iso 文件，然后使用 VMware Workstation 开始安装 CentOS 7 操作系统。

当出现如图 A-11 所示的 CentOS 7 欢迎画面，一定要选择美国英语作为安装过程中使用的提示语言。

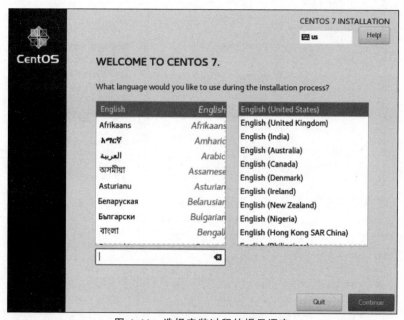

图 A-11　选择安装过程的提示语言

选择完安装过程的提示语言后，将出现如图 A-12 所示的 CentOS 7 安装配置的主画面。在 CentOS 7 安装配置的主画面中，需要配置日期和时间所在的时区、支持的语言、要安装的软件包、安装目标位置、网络配置等。

在 SOFTWARE SELECTION 对话框中，如图 A-13 所示，在 Base Environment 选项区域中选中 Server with GUI 单选按钮，在 Add-Ons for Selected Environment 选项区域中选中所有复选框。这对初学者来说很简单，等熟悉 CentOS Linux 系统，可以自行选择安装哪些软件包。完成软件选择后，单击 Done 按钮，返回 CentOS 安装配置主画面。

在 INSTALLATION DESTINATION 对话框中，选择将 CentOS 操作系统安装在磁盘 sda 上，并且由用户自己来定义分区大小，如图 A-14 所示。

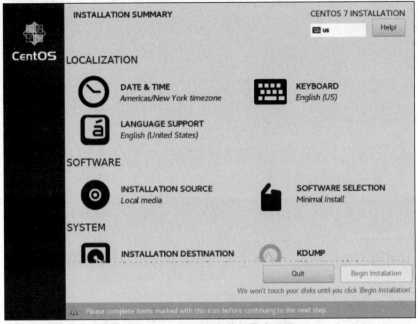

图 A-12　CentOS 安装配置主画面

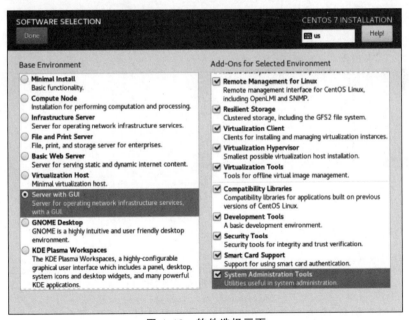

图 A-13　软件选择画面

在接下来显示的 MANUAL PARTITIONING 对话框中，选择使用 LVM 方法来管理分区，如图 A-15 所示。

在手动分区对话框中，单击"＋"按钮，将开始配置 CentOS 操作系统分区的大小。依次

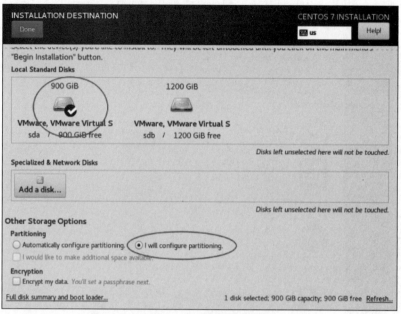

图 A-14　选择 CentOS 的安装盘和分区方法

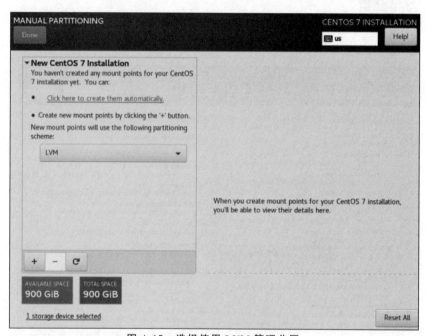

图 A-15　选择使用 LVM 管理分区

创建分区/boot,大小是 10GiB;分区/,大小是 10GiB;分区 swap,大小是 64GiB;分区/usr, 大小是 20GiB;分区/home,大小是 40GiB;分区/opt,大小是 40GiB;分区/tmp,大小是 10GiB;分区/var,大小是 40GiB;分区/toBeDeleted,大小是 665.99GiB,如图 A-16 所示。

在 NETWORK & HOST NAME 对话框中,如图 A-17～图 A-20 所示,配置网卡 ens33。

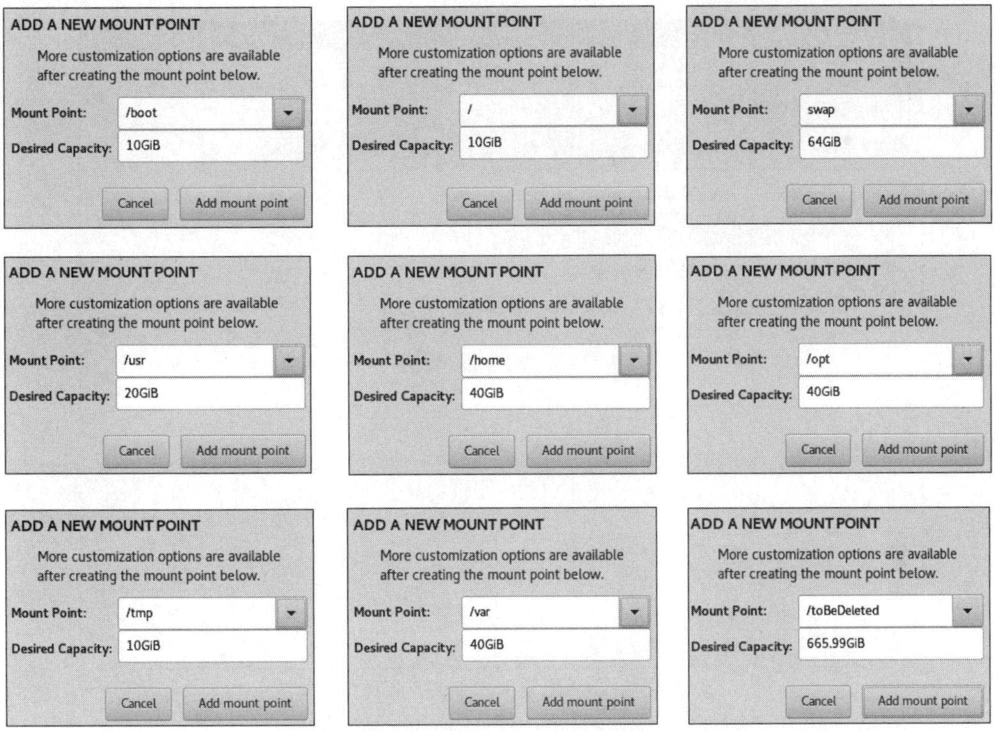

图 A-16　CentOS 分区大小

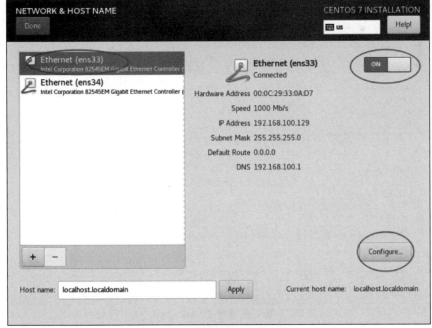

图 A-17　配置网卡 ens33 画面 1

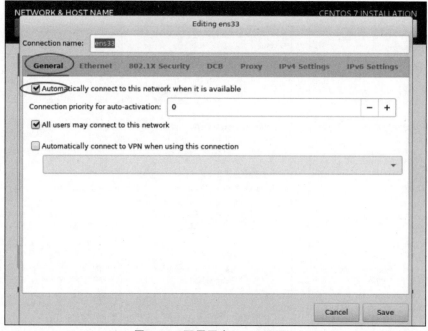

图 A-18　配置网卡 ens33 画面 2

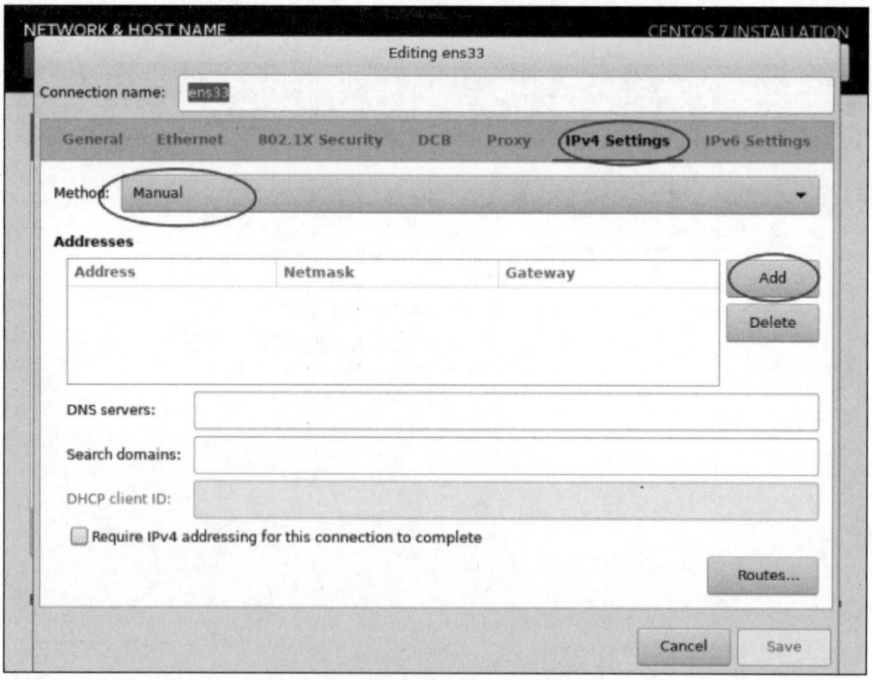

图 A-19　配置网卡 ens33 画面 3

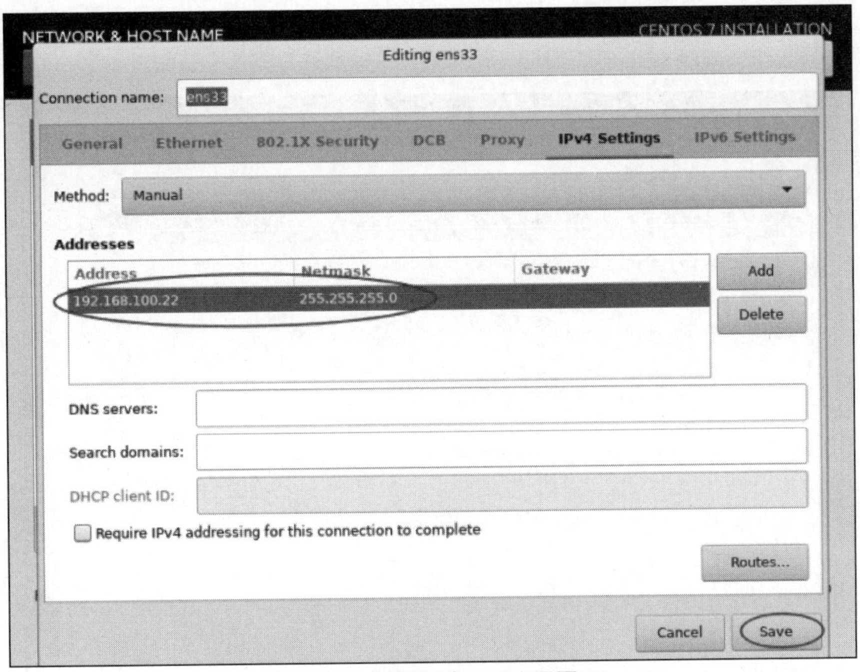

图 A-20 配置网卡 ens33 画面 4

如图 A-21 和图 A-22 所示，配置网卡 ens34。

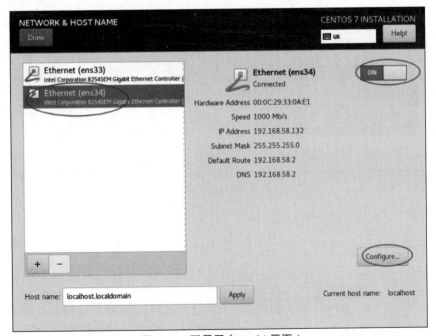

图 A-21 配置网卡 ens34 画面 1

如图 A-23 所示，配置服务器的主机名为 dbsvr。

在 DATE & TIME 对话框中，如图 A-24 所示，配置 CentOS 服务器的时区为 Asia 和 Shanghai。

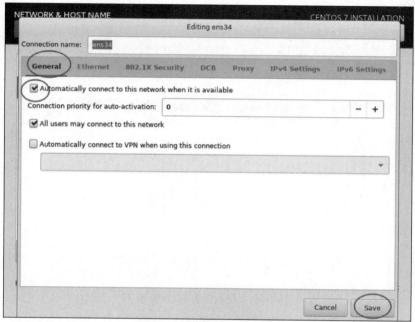

图 A-22 配置网卡 ens34 画面 2

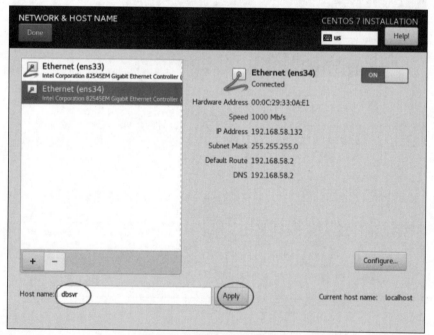

图 A-23 配置 CentOS 主机名

在 LANGUAGE SUPPORT 对话框中,如图 A-24 所示,设置 CentOS 支持的语言包含所有的中文。

接下来,如图 A-25 和图 A-26 所示,可以开始安装 CentOS 操作系统了。

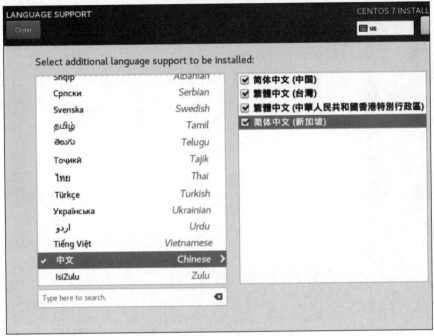

图 A-24　配置 CentOS 支持的语言

图 A-25　开始安装 CentOS 操作系统

在安装 CentOS 操作系统的过程中，如图 A-26 所示，选择 ROOT PASSWORD 选项，为用户 root 设置密码（简单起见，root 密码设置为 root123），如图 A-27 所示；选择 USER CREATION 选项，创建一个名为 test，密码为 test123 的新用户，如图 A-28 所示。单击 Done 按钮完成创建。

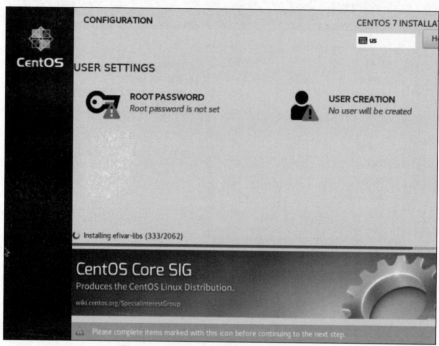

图 A-26　安装 CentOS 操作系统过程中

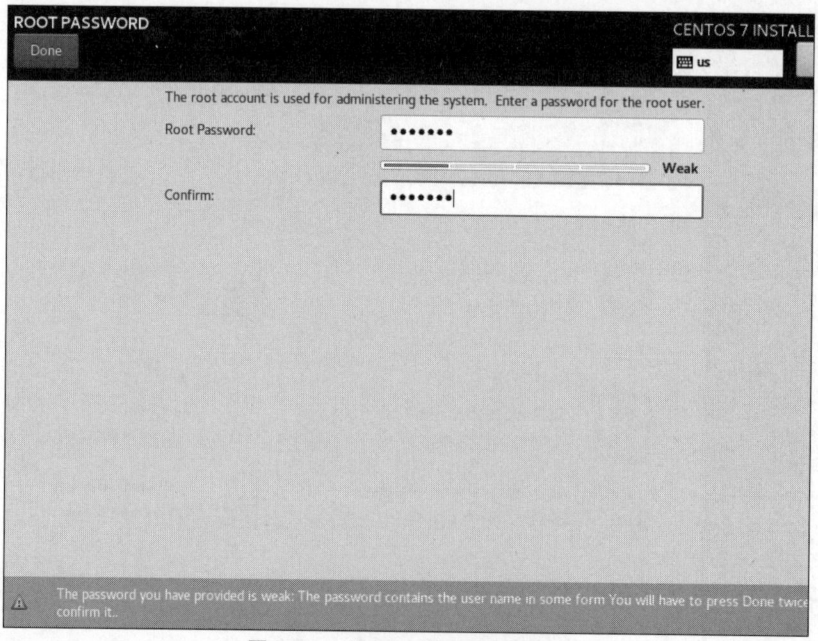

图 A-27　为 root 用户设置密码

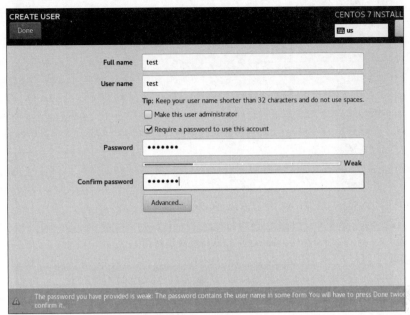

图 A-28　创建用户 test

耐心等待一会，出现如图 A-29 所示的画面，提示用户安装完成 CentOS 7，需要重新启动系统。

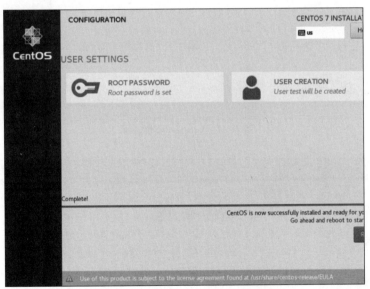

图 A-29　CentOS 安装结束提示重新启动

系统重新启动后，将出现如图 A-30 所示的画面，提示用户现在还没有接受 CentOS 软件许可证协议。选择 LICENSE INFORMATION 选项，出现如图 A-31 所示的画面。

勾选 I accept the license agreement 复选框后，单击 Done 按钮，出现如图 A-32 所示的画面，完成 CentOS 操作系统的配置。在图 A-32 中，单击 FINISH CONFIGURATION 按钮，稍等一会，出现如图 A-33 所示的画面，CentOS 7 已经安装完毕并且开始运行了。

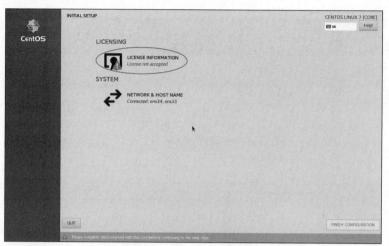

图 A-30 License 协议画面 1

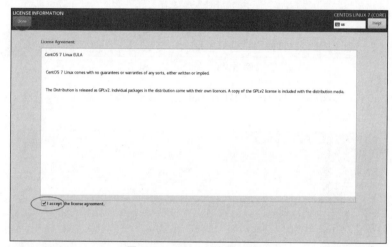

图 A-31 License 协议画面 2

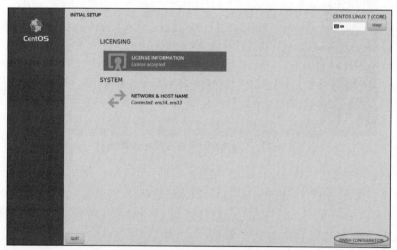

图 A-32 完成 CentOS 配置

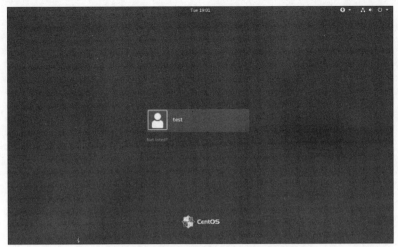

图 A-33　第 1 次启动 CentOS 的画面

A.4　安装 Google Chrome 浏览器

在 KingbaseES 数据库实战过程中，经常要用到浏览器查看一些文件。有时候，CentOS 7 上的 Firefox 浏览器不能正常工作。为了解决这个问题，请使用用户 root，执行下面的命令在 CentOS 7 上安装 Google Chrome 浏览器：

```
[root@dbsvr ~]# wget https://dl.google.com/linux/direct/google-chrome-stable_current_x86_64.rpm
[root@dbsvr ~]# yum localinstall -y google-chrome-stable_current_x86_64.rpm
[root@dbsvr ~]# rm -f google-chrome-stable_current_x86_64.rpm
```

安装完成后，可以使用普通用户（如用户 test，密码 test123），登录 Gnome 图形界面，在终端上运行命令 google-chrome，打开 Google Chrome 浏览器：

```
[test@dbsvr ~]$ google-chrome
```

A.5　删除逻辑卷 centos/toBeDeleted

安装完 CentOS 之后，执行下面的命令可以查看逻辑卷和文件系统情况：

```
[root@dbsvr ~]# vgs
  VG     #PV #LV #SN Attr   VSize    VFree
  centos   1   8   0 wz--n- <890.00g 4.00m
[root@dbsvr ~]# lvs
  LV              VG     Attr       LSize  Pool Origin Data%  Meta%  Move Log Cpy%Sync Convert
  home            centos -wi-ao---- 40.00g
  opt             centos -wi-ao---- 40.00g
  root            centos -wi-ao---- 10.00g
  swap            centos -wi-ao---- 64.00g
```

```
  tmp            centos -wi-ao----   10.00g
  toBeDeleted    centos -wi-ao----  665.99g
  usr            centos -wi-ao----   20.00g
  var            centos -wi-ao----   40.00g
[root@dbsvr ~]# df -h
Filesystem                    Size  Used  Avail  Use%  Mounted on
devtmpfs                      7.8G     0   7.8G    0%  /dev
tmpfs                         7.8G     0   7.8G    0%  /dev/shm
tmpfs                         7.8G   13M   7.8G    1%  /run
tmpfs                         7.8G     0   7.8G    0%  /sys/fs/cgroup
/dev/mapper/centos-root        10G   84M    10G    1%  /
/dev/mapper/centos-usr         20G  5.8G    15G   29%  /usr
/dev/mapper/centos-tmp         10G   33M    10G    1%  /tmp
/dev/mapper/centos-var         40G  934M    40G    3%  /var
/dev/mapper/centos-opt         40G   33M    40G    1%  /opt
/dev/mapper/centos-toBeDeleted 666G  33M   666G    1%  /toBeDeleted
/dev/mapper/centos-home        40G   33M    40G    1%  /home
/dev/sda1                      10G  187M   9.9G    2%  /boot
tmpfs                         1.6G   20K   1.6G    1%  /run/user/0
[root@dbsvr ~]# vgdisplay
  --- Volume group ---
  VG Name               centos
--删除了一些输出
  VG Size               <890.00 GiB
  PE Size               4.00 MiB
  Total PE              227839
  Alloc PE / Size       227838 / 889.99 GiB
  Free  PE / Size       1 / 4.00 MiB
  VG UUID               EE7MOx-DHe8-4OWy-ctLH-xcDu-melC-Jqpv5i
[root@dbsvr ~]#
```

可以看到，当前操作系统驻留的逻辑卷组 centos 已经没有空闲空间，因此需要执行以下步骤，删除逻辑卷 centos/toBeDeleted。

（1）执行下面的命令卸载 /toBeDeleted 文件系统：

```
[root@dbsvr ~]# umount /toBeDeleted
```

（2）使用 vi 编辑器编辑文件 /etc/fstab：

```
[kingbase@dbsvr ~]$ vi /etc/fstab
```

删除文件中的这一行：

```
/dev/mapper/centos-toBeDeleted /toBeDeleted     xfs    defaults    0 0
```

（3）执行下面的命令删除逻辑卷 centos/toBeDeleted：

```
[root@dbsvr ~]# lvremove centos/toBeDeleted
Do you really want to remove active logical volume centos/toBeDeleted? [y/n]: y
  Logical volume "toBeDeleted" successfully removed
[root@dbsvr ~]#
```

(4)执行下面的命令查看当前逻辑卷组 centos 的情况:

```
[root@dbsvr ~]# vgdisplay
  --- Volume group ---
  VG Name                centos
--删除了一些输出
  VG Size                <890.00 GiB
  PE Size                4.00 MiB
  Total PE               227839
  Alloc PE / Size        57344 / 224.00 GiB
  Free  PE / Size        170495 / <666.00 GiB
  VG UUID                EE7MOx-DHe8-4OWy-ctLH-xcDu-melC-Jqpv5i
[root@dbsvr ~]#
```

可以看到,当前逻辑卷组 centos 还有 666GiB 的空闲空间。这些空间在将来可用来创建新的逻辑卷,或者用来扩大某些现有文件系统的大小。

A.6 备份 CentOS 虚拟机文件

执行下面的命令关闭这个安装好 CentOS 7 操作系统的 VMware 虚拟机:

```
[root@dbsvr ~]# shutdown -h now
```

关闭后,在 VMware Workstation 中,为这个 CentOS 虚拟机添加 **3 块大小都是 1200GB 的 SCSI 磁盘**。

接下来使用 WinRAR 压缩软件,备份这个 CentOS 虚拟机所驻留的目录(本书目录为 G:\dbserver),将其压缩为 dbserver.rar,压缩完成后,重命名为 dbserver-1OS.rar。将来可以使用这个备份的 CentOS 虚拟机,按照 1.3 节安装 KingbaseES 数据库。

附录 B

安装 KingbaseES 单机数据库的最佳实践

在学习完本书前 7 章,以及已经深入理解了 KingbaseES 数据库体系结构之后,再按照本附录的步骤,完成安装 KingbaseES 单机数据库服务器的最佳实践。

 B.1 最佳实践安装规划

按如下方案来安装 KingbaseES 数据库服务器。
(1) KingbaseES 数据库的二进制程序将被安装在目录/u00/Kingbase/ES/V9 下。
(2) KingbaseES 数据库的数据目录位于目录/u00/Kingbase/ES/V9/data 下。
(3) KingbaseES 数据库的用户表空间位于以下目录。
① /u01/Kingbase/ES/V9/data/userts。
② /u02/Kingbase/ES/V9/data/userts。
③ /u03/Kingbase/ES/V9/data/userts。
(4) KingbaseES 数据库的临时表空间为 temp_ts,位于/u03/Kingbase/ES/V9/data/userts。
(5) KingbaseES 数据库的默认表空间为 user_ts,位于/u02/Kingbase/ES/V9/data/userts。
(6) KingbaseES 数据库的控制文件多元化 3 个复制,位于以下目录。
① /u00/Kingbase/ES/V9/data/global/sys_control。
② /u01/Kingbase/ES/V9/data/global/sys_control。
③ /u02/Kingbase/ES/V9/data/global/sys_control。
(7) KingbaseES 数据库运行在归档日志模式下,在目录/u04/Kingbase/ES/V9/archivelog 下保存归档日志文件。
(8) KingbaseES 数据库的备份保存在目录/dbbak 下。
① 物理备份保存在目录/dbbak/pbak 下。
② 逻辑备份保存在目录/dbbak/lbak 下。

 B.2 基础安装

按照附录 A 安装一个 CentOS 7 操作系统,然后按照 1.3 节安装一个 KingbaseES 单机数据库。在此基础上执行本附录后面的优化配置。

也可以直接使用本书提供的资源文件 dbserver-2DB-1nodata.rar，在此基础上执行本附录后面的优化配置。

B.3　KingbaseES 数据库优化配置

B.3.1　创建数据库用户 kingbase

首先创建数据库用户 kingbase，然后将操作系统用户 kingbase 映射为数据库用户 kingbase，这样使用用户 kingbase 管理 KingbaseES 数据库将会更方便。

（1）执行下面的 ksql 命令和 SQL 语句创建用户 kingbase：

```
[kingbase@dbsvr ~]$ ksql -d test -U system
system@test=# CREATE USER kingbase WITH SUPERUSER PASSWORD 'Passw0rd';
```

（2）使用 vi 编辑器编辑文件/u00/Kingbase/ES/V9/data/sys_ident.conf，在文件末尾添加如下的行：

```
kingbase_map    kingbase      kingbase
```

（3）使用 vi 编辑器编辑文件/u00/Kingbase/ES/V9/data/sys_haba.conf，找到如下的行：

```
# "local" 只能用于 UNIX 域套接字
local   all            all                        scram-sha-256
```

在这两行之间添加一行，添加完成后内容如下：

```
# "local" 只能用于 UNIX 域套接字
local   all            kingbase                   peer map=kingbase_map
local   all            all                        scram-sha-256
```

修改后的配置含义如下。

① 使用数据库用户 kingbase，通过 UNIX 本地套接字访问所有数据库时，将使用操作系统用户进行身份认证，映射名为 kingbase_map。

② 所有数据库用户，通过 UNIX 本地套接字访问任何数据库时，都需要提供经过 scram-sha-256 加密后的用户密码进行身份认证。

（4）执行下面的 SQL 语句重新装载数据库，让上面步骤修改的配置文件生效：

```
system@test=# SELECT sys_reload_conf();
 sys_reload_conf
-----------------
 t
(1 row)
system@test=# \q
[kingbase@dbsvr ~]$
```

（5）使用用户 kingbase 执行下面的命令：

```
[kingbase@dbsvr ~]$ ksql -d test -U kingbase
```

```
Type "help" for help.
kingbase@test=# \q
[kingbase@dbsvr ~]$
```

可以看到,用户 kingbase 登录操作系统后,以数据库用户 kingbase 的身份访问数据库 test(任意一个数据库),不再需要提供数据库用户 kingbase 的密码了。

B.3.2 优化 WAL 日志

KingbaseES 数据库有大约 16 个 WAL 日志文件,并且每个 WAL 日志文件的大小为 64MB,WAL 日志缓冲区的大小也设置为 64MB。执行下面的步骤来完成这个任务:

(1) 使用用户 kingbase,执行下面的命令关闭 KingbaseES 数据库。

```
[kingbase@dbsvr ~]$ sys_ctl stop
```

(2) 使用 vi 编辑器编辑文件/u00/Kingbase/ES/V9/data/kingbase.conf,设置如下 3 个参数值为:

```
max_wal_size = 1GB
min_wal_size = 256MB
wal_buffers = 64MB             # 将 wal_buffers 的值设置为 WAL 日志段的大小
```

(3) 使用用户 kingbase,执行下面的命令修改 KingbaseES 的 WAL 日志段大小为 64MB。

```
[kingbase@dbsvr ~]$ sys_resetwal --wal-segsize=64 -D /u00/Kingbase/ES/V9/data
Write-ahead log reset
[kingbase@dbsvr ~]$
```

(4) 使用用户 kingbase,执行下面的命令重新启动 KingbaseES 数据库。

```
[kingbase@dbsvr ~]$ sys_ctl start
```

B.3.3 配置 KingbaseES 数据库工作在归档模式

KingbaseES 生产数据库通常设置运行在归档日志模式下。执行下面的步骤来完成这个任务。

(1) 执行下面的 ksql 命令查看当前 KingbaseES 数据库的归档设置:

```
[kingbase@dbsvr ~]$ ksql -d test -U kingbase -c "SELECT name,setting FROM pg_
settings WHERE name like 'archive%' or name = 'wal_level';"
          name            |  setting
--------------------------+------------
 archive_cleanup_command  |
 archive_command          | (disabled)
 archive_dest             | (disabled)
 archive_mode             | off
 archive_timeout          | 0
 wal_level                | replica
(6 rows)
[kingbase@dbsvr ~]$
```

可以看到,KingbaseES 数据库默认工作在非归档方式(archive_mode 的值为 off)。
(2) 执行下面的命令关闭 KingbaseES 数据库:

```
[kingbase@dbsvr ~]$ sys_ctl stop
```

(3) 执行下面的命令创建用于保存归档日志的目录:

```
[kingbase@dbsvr ~]$ mkdir -p  /u04/Kingbase/ES/V9/archivelog
```

(4) 使用 vi 编辑器编辑文件/u00/Kingbase/ES/V9/data/kingbase.conf,修改配置文件中关于 WAL 和归档设置的参数:

```
# wal_level可以取以下的值:minimal、replica、logical
# 修改 wal_level 的值需要重新启动数据库
# 工作在归档模式下不能设置为 minimal,可以设置为除 minimal 之外的其他参数
wal_level = replica
# 修改 archive_mode 的值需要重新启动数据库
archive_mode=on
# 修改 archive_command 的值不需要重新启动数据库,只需要 reload
archive_command = 'cp %p /u04/Kingbase/ES/V9/archivelog/%f'
# 修改 archive_time:归档周期,900 表示每 900s(15min)切换一次
archive_timeout = 900
```

(5) 使用用户 kingbase,执行下面的命令启动 KingbaseES 数据库:

```
[kingbase@dbsvr ~]$ sys_ctl start
```

(6) 执行下面的命令查看当前 KingbaseES 数据库的归档设置:

```
[kingbase@dbsvr ~]$ ksql -d test -U kingbase -c "SELECT name,setting FROM pg_
settings WHERE name like 'archive%' or name = 'wal_level';"
          name           |                 setting
-------------------------+-----------------------------------------
 archive_cleanup_command |
 archive_command         | cp %p /u04/Kingbase/ES/V9/data/archivelog/%f
 archive_dest            |
 archive_mode            | on
 archive_timeout         | 900
 wal_level               | replica
(6 rows)
[kingbase@dbsvr ~]$
```

可以看到,KingbaseES 数据库已经工作在归档方式下了(archive_mode 的值为 on)。

B.3.4 初步优化 KingbaseES 数据库的系统参数

KingbaseES DBA 在服务器上安装完 KingbaseES 数据库后,需要对一些系统参数进行初始优化。执行下面的步骤来完成这个任务:

(1) 使用 vi 编辑器编辑文件/u00/Kingbase/ES/V9/data/kingbase.conf,修改配置文件的下列参数(基于当前 KingbaseES 数据库服务器的内存是 16GB):

```
max_connections = 300                 # 根据需要配置最大连接的数量
```

```
shared_buffers= 4096MB              # 不超过内存的 40%
work_mem= 8MB
maintenance_work_mem=1024MB         # 最大不超过 1024MB
autovacuum_work_mem = -1
superuser_reserved_connections=3    # 为 KingbaseES 数据库的超级用户保留紧急连接
```

（2）修改完成后，使用用户 kingbase，执行下面的命令重新启动 KingbaseES 数据库：

```
[kingbase@dbsvr ~]$ sys_ctl restart
```

B.3.5 控制文件多元化

控制文件对于 KingbaseES 数据库的运行至关重要，不过安装完 KingbaseES 数据库后，默认只有一个控制文件。KingbaseES 数据库支持多个控制文件。执行下面的步骤可以为 KingbaseES 数据库配置 3 个控制文件副本。

（1）使用用户 kingbase，执行下面的命令关闭 KingbaseES 数据库：

```
[kingbase@dbsvr ~]$ sys_ctl stop
```

（2）使用用户 kingbase，执行下面的命令存放创建控制文件的目录：

```
[kingbase@dbsvr ~]$ mkdir -p /u01/Kingbase/ES/V9/data/global/
[kingbase@dbsvr ~]$ mkdir -p /u02/Kingbase/ES/V9/data/global/
```

（3）使用 vi 编辑器编辑文件 /u00/Kingbase/ES/V9/data/kingbase.conf，修改配置文件中关于控制文件多元化的参数（注意下面 3 行的内容是 1 行）：

```
control_file_copy= '/u00/Kingbase/ES/V9/data/global/sys_control;/u01/Kingbase/
ES/V9/data/global/sys_control;/u02/Kingbase/ES/V9/data/global/sys_control'
```

（4）使用用户 kingbase，执行下面的命令启动 KingbaseES 数据库：

```
[kingbase@dbsvr ~]$ sys_ctl start
```

（5）执行下面的命令查看 KingbaseES 数据库有几个控制文件的副本：

```
[kingbase@dbsvr ~]$ ksql -U kingbase -d test -c "show control_file_copy;"
                       control_file_copy
---------------------------------------------------------------
 /u00/Kingbase/ES/V9/data/global/sys_control;
 /u01/Kingbase/ES/V9/data/global/sys_control;
 /u02/Kingbase/ES/V9/data/global/sys_control
(1 row)
[kingbase@dbsvr ~]$
```

可以看到，KingbaseES 数据库现在有 3 个控制文件副本。

B.3.6 设置默认表空间和临时表空间

首先执行下面的命令和 SQL 语句创建 user_ts 表空间和 temp_ts：

```
[kingbase@dbsvr ~]$ ksql -d test -U kingbase
```

```
kingbase@test=# \! mkdir -p /u02/Kingbase/ES/V9/data/userts/user_ts
kingbase@test=# CREATE TABLESPACE user_ts LOCATION '/u02/Kingbase/ES/V9/data/
userts/user_ts';
kingbase@test=# \! mkdir -p /u03/Kingbase/ES/V9/data/userts/temp_ts
kingbase@test=# CREATE TABLESPACE temp_ts LOCATION '/u03/Kingbase/ES/V9/data/
userts/temp_ts';
kingbase@test=# \q
[kingbase@dbsvr ~]$
```

然后使用 vi 编辑器编辑文件 /u00/Kingbase/ES/V9/data/kingbase.conf，修改以下两个参数的值：

```
default_tablespace = 'user_ts'
temp_tablespaces = 'temp_ts'
```

最后执行 sys_ctl restart 命令重新启动数据库：

```
[kingbase@dbsvr ~]$ sys_ctl restart
```

B.3.7 配置服务名

在 KingbaseES 服务器上，以用户 kingbase 的身份，使用 vi 编辑器编辑服务名配置文件 /home/kingbase/.sys_service.conf，添加如下内容：

```
[sntest]
host=192.168.100.22
port=54321
dbname=test
[snkingbase]
host=192.168.100.22
port=54321
dbname=kingbase
[sntestdb]
host=192.168.100.22
port=54321
dbname=testdb
```

B.4 导入测试数据集

按照 1.4.2 节，导入 TPC-H 测试数据集。

B.5 CentOS 7 操作系统安全加固

按照附录 A 安装的 CentOS 7 操作系统，存在某些方面的安全隐患，因此需要进行以下的安全加固。

B.5.1 隐藏 GNOME 登录界面中的用户名

GNOME 登录界面会显示 CentOS 的用户名，将用户名隐藏起来可以增加系统的安全性。使用用户 root，执行下面的命令配置 GNOME 在登录界面中隐藏用户名：

```
cat> /etc/dconf/profile/gdm <<EOF
user-db:user
system-db:gdm
file-db:/usr/share/gdm/greeter-dconf-defaults
EOF

cat> /etc/dconf/db/gdm.d/00-login-screen<<EOF
[org/gnome/login-screen]
# Do not show the user list
disable-user-list=true
EOF
dconf update
```

B.5.2 在 GNOME 登录界面显示文本信息

使用用户 root，执行下面的命令配置 GNOME，在登录界面显示文本信息 Hello，KingbaseES Database Server！：

```
cat>/etc/dconf/db/gdm.d/01-banner-message<<EOF
[org/gnome/login-screen]
banner-message-enable=true
banner-message-text='Hello,KingbaseES Database Server!'
EOF
dconf update
```

B.5.3 禁止普通用户关机

使用用户 root，执行下面的命令将配置 GNOME 隐藏普通用户的关机按钮：

```
cat >/etc/polkit-1/rules.d/55-inhibit-shutdown.rules <<EOF
polkit.addRule(function(action, subject) {
    if ((action.id == "org.freedesktop.consolekit.system.stop" || action.id == "org.freedesktop.consolekit.system.restart") && subject.isInGroup("admin")) {
        return polkit.Result.YES;
    }
    else {
        return polkit.Result.NO;
    }
});
EOF
```

B.5.4 禁用 Ctrl+Alt+Del 组合键

使用用户 root，执行下面的命令配置在字符终端中禁用 Ctrl＋Alt＋Delete 组合键：

```
systemctl mask ctrl-alt-del.target
```

使用用户 root，执行下面的命令配置在 GNOME 的终端中禁用 Ctrl＋Alt＋Delete 组合键：

```
cat>/etc/dconf/db/local.d/00-disable-CAD<<EOF
[org/gnome/settings-daemon/plugins/media-keys]
logout=''
EOF
dconf update
```

B.5.5 禁止系统休眠

使用用户 root，执行下面的命令配置 CentOS 7 禁止系统休眠：

```
systemctl mask sleep.target suspend.target hibernate.target hybrid-sleep.target.
```

B.6 备份 CentOS 虚拟机文件

执行下面的步骤，备份这个按照最佳实践安装好了的、导入了测试数据集的 KingbaseES 数据库虚拟机：

```
[kingbase@dbsvr ~]$ sys_ctl stop
waiting for server to shut down.... done
server stopped
[kingbase@dbsvr ~]$ su -
Password: #输入超级用户 root 的密码 root123
Last login: Fri Feb 16 18:56:43 CST 2024 on pts/1
[root@dbsvr ~]# shutdown -h now
```

使用 WinRAR 备份安装好 KingbaseES 数据库的 CentOS 虚拟机所驻留的目录（本书目录为 G:\dbserver），将其压缩为 dbserver.rar，压缩完成后，重命名为 dbserver-3DB-BestPractice.rar。

附录 C

生成 TPC-H 测试数据集

本附录主要介绍如何生成 TPC-H 测试数据集。执行如下步骤生成 TPC-H 测试数据集。

（1）下载和编译 TPC-H 测试工具。

使用用户 kingbase，执行下面的命令下载 TPCH 介质，然后编译对应的造数工具 dbgen：

```
[kingbase@dbsvr ~]$ git clone https://github.com/gregrahn/tpch-kit.git
[kingbase@dbsvr ~]$ cd tpch-kit/dbgen
[kingbase@dbsvr dbgen]$ make
```

（2）生成 TPC-H 测试数据集。

使用用户 kingbase，执行命令生成 1GB TPC-H 测试数据：

```
[kingbase@dbsvr dbgen]$ ./dbgen  -vfF -s 1
```

其中参数-s 1 是指定创建的数据大小是 1GB。

（3）将 TPC-H 测试数据集导入 test 数据库下的模式 tpch。

执行下面的 ksql 命令，使用用户 system 登录数据库 test，在数据库 test 中创建模式 tpch，并将其设置为默认模式：

```
[kingbase@dbsvr dbgen]$ ksql -d test -U system
system@test=# CREATE SCHEMA tpch;
system@test=# SET search_path TO tpch;
```

在 ksql 中运行脚本 dss.ddl，创建 tpch 测试表：

```
system@test=# \i dss.ddl
```

在 ksql 中，继续执行下面的 COPY 命令，导入 TPC-H 测试数据集：

```
COPY customer(C_CUSTKEY, C_NAME, C_ADDRESS, C_NATIONKEY, C_PHONE, C_ACCTBAL, C_MKTSEGMENT, C_COMMENT ) FROM '/home/kingbase/tpch - kit/dbgen/customer.tbl ' delimiter '|';
COPY lineitem(L_ORDERKEY, L_PARTKEY, L_SUPPKEY, L_LINENUMBER, L_QUANTITY, L_EXTENDEDPRICE, L_DISCOUNT, L_TAX, L_RETURNFLAG, L_LINESTATUS, L_SHIPDATE, L_COMMITDATE, L_RECEIPTDATE, L_SHIPINSTRUCT, L_SHIPMODE, L_COMMENT) FROM '/home/kingbase/tpch-kit/dbgen/lineitem.tbl' delimiter '|';
```

```
COPY nation(N_NATIONKEY, N_NAME, N_REGIONKEY, N_COMMENT) FROM '/home/kingbase/
tpch-kit/dbgen/nation.tbl' delimiter '|';
COPY orders(O_ORDERKEY, O_CUSTKEY, O_ORDERSTATUS, O_TOTALPRICE, O_ORDERDATE, O_
ORDERPRIORITY, O_CLERK, O_SHIPPRIORITY, O_COMMENT) FROM '/home/kingbase/tpch-
kit/dbgen/orders.tbl' delimiter '|';
COPY partsupp(PS_PARTKEY, PS_SUPPKEY, PS_AVAILQTY, PS_SUPPLYCOST, PS_COMMENT) FROM
'/home/kingbase/tpch-kit/dbgen/partsupp.tbl' delimiter '|';
COPY part (P_PARTKEY, P_NAME, P_MFGR, P_BRAND, P_TYPE, P_SIZE, P_CONTAINER, P_
RETAILPRICE, P_COMMENT) FROM '/home/kingbase/tpch-kit/dbgen/part.tbl'
delimiter '|';
COPY region(R_REGIONKEY, R_NAME, R_COMMENT) FROM '/home/kingbase/tpch-kit/
dbgen/region.tbl' delimiter '|';
COPY supplier(S_SUPPKEY, S_NAME, S_ADDRESS, S_NATIONKEY, S_PHONE, S_ACCTBAL, S_
COMMENT) FROM '/home/kingbase/tpch-kit/dbgen/supplier.tbl' delimiter '|';
```

在 ksql 中,继续执行下面的 ALTER TABLE 语句,为测试表添加主键和外键约束:

```
--PRIMARY KEY
ALTER TABLE PART ADD PRIMARY KEY (P_PARTKEY);
ALTER TABLE SUPPLIER ADD PRIMARY KEY (S_SUPPKEY);
ALTER TABLE PARTSUPP ADD PRIMARY KEY (PS_PARTKEY, PS_SUPPKEY);
ALTER TABLE CUSTOMER ADD PRIMARY KEY (C_CUSTKEY);
ALTER TABLE ORDERS ADD PRIMARY KEY (O_ORDERKEY);
ALTER TABLE LINEITEM ADD PRIMARY KEY (L_ORDERKEY, L_LINENUMBER);
ALTER TABLE NATION ADD PRIMARY KEY (N_NATIONKEY);
ALTER TABLE REGION ADD PRIMARY KEY (R_REGIONKEY);
-- FOREIGN KEY
ALTER TABLE SUPPLIER ADD FOREIGN KEY (S_NATIONKEY) REFERENCES NATION (N_
NATIONKEY);
ALTER TABLE PARTSUPP ADD FOREIGN KEY (PS_PARTKEY) REFERENCES PART(P_PARTKEY);
ALTER TABLE PARTSUPP ADD FOREIGN KEY (PS_SUPPKEY) REFERENCES SUPPLIER(S_SUPPKEY);
ALTER TABLE CUSTOMER ADD FOREIGN KEY (C_NATIONKEY) REFERENCES NATION (N_
NATIONKEY);
ALTER TABLE ORDERS ADD FOREIGN KEY (O_CUSTKEY) REFERENCES CUSTOMER(C_CUSTKEY);
ALTER TABLE LINEITEM ADD FOREIGN KEY (L_ORDERKEY) REFERENCES ORDERS(O_ORDERKEY);
ALTER TABLE LINEITEM ADD FOREIGN KEY (L_PARTKEY, L_SUPPKEY) REFERENCES PARTSUPP(PS
_PARTKEY, PS_SUPPKEY);
ALTER TABLE NATION ADD FOREIGN KEY (N_REGIONKEY) REFERENCES REGION(R_REGIONKEY);
```

在 ksql 中,继续执行下面的 SQL 语句,创建、重建索引并收集统计数据:

```
system@test=# REINDEX TABLE supplier;
system@test=# REINDEX TABLE orders;
system@test=# REINDEX TABLE part;
system@test=# REINDEX TABLE partsupp;
system@test=# REINDEX TABLE customer;
system@test=# REINDEX TABLE lineitem;
system@test=# CREATE INDEX i_l_orderkey ON lineitem (l_orderkey);
system@test=# CREATE INDEX i_ps_partkey ON partsupp(ps_partkey);
system@test=# CREATE INDEX i_p_name ON part(p_name);
system@test=# CREATE INDEX i_s_nationkey ON supplier(s_nationkey);
system@test=# CREATE INDEX idx_orders_o_totalprince ON orders(o_totalprice);
```

```
system@test=# CREATE INDEX i_c_nationkey ON customer(c_nationkey);
system@test=# CREATE INDEX lineitem_l_partkey_idx  ON lineitem(l_partkey);
system@test=# CREATE INDEX partsupp_ps_partkey_idx ON partsupp(ps_partkey);
system@test=# VACUUM ANALYZE;
```

至此,在数据库 test 的模式 tpch 中,准备好了用于 TPC-H 测试数据集。

(4) 运行 TPC-H 测试语句。

将本书的资源文件 tpch_sql.2024ok.tar 上传到 /home/kingbase 目录下,使用用户 kingbase,执行下面的命令解开 tar 文件包 tpch_sql.2024ok.tar:

```
[kingbase@dbsvr ~]$ tar xf tpch_sql.2024ok.tar
[kingbase@dbsvr ~]$ cd /home/kingbase/tpch_sql/
[kingbase@dbsvr tpch_sql]$ ls
stream_0_10.sql   stream_0_14.sql   stream_0_18.sql   stream_0_21.sql   stream_0_4.sql   stream_0_8.sql
stream_0_11.sql   stream_0_15.sql   stream_0_19.sql   stream_0_22.sql   stream_0_5.sql   stream_0_9.sql
stream_0_12.sql   stream_0_16.sql   stream_0_1.sql    stream_0_2.sql    stream_0_6.sql   tpch.sql
stream_0_13.sql   stream_0_17.sql   stream_0_20.sql   stream_0_3.sql    stream_0_7.sql   tpchtest
[kingbase@dbsvr tpch_sql]$
```

解开 tar 包后,可以看到一些 TPC-H 的 SQL 测试语句和 shell 测试脚本。

文件 tpch.sql 是一个 ksql 脚本文件,内容如下:

```
[kingbase@dbsvr tpch_sql]$ cat tpch.sql
SET search_path TO tpch;
SELECT ceil(random() * 22) as num;
\gset
\i /home/kingbase/tpch_sql/stream_0_:num.sql;
[kingbase@dbsvr tpch_sql]$
```

它首先产生一个 1~22 的随机数 num,然后执行第 num 条 SQL 测试语句。

文件 tpchtest 是一个 bash shell 脚本文件,内容如下:

```
[kingbase@dbsvr tpch_sql]$ cat tpchtest
#!/bin/bash
export KINGBASE_PASSWORD=Passw0rd
for (( i=1;i<=1000000;i++ ))
do
    ksql -d test -U system -f tpch.sql
done
[kingbase@dbsvr tpch_sql]$
```

在这个文件中,设置了用户 system 的密码,脚本将循环运行 1 000 000 次,每次执行 1 条 tpch 测试语句。

要运行 TPC-H 测试,请用用户 kingbase,执行下面的脚本:

```
[kingbase@dbsvr ~]$ cd
[kingbase@dbsvr ~]$ cd tpch_sql/
```

```
[kingbase@dbsvr tpch_sql]$ sh tpchtest
```

可以看到,执行该脚本会持续不断地运行 tpch 测试语句。

要停止 TPC-H 测试,可以持续按 Ctrl+C 组合键,停止脚本的运行。

(5) 使用 KingbaseES 数据库的 exp 命令,导出运行 TPC-H 测试数据集。

首先需要执行下面的命令创建服务名 sntest:

```
[kingbase@dbsvr ~]$ cat>/home/kingbase/.sys_service.conf <<EOF
[sntest]
host=192.168.100.22
port=54321
dbname=test
EOF
```

服务名访问的是 KingbaseES 数据库服务器上的数据库 test。

然后执行下面的 exp 命令导出数据库 test 的模式 tpch:

```
[kingbase@dbsvr ~]$ exp system/Passw0rd@sntest file=tpch.dmp owner=tpch
```

exp 命令导出数据库 test 的模式 tpch,生成了逻辑备份文件 tpch.dmp。在本书的 1.4.2 节,使用这个文件导入 TPC-H 测试数据集。

参考文献

［1］北京电科金仓公司信息技术股份有限公司. KingbaseES V9 产品手册［EB/OL］. (2023-03-06)
　　［2023］. http：//help.kingbase.com.cn/v9/index.html.

［2］王珊，杜小勇，陈红. 数据库系统概论［M］. 6 版. 北京：高等教育出版社，2023.

［3］Abraham Silberschatz，Henry F.Korth，S.Sudarshan. 数据库系统概念（原书第 7 版）［M］. 杨冬青，
　　李红燕，唐世渭，译. 北京：机械工业出版社，2021.

［4］PostgreSQL 全球开发组. PostgreSQL 12.2 手册［M］. 彭煜玮，PostgreSQL 中文社区文档翻译组，
　　译.［2023-03-06］. http：//www.postgres.cn/docs/12/.

［5］谭峰，张文升. PostgreSQL 实战［M］. 北京：机械工业出版社，2018.

［6］屠要峰，陈河堆. 深入浅出 PostgreSQL［M］. 北京：电子工业出版社，2020.

［7］Simon Roggs，Gianni Ciolli. PostgreSQL 14 Administration Cookbook［M］. Packt Publishing，2022.

［8］Rogov, E. PostgreSQL 14 Internals［EB/OL］. (2023)［2023］. https://postgrespro.com/community/
　　books/internals.

［9］TPC-H Standard Specification［S］. Transaction Processing Performance Council（TPC）.